FARMLAND OR WASTELAND

FARMLAND OR WASTELAND

A TIME TO CHOOSE

Overcoming the threat to America's
farm and food future

R. Neil Sampson

Rodale Press  Emmaus, Pennsylvania

Book design by Linda M. Jacopetti.

Printed in the United States of America on recycled paper containing a high percentage of de-inked fiber.

Earlier versions of the material in Chapters 6 and 7 were published by the National Agricultural Lands Study as Interim Report No. 4, *Soil Degradation: Effects on Agricultural Productivity.*
A more technical version of Chapter 6 was published in the "Working Group Background Papers" for the National Conference on Renewable Natural Resources sponsored by the American Forestry Association and others. Portions reprinted by permission of the American Forestry Association.
An early version of Chapters 10 and 11 was presented at the 35th Annual Meeting of the Soil Conservation Society of America and appears as "Energy: New Kinds of Competition for Land," in *Economics, Ethics, Ecology: Roots of Productive Conservation* (Ankeny, Iowa: SCSA, 1981). An early version of Chapter 14 appeared as "The Ethical Dimension of Farmland Protection," in *Farmland, Food and the Future* (Ankeny, Iowa: SCSA, 1979). Rewritten material has been used with permission of SCSA.

Library of Congress Cataloging in Publication Data
Sampson, R. Neil.
Farmland or wasteland.

Includes bibliographical references and index.
1. Soil conservation—United States.
2. Agriculture—United States. 3. Water-supply, Agricultural—United States.
4. Agricultural conservation—United States.
5. Agriculture and state—United States.
6. Land use, Rural—United States. 7. Water conservation—United States. I. Title.
S624.A1S22   333.76'0973   81-11894
ISBN 0-87857-366-6 hardcover   AACR2
2 4 6 8 10 9 7 5 3 1 hardcover

*This book
is dedicated to
Rob, Eric, Chris, and Heidi,
whose future will be affected in
many ways by the willingness and skill
of today's adults and leaders in
addressing the emerging soil
and water conservation
crisis.*

Table of Contents

List of Figures and Tables

Preface

In early 1980, the results of several new resource studies began to come together to paint a startling picture of an emerging national problem: the rapid approach of limits to our ability to produce agricultural crops. It seemed that all of the recent studies pointed in the same direction: The water studies showed growing competition for water; the farmland studies showed a high rate of loss of good land to other uses; forestland and rangeland studies indicated an emerging shortage of land for those uses; and soil erosion measurements were both high and significantly different in their effect from ever before. All told, not a very soothing picture.

There was, however, no single book or reference work that attempted to tie all these various situations together. In the U.S. Department of Agriculture (USDA), work on the Resources Conservation Act (RCA) had fairly well identified the land and water facts, but these had not been correlated with a wide variety of issues such as the demand for forest products, new crops for energy, or the National Water Assessment done by the Water Resources Council.

For those of us who were attempting to keep track of all these various pieces of information, the fact that they were in many locations, with little relation to each other, was a real problem. For concerned and thinking Americans outside the professional circles where those facts circulated, getting a clear picture of the situation was even tougher.

As a result, I set off, with the assistance of many skilled

friends within the USDA who were willing to share their knowledge and insight, to draw the information together and interpret what it might mean. Within the National Association of Conservation Districts (NACD), the Special Committee on District Outlook was studying what these facts might mean to the nation's future, and particularly the future demands that might be placed on the soil conservation districts. I shared some of our ideas and concerns with Bob Rodale of Rodale Press, who suggested that a book on the subject would be useful. I agreed, as much out of the desire to study the situation fully as any other thing, and the project was launched.

What became evident early in the research is that the story of resource scarcity has been told many times before, but time and events have always proven it wrong. From Malthus onward, people have been forecasting a limit to agricultural production. The *1923 Yearbook of Agriculture* contains this as a major theme, and the late 1930's were marked by many articulate and well-reasoned arguments for conservation, based on a growing shortage of land for food production. What followed, instead, was a period of farm surpluses.

The first question that needed answering, then, was, "what is different now, that might make the trends shown by the current data true when those of the past were not?" I hope the answer to that question emerges satisfactorily from the pages that follow. It is not an easy question to answer, but my search for a reason why we should be comforted in spite of the current trends turned up nothing of significant merit. I hesitate to call what we now face a crisis, but it seems clear that we are bearing down on a crisis situation at a more rapid rate than most Americans realize.

This situation will challenge all of us—farmers, ranchers, professional conservationists, researchers, educators, politicians, and ordinary citizens, to face new demands on our skill, patience, and resourcefulness. Most challenged will be those at the center of the issues—farmers, ranchers, conservation district officials, and professional conservationists. Farmers will be under pressure from citizens and policymakers to do two things simultaneously that may, in the final analysis, be impossible: produce more food at lower costs and protect the resource base in the process.

Both farmers and conservation planners will be blamed for getting us into the mess that we are in and hard-pressed to explain why the soil and water suffered such damage in spite of the existence of a national soil conservation program that has used a great deal of public funds over the years. Rather than wait for the situation to worsen, then look for someone to blame, we need to understand what is happening right now and what has led up to it and get a sense for the options that exist.

In trying to set out what has happened, and why, it is easy to lose a sense of perspective. The land data of 1977 tells us that too much damage is occurring; what it fails to tell us is how much damage was *not* taking place because of the ongoing conservation work on farms and ranches. It tells us that we need new emphasis on soil conservation programs; it does not tell us precisely which programs are working best, or which need to be changed, or in what direction they must be altered. Those are decisions that will require reasoned judgment by well-informed people, not more and more numbers spat out of bigger and bigger computers.

My work with the NACD has provided me with the opportunity to discuss these ideas with hundreds of professional and citizen conservationists who live and work with the American land every day. The soil conservation movement in America, despite the fact that it has been largely ignored for several decades, is alive and well, with dedicated and skilled partisans in every corner of the nation. These people, each telling a conservation story in their own way and expressing their concerns for the future by their actions as well as their statements, have shaped, both directly and indirectly, most of my attitudes about conservation. I am proud to have had the opportunity to learn from them.

The resulting conclusions, of course, are mine, not theirs. Nothing said herein has been "approved" by NACD, and it is not a statement of the policies or views held by that organization. That the officers of the organization were willing to allow me to proceed on my own, expressing personal points of view and running the risk of making statements that might be counter to their own personal positions or to those historically taken by NACD, is an indication of both their political courage and their commitment to furthering the conservation ethic in America.

The errors that will ultimately emerge are, of course, my responsibility. After having manipulated so much data, I can only hope that I have treated it as fairly and accurately as possible. But I make no claim to objectivity. At one point, the Soil Conservation Service (SCS) analysts working on the RCA study told me that their computers had produced over 100,000 pages of data printouts. Obviously, 99.9 percent of it does *not* appear in this book and probably will never appear anywhere. There are many subjective judgments when it comes to selecting those few facts that can be captured and interpreted, and I know of no way to guarantee that some important items have not been overlooked.

In trying to sort out the most reliable data from a multitude of sources, one of the problems is that some are in direct conflict, so some of the information I have captured may prove, in hindsight some day, to have been inaccurate. In addition, I have made projections and forecasts into the future that are based mainly on a "sense" of where the current trends seem to be leading. Those may be proven wrong by events. For those kinds of errors I do not apologize—I simply hope others can improve upon my mistakes. The soil and water conservation challenge facing the United States is so compelling that it will take the best that all of us can contribute to meet it. The mistakes we make along the way will pale beside the penalties that could result from either ignoring the situation or being too timid to address it.

Acknowledgments

In writing any book that attempts to cover a wide array of complex data and ideas, one must depend on many people for help. So many people helped, in so many ways, that attempting to single a few of them out for special recognition is hazardous. The people who willingly gave time and effort to make this book more accurate were always busy with other work but never failed to make time to read portions of manuscript and contribute ideas and suggestions. One scarcely knows how to say thanks to such friends.

Those who read portions of the manuscript and offered suggestions included Lyle Bauer, Norm Berg, Charlie Boothby, Steve Brunson, Wendell Fletcher, Neil Hazlett, Walt Jeske, Pat Jordan, Gene Lamb, Bill Larsen, Jerry Lee, Bud Mekelburg, Howard Tankersley, Burtt Trueblood, and Bob Williams. Some are farmers, some are scientists; all are conservationists at heart. Their ideas often sparked a new line of inquiry, cleared up a confused line of reasoning, or sharpened up the facts. To say that much of what has finally been captured is because of their efforts is inadequate; to acknowledge that I am in their debt is insufficient. They contributed time, insight, and wisdom that can never be fully recognized or repaid.

Iowa conservation farmers Chuck and Helen McLaughlin read the entire manuscript and made incisive, painstaking comments that sharpened the editorial style as well as the intellectual focus.

Until you write a book, you may not appreciate the degree to which you must depend on the skills of the editor. Sara Ebenreck read reams of draft chapters, helped untangle lines of logic that were not always clear, demonstrated how to clarify some of the technical data, chased down dozens of small details such as lost photographs, and worked as a partner in molding the whole text into final form. I am deeply indebted for her insight, hard work, and resourcefulness.

Finally, I need to recognize two sources of support without which no person would dare undertake a project so demanding. My wife, Jeanne, and our four children were not only willing to put up with a word processor and an incredible pile of reference books cluttering up a corner of the house but also read portions of the manuscript to help me understand how lay people and youth might react to the soil conservation story.

The officers and staff of the National Association of Conservation Districts gave their strong support throughout the project. Even at times when deadlines required that I "disappear" for several days, leaving them to cover my job as well as their own, they were supportive and helpful. Most important, I owe them a tremendous intellectual debt for their constant efforts, both as individuals and as a group, to improve our understanding of the land and our role as stewards.

1.

People, Food, Land: Equation for Survival

The importance of topsoil to humans and to the quality of our life is difficult to overstate. It produces our food as well as most of our clothing and shelter. It is an absolutely essential natural resource but one that is both limited and fragile. We stand, in most places on earth, only six inches from desolation, for that is the thickness of the topsoil layer upon which the entire life of the planet depends.

In spite of that, the reluctance of humans to protect their soil resources can be traced through history. The progress of civilization has been marked by a trail of wind-blown or water-washed soils that, in all too many cases, are totally barren as a result. Even today, as sophisticated as we profess ourselves to be, we still have not arrived at the point where we willingly pay the full price of the products that we eat, wear, and use in our daily lives. That price includes not only the out-of-pocket costs to the producer but also an adequate investment for maintaining the vigor of the land. Continuing to grow food year after year without appropriate soil conservation management wears out the soil as surely as though you ran a factory at full production without investing in repairs or maintenance. The factory would soon wear out; the land is no different.

There are places in America where the entire layer of topsoil has been lost in only 50-100 years of cultivation. Some of that land can still be farmed by tilling the less-fertile subsoil, but as an efficient and highly productive "food factory," the land is gone forever.

It will always require heavy subsidies of energy, fertilizer, irrigation water, and management if it is to produce at all.

In letting topsoil erode away, we lose the bounty of a natural soil-building process that took millenia. Topsoil is created primarily by weathering processes that break down inert minerals into more chemically and biologically active soil materials. These processes occur very slowly, taking centuries to transform dead rock into living topsoil.

Protecting the quality of our nation's topsoil—and, therefore, much of the quality of our lives—is largely within human control. Good management can improve topsoil quality, but bad management can destroy it within a few years. A cropping system that takes out more nutrients and organic matter than it replaces is essentially "mining" the land, making it susceptible to the main agent of destruction: accelerated erosion. Topsoil lost in this manner is the victim of human neglect and greed.

To understand what is happening to the land, we must also understand the people on the land. The social, economic, and cultural pressures that shape any society affect the way the members of that society use and treat their land. In return, the way people treat the land determines whether or not their society can thrive, or even survive. As improbable as it seems, the "people on the land" are not just the farmers, ranchers, foresters, and others who make their daily living from the soil. We all depend on the products of the soil, whether we drive a cab, teach school, run an office, or work in a factory. That is easy to forget when you live in a city and seldom leave concrete or asphalt, when your food comes from a supermarket, or when your weekly paycheck comes from a source totally unrelated to agriculture. But it is true, nonetheless. In addition, we all contribute attitudes to that elusive factor called public opinion that establishes, in many important ways, what are the "right" and "wrong" things to do in our society.

Farmers either protect or waste land in response to the economic, political, and social signals they receive from all of us. We, in turn, will find the price of the products we must buy, the quality of the environment around us, and the range of options facing our children dramatically affected by the choices those farmers make. So wasting farmland is not just an agricultural issue, it is a danger threatening all of us. The solutions demand our participation as well. A solid majority of Americans need to know what is wrong, how serious it is, and what must and can be done to improve the situation. Only then will our political and economic system devise the kind of support and incentives needed for a sustainable, permanently productive agriculture to emerge.

A sustainable agriculture is not what is developing today. American agriculture is growing larger and producing more, but also becoming more and more vulnerable to a variety of disrupting factors such as inflation, Arab oil embargoes, wars, and even weather changes. Such vulnerability leads to great uncertainty about the future, because most of the important trends affecting agriculture's future are going in the wrong direction. That is, in itself, sobering. Having a few pieces of "bad news" is not too unsettling, providing there are offsetting positive signs that could help balance the situation. But when everything seems headed the wrong way and many of the most perverse trends seem to be gaining speed, while there are few, if any, signs that effective action to reverse them is under-way, it is hard to avoid pessimism.

Demands for agricultural products are accelerating at an ever-increasing pace, while, simultaneously, the ability of the U.S. land to produce is being wasted faster than at any point in our history. It is clear that those two trends cannot continue indefinitely. Former Secretary of Agriculture Bob Bergland called them a "collision course with disaster." I do not think he overstated the case; I think he understood it.

Lessons from the Past

History clearly demonstrates the risk of ignoring the relationships of people to land and land to people. People who waste their land vanish. Entire civilizations have succumbed to the ravages of soil erosion. Their record, written on the land, was described graphically by conservationist Walter C. Lowdermilk, in *Conquest of the Land Through 7000 Years*.[1] In that classic study, Lowdermilk found that neglect of the land had often been a major factor in the fall of empires and the barren nature of much of today's world.

Agriculture developed at least 7,000 years ago in the fertile alluvial valleys of the Tigris, Euphrates, and Nile rivers. In Mesopotamia (now Iraq), the traditional site of the Garden of Eden, the waters of the Tigris and Euphrates were developed for irrigation with an elaborate system of canals, and the productivity of that agriculture freed many people from the drudgery of subsistence farming to become artisans, tradesmen, and leaders. But the river waters were laden with silt from the overgrazed highlands, and as the water entered the more placid, slow-moving canals, the silt settled out, clogging the canals and cutting off the lifeblood of the agriculture. Continuous hand labor was employed to keep the silt removed, but periodic wars would interrupt that task, bringing agriculture, and the empire, to a halt.

As a result, Lowdermilk estimated that at least 11 empires have risen and fallen in the 7,000-year history of Iraq.[2] Today, the land supports less than one-sixth of the people that lived there during its historic peaks, and Iraq is dependent on oil exports, not agricultural abundance, for its wealth. The recent war between Iraq and Iran has halted dredging in the waterway that separates the two countries, and silt buildup is sealing scores of ships behind rapidly growing mud bars in the channel. Soil erosion is still exacting its toll.

In Syria, Lowdermilk found over 100 dead villages and market towns in the region between Hama, Alleppo, and Antioch. The ruins of these towns, many of them no doubt thriving when the Apostle Paul made his journeys to spread the new Christian faith, were not buried. They stand stark on bare rock, the soils completely washed or blown away. Lowdermilk's verdict: "If the soils had

Its former glory can be glimpsed in the ruins of the city of Timgad, North Africa. This 1939 photo shows the long-deserted city, surrounded by an eroded, desolate countryside which could no longer provide food. USDA photo.

remained, even though the cities were destroyed and the populations dispersed, the area might have been repeopled again and the cities rebuilt. But now that the soils are gone, all is gone."[3]

Recent Concern in the United States

Those are the kind of stark realities that have spurred new study and debate on the future of farmland in the United States. In some ways, the concern might be seen as an outgrowth of the environmental movement of the 1960's and 1970's, since it focuses heavily on the protection of the land. But it is radically different from the majority of the early environmental concerns, which usually centered around the protection or preservation of landscapes, often in pristine condition. Productivity, at least in the economic sense, was seldom much of a factor in those environmental issues. In farmland protection, however, productivity and its protection are a major concern. Trying to "preserve" farmland is not to prevent its effective economic use in the future but to guarantee it.

Spurred by widespread indications that U.S. farmland was not faring very well under modern agricultural management, Congress has increasingly turned its attention to this issue. In a major policy move, the Soil and Water Resources Conservation Act (RCA) was enacted into law in 1977. This law required the U.S. Department of Agriculture (USDA) to assess the quantity and quality of the land and water of the United States, study the trends affecting these resources, and report its findings to the American people and the Congress.

By early 1981, USDA had completed an intensive appraisal of the nation's soil and water resources. This appraisal was based on the National Resource Inventories of 1977, conducted by the Soil Conservation Service (SCS). The *RCA Appraisal* paints a stark picture: The soil and water resources of the nation are being wasted at a rate unparalleled in recent times.

With a similar sense of urgency, the Carter administration established the National Agricultural Lands Study (NALS) in mid-1979 as an 18-month effort to answer several important questions about the nation's farmland supply. What are the nature, rate, extent, and causes of conversion of agricultural land to non-agricultural uses? What are the economic, environmental, and social consequences of that conversion? What is being done to try to slow farmland losses, and how well are existing efforts working? What else should be done?

The study, too short-lived to develop new data, utilized the existing facts from USDA—again, largely derived from the RCA study—as well as information from other federal agencies, states, universities, counties, and other sources. Its final report, issued in 1981, found that farmland conversion is a pressing concern in many localities and that it is a problem of large and growing proportions when viewed in the national perspective.

But even these new efforts fail to provide a definitive answer to some of the most critical questions being asked about the future of America's farmland. The reason is simple—government studies don't forecast the future any better than other mortal efforts. So there is still debate, and a great deal of apathy, over the wasting of America's farmlands. Is it a real problem or isn't it? When will it reach the point where significant ill effects will be felt? What will we need to do then? I will review some of those arguments, but the truth is that most of them miss the point completely. Scientists and economists can argue endlessly over the precise details of the rate of land loss, the amount of yield increases likely to occur in the future, or the quantity of food liable to be needed at some future year. Their debate can grow heated over minor differences in interpretation or projections of the data.

When a gully gets started, a great deal of soil can be moved in a short time. Not only the gullied land, but also the bottomland where all the washed-out rocks and soil were deposited, were damaged in this Massachusetts pasture. SCS photo by Arthur Verdi.

In the larger sense, none of that matters much. What is true is this: The more people there are in the world, the more demand there will be for food, fiber, wood, water, fuel, and every other product of the land. Populations are growing—rapidly—all over the world, including in the United States. The productivity of agricultural soils is not rising as rapidly as it has in the past and in many cases is actually dropping. In addition, good farmland is being wasted by conversion to non-farm uses, by topsoil erosion, and by a variety of other factors. These changes, too, are occurring all over the world, including the United States. To argue about whether such trends will ultimately reach the stage where they result in problems in 1990—or 2000—or 2030—is to miss the whole point. What does 10 years—or 50—mean in the scale of human time? Will today's generation feel good because it did not starve, but in the process of producing food, guaranteed privation for its children?

The real point is that wasting good farmland is senseless, for both current well-being and future security. It cannot be defended

This cornfield was badly gullied when an irrigation canal overtopped during a sudden rainstorm in Sedgewick County, Colorado. USDA photo by R. Moreland.

on any grounds—legal, moral, or economic. To the extent that the United States is wasting the greatest body of prime farmland on the face of the earth, it is tragic. To the extent that we do not understand how and why we are wasting it, it is culpable ignorance. To the extent that we realize the problem, but fail to act to solve it, our negligence is criminal. We are rapidly, as a national population, moving from the point of ignorance about the problem to knowing about it, but not acting. Part of the reason for that is that we have not yet defined a rational relationship between people and land that makes sense in today's world.

Our Soil Savings Account

Our relationship to the soil can be compared to a person with a savings account in the bank, whose interest income each year is not enough to meet his consumption needs (or desires, which are often very different). Such a person has two choices. The first is to cut back his consumption to a level that the interest income from the account would support. If he wants more money in the future, he must invest something from current income to build the savings account for future yields. This is tough and requires that a person be willing to forego current pleasure for future security. It is no fun, at least in the short run.

The second alternative is to keep taking money out of the account for spending. That gratifies the need for immediate consumption, even though the consumer knows that it also reduces the savings account and assures that it will yield even less in future years.

The land, and its productivity, are our national savings accounts. But private economic pressures are driving farmers to extract as much from those accounts as possible, even to the point of depleting the soil permanently. And it is not just the "soil account" that is depleted in the process. As farmland is wasted, it must be replaced with new land that often requires the drainage of wetlands, cutting down of bottomland hardwood forests, plowing of productive rangelands and pasturelands, mining of limited groundwater supplies, and consumption of limited mineral and petroleum resources. Each time we must resort to these efforts, we deplete yet another resource "savings account," so that the net effect of wasting farmland is to compound the nation's vulnerability by reducing productivity virtually everywhere. This is the current course of action; one that would be called totally irresponsible if it were applied to the management of a bank account. But this course of action is strongly supported, and, in many cases, vigorously promoted by private citizens, corporations, and politicians.

In the face of growing economic strife, much of which is caused by resource shortages, admonitions to conserve soil — or any other natural resource, for that matter — are not very palatable, even though they make the best of sense. Exhortations to "produce more," on the other hand, find a receptive audience, even when they are clearly irresponsible. The nation's current preoccupation with petroleum fuels is an example. Conservation is clearly the most cost-effective near-term method of reducing the country's dependence on imported foreign oil, but it is a politically hazardous course. Added production from our ever-dropping reservoirs of domestic oil is a much more popular prescription for overcoming the stress caused by declining supply and increasing cost, even when it is patently apparent that such a strategy simply hastens the depletion process.

Similar prescriptions for food have been common. Food prices went up in the early 1970's as commodity supplies were drawn down. "Plant fence row to fence row," exhorted Agriculture Secretary Earl Butz. Farmers followed his advice, for it was certainly to their personal financial advantage to do so, and Butz was hailed as a hero in farm country. But that surge in demand for farm products was short-lived, and Butz's successor at USDA, Bob Bergland, was forced to tell farmers the bad news: They had overproduced their market and would have to cut back production again or face disastrously low prices.

In addition, Bergland reminded farmers, the topsoil they were losing and the soil conservation practices that had been destroyed in the "great plow-out" would have to be rebuilt. Farmers were angered, of course, and some of the more militant marched in rumbling tractorcades to Washington, D.C., where they spent most of the winter of 1978-1979 racing their tractors on the Mall and demanding higher prices for farm products "or else."

That approach, while it failed to solve anything, clearly pointed out the basic problems in U.S. agriculture: Many farmers do not have the capacity to be flexible to changing circumstances. The "factory farm" so highly touted as the wave of the future has its weaknesses, one of which lies in its apparent need to continuously expand and produce more in order to be profitable. Most observers agree that the total demand for food, fiber, and the other products of the land will dramatically increase in the future. But will the nation's land base — and its productivity — be able to keep up with the increased demands created by continued population growth, per-capita consumption, and economic pressures? Are there limits to our capability to produce?

Will we be able to produce more and more products from less and less land—land that is not only reduced in quantity as the forces of "progress" bury millions of acres under concrete and asphalt, but land that is also losing vitality, resilience, and productivity as its precious topsoil erodes away? My view is that we must finally face the fact that getting more from less will be difficult, if not impossible.

Entering a New World

We are, as Richard Barnet points out in his book *The Lean Years,* entering a new age of resource scarcity that is bound to be marked by fierce competition over the control of resources.[4] That struggle is already taking place, and the United States will, in the coming years, become more and more embroiled in the vital need to control resources. As certainly as the OPEC nations control petroleum today, the United States controls something even more valuable: the ability to produce extra food. But just as those oil resources can be destroyed by war, or depletion, or waste, the agricultural strength of the United States is vulnerable.

Part of that vulnerability, of course, is due to danger from outside aggressors. This fear fuels a national defense budget of large and growing proportions, along with political pressures to increase it still more. But another, more insidious danger is posed from within. Control over resources and their productivity is more than simply the occupation of territory. The nation's current political tendencies to strengthen national defense and "reindustrialize" America must also include an intensive effort to conserve, improve, and "revitalize" agriculture and agricultural land. Without such an effort, the other attempts to curb inflation, improve defense, and rebuild American strength at home and abroad will be only temporary.

We must look to the land that sustains us, diagnose its ills, and move to implement both preventative and curative measures for the steady decline in productivity that now threatens it. The fact that today's predictions indicate reasonably good food security in the United States until the year 2000 (barring unforeseen difficulties) is no particular cause for rejoicing. If you take comfort in that fact, recognize that the corollary of that argument is to say that serious shortages are likely some time shortly after 2000 unless major technological breakthroughs continue. Ask someone who is 16 how they feel about that kind of future. If you listen closely to their reaction, you will learn some of the real implications of today's resource waste.

A Time to Choose

Even the most gloomy combination of predictions and forecasts about America's future should not lead to inactivity or despair, because options exist today, and we should take advantage of them. But we should not count on their permanence. We have let many opportunities slip by and wasted too many resources in the process. A million acres of farmland wasted this year are a million opportunities lost for next year, and that is no small matter.

Our options are of two different types. First, we can prepare ourselves for the inevitable. The fact that there are limits to some resources, or perhaps to our productive capacity, is not, of itself, a surprise. That is simply a statement of the human condition throughout history. The idea that our social and economic system might not be able to adjust to those realities is serious, however. We all know that some day the oil wells are going to run dry if we keep pumping out the oil and burning it up. The faster we pump, the sooner they dry up. Well before that time, however, rising prices will mean that only the most high-value users can afford to burn oil.

That does not mean the end of civilization. The Greeks and Incas both developed a fairly sophisticated civilization without a single automobile! We can't imagine a world with limited mobility, but one may be coming. It most certainly will be coming if we don't develop a technology to provide new sources of energy for transportation after petroleum becomes too scarce and expensive. Yet the transition to such a new world is not impossible nor does it need to be disruptive. In fact, it may already be under way. Conservation and rationing of the remaining petroleum supplies is pretty well started already, simply because of the rising prices. As prices go still higher, consumption will drop more. Despite short-term lapses due to momentary oil gluts caused mainly by market aberrations, the next generation will be thoroughly conditioned to use petroleum sparingly and to provide money for research and development of post-petroleum technologies, for farming as well as for transportation. The beginning of the necessary accommodation to a resource-constrained world will be in effect.

Other opportunities lie in altering current trends. Few trends need to be destiny if people do not wish them to continue. We waste far too much farmland today, but nowhere is it written in stone that we must do so. Millions of acres have been buried needlessly under concrete and asphalt due solely to the construction of federally funded projects. That is *our* money, being spent to do the things *we* want done. But what if we, as the public, were to say "Enough! Build elsewhere, but save our prime farmland." Not all farmland conversion would be halted, to be sure, but the rate of loss would be dramatically slowed. Will we be able to gain the public consensus required to develop such a public policy? I would hope so, and there are significant efforts in the Congress today to do so. Is it possible? Of course.

But moving into the new realities created by growing populations pressing on shrinking resources will require a new set of individual actions and collective, or public, policies geared not to the past but to the present and the future. In order to agree on those actions and policies, people must get a clear understanding of where current trends seem to be leading us. Then, if we agree that we are uneasy about those trends, we should devise and carry out strategies to alter their course.

Such an opportunity exists today in regard to our waste of farmland. Because of the intensive studies carried out by the USDA and the NALS, we have more facts at hand than ever before. Studies ranging from the RCA to the *Global 2000* study done for President Carter, as well as many pieces of public and private research, point to similar conclusions. There is no longer any reason to question seriously the need to stop farmland waste. That question, as the data gathered in this book will demonstrate, can be put to rest. What we need now is a plan of action.

Our current situation is serious, but far from hopeless. The technology of soil and water conservation available today is by far the most sophisticated ever known. The means to avert waste are at hand, but the clock is running. The forces that are gnawing away at the productive land base of the United States are symptomatic of very powerful economic, social, and environmental factors. The time to start turning the current trends toward more constructive, productive directions is now, while there is still time, and while today's options remain open. With each acre of prime farmland we lose and each ton of topsoil we allow to wash away, another option is closed for our future and the future of those who follow us. So what is needed—and soon—is enlightened action by people working to improve the world for themselves and their children.

For people are the real crux of this issue. Saving a ton of topsoil so that the land may somehow be more beautiful, or productive, or "whole" is of little interest to most people. We would all agree that it is a "good" thing to do, but when the inevitable time comes to count the cost and make hard choices as to how much we would willingly spend to protect farmland, we will opt for very little effort if our only hope is somehow to benefit an abstract concept of "environmental integrity." But the issue is far more compelling than that, for it involves the right of humans to a life where they can meet their basic needs and have the opportunity to achieve a high quality of life within cost ranges that they can afford.

I do not agree with economists such as Julian Simon of the University of Illinois, who argues that there can be no such thing as resource shortages because "the only meaningful measure of scarcity in peacetime is the cost of the good in question."[5] Such a view suggests that cost levels don't really matter; that it really makes little difference to society as a whole where the equilibrium is achieved in the economic equation.

That is simply not true. Costs are only neutral to economic technicians who use them as bloodless numbers fed into equally bloodless equations. To real people, in the real world, costs are important, often critical. The trends that are affecting U.S. farmland today could cause a doubling of the real cost of food over the next two decades. That is not an inconsequential change, particularly to the millions of people, both in the United States and around the world, who live near the margin of economic survival.

The real implications of food costs, and the price one must pay when he lacks the resources to meet them, were brought forcefully to my attention on a hot, dusty day in a rural farming village in the Gambia, a small African nation on the western horn of that vast continent. It was 1975, and the region was recovering from a serious three-year drought that had brought famine to much of the Sahelian region of Africa. The Gambia, on the southern edge of the Sahel, had not been as critically affected as the countries to the north, and food supplies were normal in the villages where we visited.

During a visit to a family compound to meet the family and take pictures, a young mother brought me a baby boy. Holding him out to me, she asked, in her native Mandinka, if I would make him well. He was perhaps 18 months old, but weighed no more than six-eight pounds, if that. He was near death, but probably not from starvation as such. He was dying from any one of the myriad diseases that take the lives of the malnourished. His breath came in labored, rasping sobs that racked a small frame that was little more than a skin-wrapped skeleton.

Helpless and shaken, I urged her to take him to the regional clinic to see the doctor. But she knew, and I knew, that there was no hope. And I understood, as never before, the awesome responsibility that we Americans hold, both as stewards of a land base that can produce bountiful supplies of food and as bearers of the technical knowledge on how to keep it productive. To waste those resources, or squander that knowledge, is not just to lose one small human, but to consign millions to an equally cruel fate.

2.

Resource Limits – A New Concept for the United States

As reasonable as it sounds on the surface, the idea that natural resources have limits, and that those limits make important differences to people and their social systems, is not easy for Americans to understand. We are a people that have been accustomed to surplus, and we have grown to take productivity for granted.

Because this nation has been blessed with the largest, most productive farmland base of any nation on earth, people have been misled by its very size and complexity to believe that it was limitless. That is like the "flat earth" theory in Columbus's time, however, only plausible until men had explored, measured, discovered. Yet even today, when we view our world from outer space and see its dimensions and limits much more clearly, there is a lack of widespread knowledge — and agreement — on those dimensions and limits. We know a great deal more about our land than ever before in history, but it is still huge, and understanding even a small portion of it is a major task.

There have been significant periods in the history of the nation when the concern for natural resources and their future spawned landmark conservation efforts. But while those efforts established new programs and protected the resource base in the past, they have not really prepared most people to face the realities of the present and the future. The protective concerns were, in the main, too short-lived, and too often countered by periods of surplus production which made the need for conservation seem insignificant. Too often, the programs were an attempt to encourage farmers to protect farmland by making it seem easy and profitable to do so.

15

In the process of making it simple, conservation efforts were made to sound trivial, so farmers paid little attention. In the meantime, non-farmers, knowing that conservation programs were in effect, assumed that the situation was under control and that the problem was solved. The story of these early efforts demonstrates the painful slowness with which Americans are willing to admit the idea of limits to our seemingly boundless land, and their willingness to forget conservation commitments quickly when the production pendulum swings to the surplus side.

Early Conservation Efforts

There were people that were aware of the value of the soil and the need to conserve it from the time that the first settlements were established in the new land, but they were few, and their voices went largely unheard. Jared Eliot, a Connecticut minister, doctor, and part-time farmer who lived from 1685 to 1763, carried out experiments on how to stop soil erosion and, in 1748, published a series of essays on his methods. He recommended the application of manure, except on sloping lands where it would be washed away by the rain. He also recognized the value of limestone and green manure. For pastures, he recommended manures, red clover and various grasses, calling red clover the most valuable crop for building up poor soils.[1] Even today, his ideas make a great deal of sense, but the fact that the farmlands of the eastern United States were essentially worked to exhaustion in little more than a century indicates that they had little impact on what happened to the land.

Solomon and William Drown of Providence, Rhode Island, were a father and son who jointly authored *The Compendium of Agriculture* in 1824. Recognizing that soil erosion was being hastened by continuous cropping accompanied by poor plowing practices, they recommended crop rotations and contour plowing as methods for keeping soils productive.[2] In the mid-1800's, the practice of horizontal plowing was a topic of wide discussion and debate, and in 1894, USDA's *Farmer's Bulletin 20* discussed eroded soils and how to reclaim them. But the problem did not gain national attention until the late 1920's. Until then, there was just too much new land to be had in the West, and it was too easy simply to wear out one farm and move to the next.

In the 1920's, a young soil scientist, Hugh Hammond Bennett of North Carolina, began writing and speaking on the menace of soil erosion. His soil survey experience had convinced him that erosion was causing serious and permanent harm to the nation. In 1928, Bennett persuaded USDA to publish Circular 33, *Soil Ero-*

sion, A National Menace. That publication drew public attention, and Congressman James P. Buchanan of Texas was moved to hold hearings out of which came, in 1929, the first federal money to begin a national survey of soil erosion damage. It was $160,000, to be used by USDA to investigate "the causes of soil erosion and the possibility of increasing the absorption of rainfall by the soil in the United States."[3] The soil erosion peril was, for the first time in the nation's history, an official concern.

In 1933, under the urging of Bennett, the Soil Erosion Service was established to carry out a study of soil erosion and to begin demonstration projects to show how erosion could be prevented. In 1934, just as the first national survey of soil erosion was being completed, dust storms hit the drought-stricken Great Plains and the term *Dust Bowl* was born. On May 12, 1934, a major storm hit the plains, later to be described by Bennett as a turning point in the battle to get public attention to the erosion problem:

> This particular dust storm blotted out the sun over the nation's capital, drove grit between the teeth of New Yorkers, and scattered dust on the decks of ships 200 miles out to sea. I suspect that when people along the seaboard of the eastern United States began to taste fresh soil from the plains 2,000 miles away, many of them realized for the first time that somewhere something had gone wrong with the land.
>
> It seems to take something like a disaster to awaken people who have been accustomed to great national prosperity, such as ours, to the presence of a national menace. Although we were slowly coming to realize that soil erosion was a major national problem, even before that great dust storm, it took that storm to awaken the nation as a whole to some realization of the menace of erosion.[4]

That "something" had gone wrong with the land was indeed evident in 1934, and the resulting public concern led to a great deal of public and private effort to remedy the situation. The Soil Conservation Service (SCS) was created as a permanent agency in the USDA by order of President Franklin D. Roosevelt on March 25, 1935. Only a month later, Roosevelt signed into law the Soil Conservation Act of 1935 (Public Law 74-46), which remains, having never been amended, as the basic statement of official soil conservation policy in the federal code:

> It is hereby recognized that the wastage of soil and moisture resources on farm, grazing, and forestlands of the nation,

resulting from soil erosion, is a menace to the national welfare and that it is hereby declared to be the policy of Congress to provide permanently for the control and prevention
of soil erosion and thereby to preserve natural resources,
control floods, prevent impairment of reservoirs, and maintain the navigability of rivers and harbors, protect public
health and public lands, and relieve unemployment.

But soil erosion didn't just go away, banished by legislative
fiat. The continuing seriousness of the situation was illustrated by
Hugh Bennett, when he brought the results of his new agency's
studies to the attention of Congress in 1939:

> In the short life of this country we have essentially destroyed
> 282,000,000 acres of land, cropland and rangeland. Erosion
> is destructively active on 775,000,000 additional acres. About
> 100,000,000 acres of cropland, much of it representing the
> best cropland that we have, is finished in this country. We
> cannot restore it. . .We are losing every day as the result of
> erosion the equivalent of 200 40-acre farms.[5]

*Not only soil, but people, too, were blown into oblivion when drought and
wind struck in southeastern Colorado during the Dust Bowl. These 1937-
1938 storms were instrumental in sparking early soil conservation programs. SCS photos by Thomas Meier (above) and B. C. McLean (right).*

The startling data produced by Bennett and his new agency convinced national leaders of the fact that protecting the physical integrity of our nation's soils is vital if we are to safeguard the national welfare. Lowdermilk said it eloquently: "If the soil is destroyed, then our liberty of choice and action is gone, comdemning this and future generations to needless privations and dangers."[6] President Franklin D. Roosevelt put it more succinctly: "The nation that destroys its soil destroys itself."[7]

Now after almost half a century of concern that has featured a national soil conservation program which far exceeds any other national effort anywhere in the world, along with a much more sophisticated data system that helps us understand the extent of the problem, soil erosion is still serious. In 1980, Secretary of Agriculture Bob Bergland estimated that more than 10 percent of the land under cultivation in the United States was eroding at unacceptably high levels of more than 14 tons of soil per acre per year. "If something is not done about that land," he said, "it will be barren."[8] Bergland's information came from the *RCA Appraisal,* which demonstrates that the United States is losing topsoil at a faster rate today than at any time in history, including the Dust Bowl era.

How could such a thing be possible, in the face of the concentrated conservation efforts ongoing since the 1930's? Where had the conservation programs gone wrong? Where had the money invested in conservation gone, and what, if anything, had it accomplished? Those were important questions being asked by members of Congress, environmentalists, farmers, and citizens alike.[9] Part of

the answer, of course, is that the conservation programs had been working, but at a very low level of funding and activity. While the task of protecting farmland had been growing more and more difficult, the amount of money directed toward the task had actually been shrinking in purchasing power. One of the main reasons for that was the widespread feeling that the problem really did not exist, or at least, was nowhere near as severe as Hugh Bennett had professed it to be. After all, there was no shortage of land or food. There was, in the minds of many, too much.

Threats of Scarcity in the Face of Abundance

There have always been, and still are, valid differences of opinion as to whether or not the people:food:land balance is a real dilemma. Neo-Malthusians think there is relevance in the admonitions of the eighteenth-century monk Thomas Malthus, who pointed out that while there is a limit to which the productivity of the land can be pushed by science, technology, and investment, there is no limit to the amount of demand that can be created by unchecked population and economic growth.[10] Malthus, in his *Essay on the Principle of Population,* published in 1798, proposed that while population grew at a geometric rate (2, 4, 8, 16. . .), agricultural production could be expected to grow only at an arithmetic rate (1, 2, 3. . .) as land was added to cropland production. The nineteenth-century economists who followed him (David Ricardo was one of the best known) used this idea as a basis for many of their economic theories.

Scientists and agriculturalists have periodically joined the chorus. In 1898, Sir William Crookes, president of the British Association for the Advancement of Science, predicted a world wheat shortage by 1931.[11] Stuart Chase, a prominent American author, wrote a classic book in 1936 entitled *Rich Land, Poor Land.*[12] Chase had been on the staff of several government agencies and national organizations concerned with food and resources, then had prepared a special study for the National Resources Planning Board. He blended the technical data available from the resource studies of that era and demonstrated the relevance of natural resources and their condition to the current national crises—the flood of 1936, the lingering Dust Bowl drought, and a heated presidential campaign. Chase argued that attending to the needs of the land would help Americans address two vitally important problems: the short-term need to provide employment and the longer-term need to protect the land for future generations. He also demonstrated graphically how the land could be rendered virtually worthless if the trends of the day were allowed to continue.

In 1967, William and Paul Paddock used their lifelong experiences in agricultural research and foreign affairs to coauthor a book called *Famine: 1975.* Their projections of world trends in agricultural productivity and population growth foresaw a "population-food collision," which they called inevitable, some time around 1975. They predicted that "ten years from now parts of the undeveloped world will be suffering from famine. In fifteen years the famine will be catastrophic and revolutions and social turmoil and economic upheaval will sweep areas of Asia, Africa, and Latin America."[13]

Overstated? Perhaps, particularly in light of the five consecutive world bumper harvests from 1975 to 1979. But sobering, at least, to observers of the 1974 world food crunch and the human misery brought on by drought in the African Sahel, war in Cambodia, and poverty in many corners of the world. Not too far from reality, one might note, in 1981, when food security in many regions of the world hangs on a slender thread that is absolutely dependent on favorable weather conditions.

The most dire of the events predicted did not occur of course—at least not in the form, magnitude, or at the time that they were foreseen. The world has been able, as Figure 2.1 shows, to keep raising agricultural productivity as rapidly as the number of people has grown. Even in the developing countries, agricultural production per person has registered slight gains, although most of the progress made in the past decade was apparently lost in 1979. The graphs in Figure 2.1 show averages, however, and tend to hide the fact that major portions of developing Africa and Asia are suffering a serious decline in food production per person. Nonetheless, there are observers who look at the trends such as those shown in Figure 2.1 and are reassured that there is no need for concern. Keith Campbell, an Australian agricultural economist, says that "the progress made. . .augers well for agriculture's ability to cope with the demands ahead. . ." and argues that additional attention to agricultural research and the application of new technology will more than offset future needs.[14]

Julian L. Simon, an economist from the University of Illinois, has been widely published in recent months with his theory that it is a shortage of people, not land, that is the real constraint to human progress. Writing in *Science* magazine, he argues that the world supply of arable land has actually been increasing, energy and food have been getting more plentiful and less costly, and the U. S. environment has been getting better, not worse. He incorporates these past trends into a computer model that demonstrates, he avows, that as population increases, so does productivity; thus the best thing that could happen would be to add to the supply of people who would, in turn, add new inventions, technological advances, and knowledge.[15]

The public, of course, is confused. Who is right? Are the scientists who warn of coming crises to be believed? Or are they, as Simon suggests, simply spreading phony information because "bad news sells books, newspapers, and magazines; good news is not half so interesting."[16] Are the folks who say there is no need to worry feeding the public false reassurances and thus delaying the initiation of preventative or corrective action? Unfortunately, that debate is not easily settled, because we all guess, to some extent, when we predict the future. To people who base their predictions, or focus their memory, on the experience in the United States in recent decades, Simon's thesis sounds reasonable.

The United States:
A History of Abundance

Throughout our entire history, this country has been a notable exception to any problem of land resource limits. Ever since the North American continent was inhabited by European settlers, the notion of a limit on land supplies has been simply unthinkable. There was always land to be had—often just for the plowing. And it was fertile land, for the most part, with rich soils and a climate well suited for growing crops.

The fact that the best soils were plowed out first as settlers spread across the nation was of little consequence; there were other lands where only slightly more investment was needed. Thomas Jefferson thought it foolish to fertilize on his Virginia plantation, because he could buy an acre of new land cheaper than he could manure an old acre.[17]

During this same period, national policies hastened an agricultural revolution. Homestead laws made land available and created family farms. Congress established land grant colleges in every state, and the USDA provided national leadership, research, and education for farmers. A transcontinental railroad system carried people west to new lands, and then brought the products of their labors back to populous eastern markets.

Figure 2.1. Changes in World Agricultural Production, 1960-1979

% of 1961-65 average

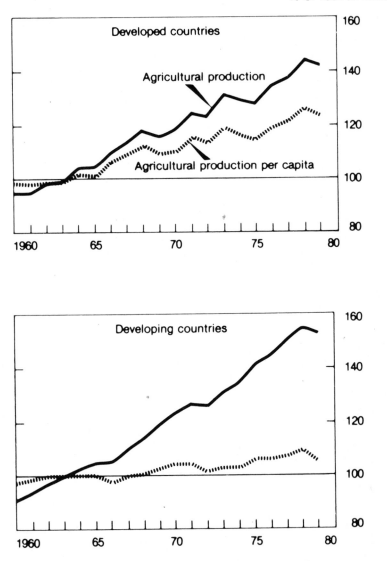

Developed countries include the United States, Canada, Europe, USSR, Japan, Republic of South Africa, Australia, and New Zealand. Developing countries include South and Central America, Africa (except Republic of South Africa), and Asia (except Japan and Communist Asia).

SOURCE: USDA

In the arid West, desert soils with virtually no agricultural value in their natural state were converted to high-yielding gardens by water storage and irrigation projects. The investment required, although significant, was readily undertaken by people whose goal was to increase productivity. Where private capital was not sufficient for the task, huge public works were undertaken to "reclaim" the desert and establish agriculture.

In the wet soils of the Corn Belt, farmers learned that drain ditches and, later, tile drains could keep the water table low enough in the soil to allow maximum crop growth. Like their western counterparts, they banded together in community drainage schemes when the task required. Again, the goal was U.S. agricultural productivity, strongly valued by Americans as both a virtue and a necessity.

World War II brought another revolution to American agriculture. Urged on by labor shortages, high prices, and the need for food to rebuild much of the world, farmers greatly accelerated the mechanization of their farms and added new techniques that would intensify the use of each acre. In addition to newer, more powerful tractors that would cover more acres, the focus shifted to methods of gaining a high yield from *each* acre. These methods included commercial fertilizers and pesticides, improved seed and livestock, and expanded use of irrigation water. As a result, U.S. crop production per acre in 1970 was 48 percent higher than it had been in 1950.[18] American agriculture was widely cited as living proof that the free enterprise system worked. In fact, the "farm problem" facing national policymakers rested not on the fact that U.S. agriculture didn't work, but that it worked too well.

Conservation was a hard case to promote. I remember one of my early attempts to link soil erosion to the loss of productivity. I could see the damage being suffered by local soils in the Idaho drylands, and knew that there was little moisture-holding capacity left in some of them. But when yield records were compiled, it turned out that crop yields had been steadily rising during the period of time when soil erosion had been steadily reducing the topsoil thickness. If that data were carried to its illogical conclusion, it would have demonstrated that maximum yields could be anticipated when the soil was completely gone! The idiocy of such a conclusion was apparent, but, in a larger sense, that is what the American people were experiencing. Soils were being wasted away, but crop yields per acre were climbing. Was soil erosion really the problem it had been made out to be? The evidence of the day was, to say the least, mixed.

Farm Policy: Fighting Surplus Production

The development of U.S. agriculture was so successful that the major agricultural policy issue in America following World War II ceased to be production; in fact, the opposite worry emerged. American farmers, spurred to expanded efforts during the war and fueled by an avalanche of technical improvements in plant genetics, fertilizers, pest control, and farm equipment, flooded the market with food products. Beginning in the early 1950's, U.S. farm policy was concerned with little else than ridding the nation of "burdensome surpluses" of commodities. As Richard Barnet says, "From the days of the New Deal when little pigs were slaughtered to keep them off the market, a major task of the Department of Agriculture has been to keep food production down in order to keep farm prices up."[19]

This federal policy had strong support, both inside and outside the farm community. The evidence was clear: American food production capacity could rise to heights far beyond any imagination. Land might not even be needed for food production before too many more years elapsed. Productivity, rather than being a virtue, was a burden. American farmers could glut any market, any time. At the very least, America would never face a land shortage in the foreseeable future. It was generally recognized that this would require a continued string of scientific discoveries and their rapid adoption by most farmers, but it was easy to see how that could happen. It had happened before and could just as easily happen again.

In 1967, Secretary of Agriculture Orville Freeman told his audiences, "Agriculture is alive with change. It is a force without bounds. . .without limits." In a series of major policy speeches entitled *"Agriculture 2000,"* Freeman told of the futurist thinking that was then taking place in USDA. Ideas like computer-controlled farms, robot harvesters, automated field work, and climate control were popular. Reviewing what all this would mean to farmers, Freeman noted that "by the year 2000, optimistic visionaries say, this interplay will have become so successful that yields of today will be doubled or tripled. . .that corn yields, for instance, could run from 200 to 400 bushels to the acre."[20] Obviously, compared to the 100-125 bushels of corn per acre that are still considered a good average, this sounded as though all the questions of production had been answered, or soon would be.

The public policy response was predictable. Federal funding support for agricultural research budgets fell off. Why spend limited public dollars finding new ways to grow more crops? We were drowning in piles of surpluses. Why spend money to encourage soil and water conservation? Why worry when cities, highways, airports gobbled up prime farmland? Land was not limited, and there was no likelihood it would be at any time soon. As agricultural economist John Timmons of Iowa State University points out, American public opinion and policy spent virtually the entire half-century from 1920 to 1970 (with the exception of the war years) trying to reduce crop surpluses and the cropland that produced them.[21]

What Freeman and his USDA experts couldn't—or didn't—see in 1967, of course, was the terrible vulnerability that was, even then, being built into American agriculture, with the enthusiastic aid of USDA policies and programs. Foreign oil, at low prices, made it easy to develop and encourage the adoption of intensive crop management techniques that were not only high-yielding, but high-risk. By 1975, however, the situation had dramatically changed. An Arab oil embargo drove the price of Middle East oil from $2 a barrel into the $30-$40 range. In 1972, the Russians purchased 28 million tons of cereal grains—18 million tons from the United States—and wiped out the American farm surplus.[22] But, even more critical, agricultural scientists had watched short crops in 1970 (due to a serious outbreak of corn pests) and 1974, when bad weather kept crop yields down.

In 1979, Secretary of Agriculture Bob Bergland told Congress that American productivity could no longer be assumed to stay high year after year. He noted that "between 1963 and 1977, there were significant shortfalls in the level of corn production during five crop years. . . ."[23] In addition, he pointed out that the long-term prognosis was for more of the same, due mainly to the widespread emergence of the oil-based agriculture that USDA and the major agricultural corporations had been so energetically promoting: ". . .the volatility of farm production levels is *increased* by intensive agricultural practices. For example, heavy use of nitrogen fertilizers to increase corn yields has also increased the amplitude of fluctuations in corn yields due to weather variations, since positive yield response to nitrogen is dependent upon favorable weather conditions."[24]

During those same years, while the vulnerability of American agriculture was being so graphically demonstrated both by whiplashing commodity prices and by angry farmers driving tractors around USDA headquarters in Washington to protest their plight, the market for U.S. farm products was steadily growing. From 1970, when around $7 billion of farm produce was sold abroad, the dollar value of exports rose five-fold, to $35 billion in 1979.[25] In 1980, in spite of a grain embargo imposed against the Russians, exports rose to over $40 billion, and American grain farmers sold 40 percent of their grain abroad.[26]

As a result, American farmers enjoy a steadily expanding market, both at home and abroad. Ironically, the market keeps growing larger and more demanding at the same time as U.S. agriculture's vulnerability to soil erosion, farmland loss, weather changes, petroleum supply disruptions, and runaway inflation has also been growing. American farmers today face a more demanding market, with a reduced capability for flexible response.

Rethinking Malthus

Although the unbroken record of technological advances in agriculture over the past centuries has demonstrated that Malthus's concerns about the people:food crunch were largely without foundation, that history may be misleading as the sole prescription for thinking about the future.

It is possible that Malthus was more right than wrong.

His timing was off and he couldn't foresee the ways in which science and technology would add a "multiplier" effect to the productivity of the land, but his basic idea bears closer scrutiny. It is time to rediscover Malthus, but apply a more complete set of elements to his gloomy prediction. We might rephrase it as follows: If population and food consumption rise on a geometric scale faster than productivity can be increased on the available agricultural land through the application of science, technology, and husbandry, then shortages of food and other essential products from the land will become constraints on consumption and, at some ultimate level, the number of people.

Such a restatement gives us four variables to calculate instead of the two used by Malthus. In addition to the number of people, we add the amount each consumes. Both tend to rise in geometric progression, but consumption can, and often does, go down sharply in response to short-term shortages or price changes. To the amount of land available for agriculture, which can only rise arithmetically as new acres are added to the inventory, we add the effect of technology which, like population, has demonstrated a tendency to rise geometrically over time.

To illustrate: If a population is growing at the rate of 2

percent per year, it will double in something like 35 years. If agricultural yields rise by a similar 2 percent per year, and if they are fairly dependable every year, no new agricultural land will be needed in order for the population to maintain its food security. But let technology cease to raise yields effectively, and the need for new land becomes important, perhaps critical. Such a growing population would, without any yield increases, need to double its cultivated land every 35 years, plus add similarly to the number of farmers and the amounts of fertilizers, farm animals, and other farm inputs. For most countries, limited supplies of arable land would preclude any such expansion.

As we have seen, the experience of recent years has been one of rapid population growth matched by equally rapid gains in agricultural productivity. So observers can look at that segment of history and suggest that it is little, if any, trouble to keep agricultural yields rising at least as fast as population. With Julian Simon, they can call the notion of an increasingly serious food situation in the developing nations a prime example of "false bad news."[27]

But we may be in a period when historic comparisons — even recent ones — no longer hold. Consider: In the first 1,500 years of our current calender, the number of people on earth grew at the rate of 2-5 percent a *century,* but in recent decades many countries have been growing at rates of 3-5 percent a *year.*[28] Figure 2.2 shows how this looks on a graph, with the rate of growth accelerating to the point where it is now climbing at an almost vertical angle.

Many observers, looking at this trend and where it is heading, come away with reservations. Keith Campbell, despite his optimism and faith in agriculture's capability, says, "Though there is evidence of overall long-term improvement, the rate of improvement in the developing countries is not great enough to give any grounds for complacency."[29] Lester Brown of Worldwatch Institute, a long-time observer of land and food issues abroad, paints a far bleaker picture. "As world population moves toward 5 billion, humanity is moving into uncharted territory in the relationship between population size and the earth's natural systems and resources. If population continues to expand rapidly, it may well lead to a decline in the level of living for a large segment of humanity."[30]

In 1975, world population was estimated at 4 billion people; 214 million of whom lived in the United States. In the United States, a declining birth rate was often cited as a signal that, at least in this and other highly industrialized nations, the "population bomb" was defused. Forecasts of serious demands on the world's resources had been proven false, it was argued.

Figure 2.2. World Population Trends

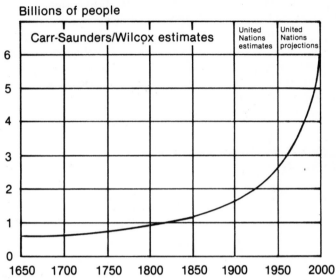

SOURCE: *Donald Bogue,* Principles of Demography *(New York: John Wiley & Sons, 1969) (Reprinted with permission of the author)*

But declining rates of population growth do not ease the pressures on land and food as much, or as rapidly, as one might think. Numbers tell a different story. Between the late 1920's and the late 1970's, about 100 million people were added to the U.S. population. Now, despite the fact that rates of growth are slowing, the Census Bureau's medium-growth projections predict that in 2030, the total U.S. population will be 300 million.[31] Thus, in the next 50 years, the U.S. population may be 75-90 million more than today, even though the *rates* of growth will have slowed.

The situation, of course, is due to the larger total number of people. A slow rate of growth in a large population creates as many new mouths to feed as a faster growth rate in a smaller group. So, despite a reduced rate of population growth, the demand for food and other products in the United States will continue to grow about as rapidly as before, since it is a demand created by *numbers* of *people* and their purchasing power, not by *rates* of *growth.*

Demands for food are created not only by numbers of people, but also by what they can afford to eat. In the United States, food consumption per person has been rising at a fairly steady rate over the past decade, as Figure 2.3 shows. USDA predicts a continuing upward trend in food consumption, but I have a hard time seeing how that will happen. Middle and upper classes of Americans seem to suffer more from obesity than from under-nutrition, and rising food prices will reduce the amount that poor people can afford. Those assumptions would suggest a lower average consumption per person in the future. My guess is that the "flattening" of the consumption line in Figure 2.3 that started in 1976 might be the start of a longer-term drop that will mean less, not more, food purchased by the average American.

Figure 2.3. Population and Food Consumption in the United States

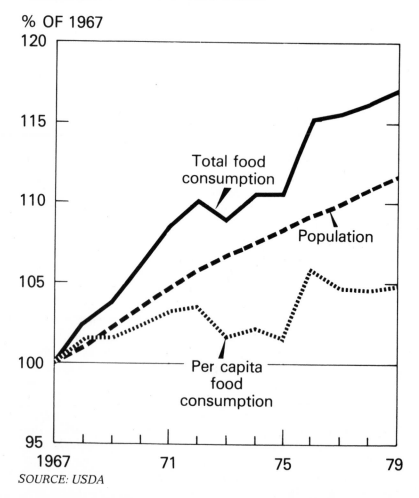

SOURCE: USDA

Debates about population growth in the United States, and its effects on land and food, should be tempered by the realization that our situation will be dwarfed by the growth in other areas of the world. There are four times more people alive today than all the humans who ever lived on earth from the time of creation until 1830.[32] In the *Global 2000 Report to the President,* a study of international trends released in 1980 by the Council on Environmental Quality and the Department of State, a sobering world picture for the future emerges.

This report looks at the effects of continuing today's trends to the year 2000. The result is a forecast for a population of 6.35 billion by 2000, up some 55 percent from the 4.1 billion of 1975. This estimate results from a "medium-growth" projection that assumes a continuation of present governmental policies, fertility trends, and economic market mechanisms, as well as the absence of a major war. Even the "low-growth" projection under these assumptions indicates a 46 percent increase by 2000 — to a world population of 5.9 billion.[33]

The implications of these projections are that, by 2000, the number of people will be increasing faster, in *absolute numbers,* than it is today. Put it into perspective this way: In 1930 there were 2 billion people on earth. We have added another 2 billion since then. In the next 20 years, we will add still another 2 billion.

Most of the problem lies in the less developed countries (LDC's), where within the next 20 years, the situation will grow much worse. The reason is, simply, that most of the population growth over the next 20 years is liable to occur in the LDC's. The *Global 2000* report estimates that 92 percent of the new people that will be added will live in those regions. By 2000, the report says, almost four out of every five people will live in LDC's.[34]

Nothing short of disastrous wars, famines, or plagues can slow this trend. The age distribution of world populations today, particularly in the LDC's, shows a very high percentage of young people with childbearing years ahead of them. Even though demographers expect worldwide fertility rates will drop over 20 percent — from 4.3 children per fertile woman to 3.3 — the fact that there were over three-fourths of a billion women under the age of 20 in the LDC's in 1975 guarantees a continued high rate of births into the immediate future.[35]

In addition to rapidly growing numbers in the LDC's, the *Global 2000* report indicates that people will be increasingly concentrated in urban areas. If present trends were to continue, Mexico City would have over 30 million people by 2000, and 400 cities would have over 1 million. Table 2.1 shows what the future may hold for some of the rapidly growing urban areas of the world. Where are the food, energy, housing, water, jobs, and other elements of a quality life for this many more? Do we have it to spare today? Can we develop it fast enough?

Table 2.1. **Estimates of Selected Future Urban Agglomerations in Developing Countries**

	1960	1970	1975	2000
	(millions of persons)			
Calcutta	5.5	6.9	8.1	19.7
Mexico City	4.9	8.6	10.9	31.6
Greater Bombay	4.1	5.8	7.1	19.1
Greater Cairo	3.7	5.7	6.9	16.4
Jakarta	2.7	4.3	5.6	16.9
Seoul	2.4	5.4	7.3	18.7
Delhi	2.3	3.5	4.5	13.2

SOURCE: *Global 2000 Technical Report*, Table 13-9.

Feeding the World

No matter how far this growth proceeds, the implications of growing urbanization in the LDC's will mean that the bulk of the world's people will be dependent on food grown in another area and transported to them. Whether that food comes from their own farm regions or from another country will be less important than the urgency of simply acquiring enough, so these urban people will push their leaders to obtain adequate food supplies from abroad as needed. That, of course, means more food imported from the United States, since we are the only source that can even begin to provide the quantities involved.

In developing a "moderate" projection of exports in the future, USDA assumed that countries overseas would continue their existing attempts to improve the quantity and quality of their people's diets, but would reduce their purchases as they developed their own capability to meet internal needs. In addition, USDA predicted that exports would slow down after 2000, as higher production costs forced higher prices and dampened demand.[36]

The problem with these projections—or any other projec-tions of world food demand, for that matter—is that they are almost hopelessly speculative. In 1981, actual exports will exceed the official USDA predictions for the year 2000—and those predictions were made as recently as 1979! The rate of growth has far exceeded expectations, and will, if it continues, rapidly outstrip the most optimistic estimate of U.S. production capability.

My reading of all this inconclusive calculating is simply this: The capacity of the people of the world to produce more people, and the ability of those people to purchase food, is rapidly approaching the functional limits of the global food system. I can't prove that conclusion with numbers, because all the numbers still seem to point to the conclusion that we should be able to feed today's world. But even a cursory look at what is happening around the world today demonstrates conclusively that we are not doing all that well. We need to take our current performance as a starting point, then see how we rate the odds of feeding the additional people that the demographers assure us are coming.

Some scientists have claimed that the earth could support many times today's population—up to 60 billion, perhaps. Who knows the correct answer? We have already passed some of the population milestones that were the basis of earlier neo-Malthusian concerns. We may pass Lester Brown's 5 billion target without noticing too much has changed. But does that mean that there are no limits, or that those limits are so far ahead as to be meaningless? I don't think so. I think the time has already come when limits are beginning to intrude in our everyday lives.

Since 1973, when petroleum started the parade, a steady chain of events has signaled the growing importance of resource issues to human well-being. Today, a trip to the gas station, a ride in a taxicab, or a swing through the supermarket provides a graphic lesson in the laws of supply and demand: As supply goes down—or demand goes up—prices go higher.

The post-1973 petroleum situation has given us $1.30-per-gallon gasoline, but it has barely begun to intrude into our food budgets. By raising real costs on the farm and in the transportation and processing industries, it has raised food prices, but only at about the same rate as the general rate of inflation.

The real Malthusian people:food:land crunch, if it comes, will exert a much more direct and dramatic multiplier on food prices. If that happens, public attention will quickly focus on the problem, and there will be a new level of concern for farmland and its future. Such a concern, when it happens, will be a new experi-ence for the people of the United States.

American farmers don't think now in terms of limits to food production any more than most consumers do. Farmers have eagerly sought, and continue to support, the all-out efforts to increase agricultural exports. One of the main factors behind farm resentment of the 1980 Russian grain embargo was the feeling that the United States would need to remain a dependable supplier in order for buyers to commit to large purchases every year. Shutting off the flow, for whatever reason, interrupts the market flow and forces customers to other buyers. American farmers are still convinced they are dealing in a "buyer's market" when it comes to food and fear that their behavior as a "seller" is the most important factor in creating good markets for the future.

With the "slack" gone from U.S. agriculture, and virtually no slack in existence anywhere else in the world, a steady rise in domestic and export demand should result in strong prices for most American farm commodities, particularly those such as wheat, corn, and soybeans, in spite of any harm caused by the embargo. The natural tendency—of both farmers and farm policymakers, will be to move aggressively to meet new market demands.

That strategy is not without its hazards, however. As Iowa State economist John Timmons has noted, "Foreign demands for our agricultural products are very fickle and unreliable, depending upon such events as a drought in Russia, the Anchoveta catch in Peru, or a political change in China. Yet our agricultural production cannot readily be turned on and off in response to such changes. Such rapid changes leave in their wake waves of inflation for consumers and recession for farmers."[37]

It appears logical that the coming resource limits will be translated into higher food prices. This will, farmers argue, bring food more nearly into line with the other basic commodities people must buy, but it is doubtful if consumers will accept that logic—and those price rises—gladly. Even if they only spend an average of 17 percent of their income on food, that cost seems high, and the average is deceiving. For many millions of Americans in the lower middle class, the percentage is much higher. Inflation on all fronts has eaten into their purchasing power and any additional price rises, particularly in food, will be bitterly fought. Those price rises are liable to pit consumers against farmers in a battle that will make reasonable discussion about long-term farm and food policies virtually impossible.

The implications for topsoil in the United States are grim. High prices provide incentives to plow up marginal land, harvest a crop or two at peak price, then get out, leaving the land to erode.

Such practices were common in 1973-1974, particularly in the Great Plains and the Corn Belt, and are starting to be seen once more, a factor we will discuss at more length in the next chapter. Farmers often argue that high prices will elicit higher levels of soil conservation because they create the profits needed to install conservation practices. That is true, but only to a certain point and for certain kinds of farmers. Conservation-minded farmers who have put off investments in conservation practices or structures may, indeed, build more of them when their profit picture brightens.

But marginal farmers trying to eke out a living from marginal or unsuitable land, or industrial farmers trying to maximize short-term profits, will respond to high prices by growing as much as possible, even when that means having to farm the land too intensively. Since there are growing numbers of farmers in these situations, there is good reason to forecast that, as growing foreign demand strengthens farm prices, it will also accelerate the wasting of America's farmland. The United States is mining its soil to generate dollars to trade for oil, and in the long run, we will get by far the worst of the bargain.

3.

Pressures on Farmers to Exploit the Land

Not all farmers are mining their topsoil, but too many are. A good number of those who are exploiting their land know that what they are doing is wrong and would like to change their methods but feel they can't. The reasons they give are many, and they are valid. If the nation is to devise cures for today's soil and water ailments, those cures need to address the real pressures causing farmers to waste land and water, not just the symptoms of the disease.

Although it is too seldom recognized in the rarified atmosphere of Washington, D.C., when national conservation programs are discussed, the fact is that public programs don't save soil or manage water. Farmers do. They manage those resources as part of the day-to-day work of their private business. In soil conservation, as in crop, livestock, or family financial management, they will do what they have the knowledge and skill to do and the equipment to carry out, and what seems, in their own private calculation of costs and benefits, to be the "right" thing to do.

This is a matter of considerable importance and some delicacy. If a hardware store owner uses all his current receipts to support his family instead of investing to keep an adequate inventory, it doesn't take a Harvard Business School graduate to tell him that he won't last long in the business. A farmer who fails to invest in keeping the soil productive is mining his business inventory just as surely. But, in general, society seldom cares whether or not a hardware store goes out of business, so there aren't any overt public pressures on store owners to conserve their capital. They can fail if they choose.

Farmers are different, for their business does not exist within four man-made walls that can be as easily used for another purpose or rebuilt at another place or time. If a farmer fails because he has ruined the land, the ability of many other persons to succeed has also been lost. If too many farmers fail, the ability of the society to survive has been lost. That means the success of the farmer's business is, in part, everybody's concern. So the farmer, who prides himself in being a rugged individualist, is not totally free from restraints on his business, and he often resents this. His anger is often directed at "government" in general and finds expression in the widespread view that "they" or "those faceless bureaucrats" or "politicians" don't understand his situation and have no right to "meddle" in his private business.

The trouble with that idea is that, while it may have worked well in the days when most, if not all of the farmer's business was conducted on the local level, it runs into serious strains now that the farm business has gone global. Today's American farmer is not just a food producer, he is an important cog in the nation's economic and foreign policy. He, and the land he farms, are expected to perform well so that the nation can meet its commitments at home and abroad.

Those kinds of demands are not likely to be met unless there is also a measure of public support and understanding for the conservation and management investments that will have to be made by farmers. That means, in short, some public support for the fact that farmers cannot simply extract from the land; they must constantly reinvest in it as well. That will take both effort and dollars, and those dollars must eventually come from either the consumers of the farmers' products or the public. Either way, public support will be critical.

Lack of Public Support

Effective support for conservation will not be easy for Americans, for we may be no better prepared to accept the discipline required to achieve a sustainable agriculture and an economy based on renewable resources than we were in 1850, or 1930, or 1970. Secretary of Agriculture John R. Block sees the conservation dilemma in this way: "Americans are . . . people of surplus. We consider it our heritage and our right. We see no limits, recognize no boundaries. Yet in taking bounty for granted in our own lifetime, we may deny our children."[1]

Most Americans have left the land, and few feel real roots back to the soil. Most of us live in cities, where food prices are far more the result of transportation costs and our own penchant for having everything trimmed, packaged, frozen, and premixed than they are a reflection of the costs of production on the farm. There may be a widespread desire on the part of city dwellers to "get back to the land," but that seems to be more a reflection of wanting to escape the pressures, high taxes, and other costs of urban living than a desire to grow food or be a farmer.

In the minds of most American farmers and urbanites, farming is no longer accompanied by a special attachment to the land. It is simply one industry that just happens to use the land. That may make us insensitive to the plight of the farmer who feels forced to waste land. As Wendell Berry points out, we have lost much of our sense of the "culture" involved in agriculture as farmers spend less time and attention on the husbandry of land, plants, and livestock, and more on the details of finance, investment, technology, marketing, commodity trading, and land speculation. The farmer today is a manager. He handles cash flows, machinery purchases and maintenance, labor relations. His time to concentrate on the subtle signs that indicate soil damage may be lessened as a result. To many observers, that is no cause for concern, but to Berry it is a sign that "the economy of money has infiltrated and subverted the economies of nature, energy, and the human spirit."[2] In other words, if farmers are not sensitive about the needs of the land, the general public will be even less so.

In recent years, many observers of farm politics have felt that farmers are losing their economic and political clout in American life, as well, and can no longer control the public policy agenda that affects their lives. When Don Paarlberg, chief economist for USDA during the Hardin-Butz era, constructed a list of the major policy issues that have dominated the recent scene in Washington, he found that most of the issues were generated by non-farmers, not farmers or USDA. The list starts with food prices, welfare programs such as food stamps, food quality, environmental protection, rural development, grain embargoes, civil rights, occupational safety and health, collective bargaining, and a host of others. As a result of his experience, Paarlberg concludes that "the agricultural establishment — the farm organizations, the agricultural committees of Congress, the Department of Agriculture, and the land grant colleges — had lost control of the farm policy agenda."[3]

One of the reasons for this turnover may be the declining political power of farm people. Back in the early 1930's, they made up slightly over one-fourth of the U.S. population. By 1978, that percentage had dropped under 4 percent and was still going down. This story, along with the current statistics about the dwindling number of Congressmen whose districts are dominated by agricultural interests (by 1980 this was something like 35, but the new census will no doubt lower it further), are often used as the basis to suggest that farmers have a difficult time getting the kinds of programs and policies they need from the national level.

Despite the fact that farmers have found themselves pitted against, at one time or another, powerful coalitions of labor, environmental, and consumer interests, the recent record may not be all that bad. John Kramer, the Associate Dean of the Georgetown University Law Center in Washington, D.C., and a regular consultant to the House Agriculture Committee, feels it is, if anything, slightly in favor of the farmers. "The hard truth is that the way the policymaking process actually has functioned in the recent past is more progressive and favorable to the farm than the farmer has been led to believe," Kramer asserts.[4]

But the hard fact still remains that farmers feel like a politically isolated and endangered group. They tend to view the initiatives brought by other groups with hesitation, if not out-and-out resistance. Years of confrontations with environmentalists, consumerists, and laborites—on issues where the farmers were largely placed on the defensive—have conditioned that response.

That kind of political confrontation will have to be avoided in establishing a new policy agenda on soil and water conservation.

Protecting farmland will, naturally, need to be done by farmers. But the costs, in large measure, must be willingly borne by non-farmers, either in the cost of their food products, in government-funded incentive programs, or in some combination of both. The kind of mutual agreement and broad consensus among many groups that will be required for the development of such an approach will certainly not come out of distrust, poor communication, and confrontation. Instead of public *pressure* to stop wasting land, farmers will need public *support.*

Today's farmer is no longer isolated in terms of his business—he is part of a global food system that affects the economic and political life of the entire world. The public will expect him to produce the food needed to keep the nation's economy healthy and its foreign policy on track. And, increasingly, that is no small matter.

Farming for a Global Market

The American wheat, corn, or soybean producer is, by any measure, the most "international" farmer in the world, and the growth of the international market is the primary cause of the production pressures facing him in the future.

Wheat, feed grains, and soybeans are his major export crops, with 60, 26, and 56 percent, respectively, going into trade channels.[5] In dollar value, soybeans are America's second largest single export, exceeded only by jet aircraft. Cotton and tobacco, two major non-food farm crops, also are heavily traded on overseas markets. In 1979, over half of U.S. cotton was sold abroad, and USDA reports that 38 percent of the 1978 tobacco crop went into export trade channels.[6]

The unique impact of U.S. agriculture on world trade is indicated by a look at the top five grain-exporting nations in the year ending June 30, 1980.

According to USDA, they were:[7]

United States	111.5 million metric tons
Canada	19.9
Australia	19.6
France	17.1
Argentina	11.4

(A metric ton is 2,205 pounds.)

As can be readily seen, the U.S. role in the world market is almost six times larger than its closest competitor, and 60 percent greater than the next four largest exporters combined!

These overseas grain markets have little to do with "feeding the world's hungry people," however. The big buyers are countries who have reached an adequate income level to be able to afford to upgrade their diets. Economist John Timmons of Iowa State University points out that, in the past two decades, it has been rising affluence, rather than rising populations, that has fueled the increased demand for imported farm products.[8]

The American producer's point of view was expressed very well at a National Farm Summit held in 1979. There, Carl C. Campbell of the National Cotton Council of America pointed out that feeding hungry people, while certainly a laudable goal, was not the basic reason why we export food and feed. "We need to export our agricultural commodities," he said, "in order to have a viable agricultural program in the United States, in order to provide work and income to our farmers and exporters, and in order to assist our country in maintaining reasonable balances of trade and payments."[9] To much of U.S. agriculture today, that truth is readily apparent, but so are the risks involved.

A weather situation or a political decision, anywhere in the world, can affect U.S. agriculture severely, as the events of 1980 show so graphically. Farmers in the Wheat Belt of the Great Plains planted the 1980 crop in September and October of 1979. Market outlook was strong, and there were no federal acreage restrictions on wheat to curb production. Wheat prices at planting time were around $3.75 per bushel, with most farmers anticipating prices to hold at about that level by harvesttime.

In December, the Soviet Union invaded Afghanistan, setting off a clamor of world rebuke, but little in the way of real retribution. In the United States, President Jimmy Carter searched for ways in which the United States could sanction the Soviets or attempt to get them to back down. Despite USDA's protests that an embargo would bring down the anger of farmers, Carter imposed an embargo on grain shipments above and beyond the 8 million metric tons contained in the five-year trade agreement between the United States and Russia. The effect, it was calculated, would be to reduce U.S. grain shipments by some 17 million metric tons. This would, foreign policy analysts hoped, be enough to dampen Russian plans for increasing meat production and bring noticeable political pressure on the Soviet leaders.

We don't know what happened in Russia, but the move caused panic in U.S. grain markets. Trading on the Chicago Board of Trade was suspended for two days to let the situation cool, but wheat prices dropped the maximum allowed (10 cents per bushel) for two days after the market reopened. By mid-May, the price of wheat had dropped to $3.40 a bushel from its January level of $3.95, and wheat farmers faced a far grimmer harvest than they had hoped for.[10]

In June, when the first of the crop began to roll in, two facts were apparent: There was going to be a good crop—perhaps a record in the early wheat regions—and the price was going to be terrible—about $3.25 a bushel. Farmers faced with payments on operating loans or other debts had little choice but to sell their crop at a loss of some 75 cents a bushel below the cost of production and curse Carter for imposing the embargo.

Much of the farm press was highly critical of the grain embargo, and farm state senators introduced a bill rescinding it. They claimed that the embargo did little to affect the Russians, since they were seemingly able to buy what they needed from Argentina, Canada, Australia, and other grain exporters.[11] The penalty, they claimed, had been unfairly laid on American farmers, who had lost a $3.5 billion market they badly needed and suffered, as a result, staggering economic losses. Senator Larry Pressler of South Dakota thought it was obvious that the embargo had "cost farmers and the American economy plenty."[12]

But the President stuck to his decision, and for those farmers who held their grain, the story wasn't finished. A serious drought had been in progress through the summer, and, while it did not affect much of the early wheat crop, it began to cut into yields in the northern plains. The drought also reduced the corn crop in large areas of the Corn Belt. This caused USDA forecasts for the corn crop to shrink and buoyed market prices for all grains.

In late October, the Russians acknowledged a major shortfall in Soviet grain production—something on the order of 50 million metric tons. For the second year in a row, Soviet planners were forced to admit publicly their inability to meet goals for improving the quality of the Russian diet through increased meat and milk production. Their need to go back into the world markets to make up at least part of the shortage had immediate impact on wheat prices. By October 27, wheat futures on the Chicago market were $5.35 per bushel for wheat to be delivered in December 1980, and the cash price offered in Denver was $4.02 for ordinary wheat and $4.27 for 13 percent protein milling grain.[13]

For the wheat farmer who sold his 1980 crop in June or July, the price was a disaster, and the embargo was the major cause of the soft market. But those who held their grain until November or longer had a strong market, the way events turned out. As usual, that difference tended to trap marginal farmers, or those who guessed wrong, or those who needed cash immediately after harvest. It favored the more affluent (or lucky) ones who did not have heavy debts or carrying costs and who could afford to store their crop and wait.

The effect, politically, was to allow challenger Ronald Reagan to campaign on the "failure" of the embargo, while forcing Carter into the defensive position of telling farmers that, despite the embargo, wheat prices were at $4 in October of 1980 instead of $2, as they had been in 1977. Carter proudly pointed out the farmer-owned grain reserve, which has "taken the government out of the grain business," and the opening of new markets such as China, calling them the high points of his administration's farm successes.[14] But farmers were convinced that their $4 wheat would have been much higher without the embargo, so the net result of the grain embargo was probably that it caused far more political damage in the United States than economic damage in the Soviet Union.

It was certainly a major factor in the smashing defeat suffered by Carter in the farm states, where it was estimated he carried scarcely one-third of the vote in the November elections. It may have contributed as well to the demise of farm-state Democrats in the Senate, notably John Culver of Iowa, George McGovern of South Dakota, and Frank Church of Idaho. The farmers were angry, and it showed.

Less than a week after the election, however, more bad news came in about the 1980 crop: corn—6.46 million bushels—down 17 percent from 1979; cotton—22 percent below 1979's record crop; peanuts—42 percent less than 1979.[15] The drought and heat of 1980 had cut into crop production far more than the embargo on grain sales had cut into exports. Howard Hjort, USDA's chief economist, predicted that the world would arrive at the 1981 harvest with no reserves of feed crops. Advisors to the Reagan administration, who had been urging Reagan to cancel the Russian grain embargo immediately, began to have second thoughts, and once the new administration was in office, new signals were heard. After several cabinet meetings on the subject, the embargo was left in place, at least for a while.

The events of 1980, in addition to polarizing farmers politically, have heightened their anxiety over farm, food, and foreign policies as they realize even more graphically how events and decisions far beyond their control can affect their lives. American farmers are the world's only significant source of purchasable grain. No other country really matters. When world markets flutter, for whatever reason, it is a significant event down on the farm.

An agriculture that depends on foreign trade to stay in business also depends on foreign political decisions that it cannot control (or, often, even guess in advance). Farmers—or Secretaries of Agriculture—or even U.S. presidents—have a hard enough time controlling the flow of economic events as they respond to changes in U.S. policy. The difficulty in controlling events and stabilizing agricultural markets on the home front is made many, many times more difficult and uncontrollable as more of the agricultural economy is dictated by foreign governments whose basic motivations may range from the simple desire to gain maximum economic advantage for their own people to the political goal of heaping maximum disadvantage on U.S. leaders and the U.S. economy.

American farmers and businessmen understand and accept the idea that each participant in a business deal seeks to maximize his own economic advantage. It is something else, however, to have an adversarial foreign government in the market who may make decisions with the basic motive of furthering their own political ambitions.

When an American farmer goes to sell his grain, he does not deal with a foreign buyer, but with a grain elevator that, if the grain is to be traded, deals in turn with one of five major grain exporters. Because they are virtually the only brokers for large international sales, the "five families of grain" effectively set the prices.[16]

When Russia, or another large grain buyer, seeks out American grain, it does not deal with the U.S. government. They deal, again, with those same five grain traders. When the companies have obtained a sale—by offering grain at the lowest price, to be delivered at some future date—they go into the futures market and buy a futures contract at a price that will allow them to hedge against any price rises that might occur before the sale is to be delivered. In this way, they have guaranteed protection against loss, because if the price of grain were to jump just before their sale were to be delivered, they could purchase the grain contained in the futures contract and still stay at a profitable margin. On the other hand, if the price of grain drops for any reason, they would be able to buy on the open market and either sell the futures contract at a loss or try to make some profit by selling it when the market price changed again. Either way, the grain company is protected.

It is a system that, in order to work, depends on low grain prices. When high prices come, says author and lecturer Ray Jergeson, the system does not work any longer.[17] When prices are climbing, too many farmers hold their grain waiting for prices to go still higher, and the flow of grain into the market is interrupted. Companies are left without the protection they get when prices go through normal up and down fluctuations so that they can use the protection of the commodities market to stabilize their purchase prices. As a result, all the forces in the market tend to push prices lower. It is no conspiracy; it is the way the system works. But farmers, who have been sold on the idea that low prices help them "compete," have a hard time taking the accompanying low profits to the bank to pay their bills.

In this segment of the "free enterprise system," the farmers take the risks, the companies make the profits, and foreign governments call the shots. In 1972, working skillfully and secretly between different companies, the Russians brought virtually our entire stock of grain, dropping U.S. supplies to low levels and ushering in the current period of rapidly fluctuating supplies and prices. A free-market-oriented Secretary of Agriculture hailed the event and took full credit for the price boom that buoyed farmer's profits (and hopes) for the next few years, but the real truth is that USDA had no idea of the real implications of the Russian move until after it was an accomplished fact. The strategy, to the extent there was one, was a Russian strategy, not a U.S. strategy.

To farmers, living in a tense world that can't decide whether a cold war or détente is the most feasible means for the super-powers to maintain an arms-length accommodation, having the most vital decisions that affect their lives made in the Kremlin cannot be too reassuring.

The rest of Americans have something to be concerned with as well. The Russian grain purchase of 1972 triggered a great "plow-out" of marginal U.S. farmland, followed by a great "wash-away" that continues to this day. The assault was not just on grain stocks; it was on the American land.

Bigger Farms, Higher Risks

Every projection for agriculture in the future predicts a continuation of the trend toward fewer, bigger farms. If farms do become fewer and larger, that will mean less farm family labor per acre. But hired labor isn't readily available, either. The move toward specialization on farms makes it more and more difficult to maintain year-round work for a hired man. This is particularly true on grain farms, where the abandonment of livestock as a portion of the operation has virtually eliminated wintertime farm jobs.

The increasing cost and vulnerability of today's sophisticated farm machines makes a farmer reluctant to hire an untrained person, so the bind becomes severe. When the farmer can't afford dependable year-round help, and can't find skilled seasonal workers, he has little choice but to try to buy the size and kind of machinery that will let him do what needs to be done with his own labor and that of his family.

In the past several decades, labor inputs per acre have fallen off sharply. That can't continue much longer, or there would be no need for farmers at all! Only a radically different future, with smaller average farm sizes, more family farms, and more time for the farmer to spend managing each acre offers any hope that increased labor intensity could become a positive factor in future agricultural productivity. While there are people who fervently believe that such a future would be vastly preferable than the one to which current trends appear to be leading, there is little evidence of any strong forces that will lead to a future of small family farms.

As labor has dropped, it has been replaced by more, faster, and more expensive farm machinery and buildings, and those represent significant capital investments. But farm machinery, while it raises labor productivity and allows each farmer to handle more land, simply represents a substitution of capital for labor, and productivity in the overall sense may change little, if any. Farmers have been heavy users of capital in recent years, but that may be getting more scarce, as well.

A yard full of costly equipment is needed to operate even a small Iowa corn and hog farm. USDA photo.

For the past several years, financial analysts have been warning of an impending shortage of capital in the U.S. economy. The *1976 Report on National Growth and Development*, prepared for President Ford by the Department of Housing and Urban Development (HUD), pointed out that the nation's need for investment capital would soon far outrun the supply.[18] With the passing of a few years, many of HUD's predictions seem to be coming true. Cities all over the country are bemoaning the fact that street, water, and sewer systems are outdated, crumbling, and in need of immediate repair. The condition of the nation's bridges is a scandal, and the railroads are a shambles.

The need for new energy-related developments ranging all the way from coal-loading facilities at the nation's ports to dozens of power plants and thousands of miles of transmission lines carries a huge price tag, as do the 23 million new houses that the National Home Builders Association predicts will be needed in the 1980's. What does all this mean to agriculture? It means plenty. Farmers compete on the same money market as shipbuilders. When interest rates bounce around between 12 and 20 percent, as they have done in 1980, very few farmers can compete for capital. They are forced into a situation where they work as hard as they can and manage as well as they can, but end up simply trading money, with little or no profit left over after having taken extremely high risks and handling large cash flows. It is a situation most non-farm businessmen would avoid like the plague.

What suffers most, under those conditions, is capital for long-term investments with high initial costs and slow paybacks. High interest rates make it almost impossible to justify investments in soil conservation and resource development on a strictly financial basis. Water resource development projects such as storage reservoirs or canals, for example, start dropping out of economic feasibility when the interest rates (or discount rates) used in planning pass the 6-8 percent mark. Planting trees today for a harvest in 30-50 years is uneconomic at interest rates above 10 percent unless one is willing to gamble on very high stumpage prices in the future. Building terraces to conserve soil is even worse, since there won't be any income from the terrace at all, only a hoped-for benefit at some far-off time when soil productivity will be better as a result of the topsoil saved.

In short, the rising competition for capital in the United States is certain to hit the farmer hard, particularly in his ability to invest in resource development. As a small businessman, he is far less prepared to compete for limited capital than a major oil company or a state or local government. When he uses capital, as every farmer must today, he must limit it to those things that provide immediate cash paybacks in order to keep the high interest rates from taking all the profit from the agricultural enterprise. As a result, the pressures will be to exploit the land, gain as much quick return as possible, and invest as little as possible in long-term, slow payback practices. That is bad news for the soil and, ultimately, bad news for agriculture, because the soil can be exploited for only so long until it is "mined" to its economic limits.

Carrying Conservation Costs

Farmers take a stronger pro-conservation position than does the public as a whole, according to the Louis Harris public opinion poll done for USDA in 1979.[19] But that still does not mean that farmers are carrying out the soil conservation practices they espouse in theory. When asked the primary cause of this, most point to the financial bind they face. One "working farmer" explained it this way:

> Loaded down with increasing indebtedness, farmers are locked into a year-to-year battle for survival. They are forced to pay ever-inflating prices for production inputs while lacking any way to pass on these increased costs. Farmers have responded to this financial squeeze by adopting farming practices that offer short-term success.[20]

Charles McLaughlin, an Iowa farmer and past president of the Iowa Association of Soil Conservation District Commissioners, contrasted the economic situation of surface mine reclamation and soil and water conservation in testimony before a Senate Agriculture Committee hearing in 1980:

> [In surface mine reclamation] there is a law that the land must be restored, by all miners; the cost can be added to the price of coal, and the public pays for reclamation as part of the cost of the fuel. In contrast, agriculture in the United States is composed of a motley crew of individual entrepeneurs who pay what is asked for supplies and equipment, gamble on the weather and on government policy, and take what is offered for their products. Although we feel that this system has irreplaceable values in our society, it

does not permit adding the cost of soil conservation to the prices of food, fiber, and fuel. The products of the land are used by all who eat or use energy. It is, therefore, indisputable logic that the public should bear a part of the investment in protecting the soil resource base.[21]

In short, *installing and maintaining soil and water conservation systems on the land creates important costs for farmers and ranchers that they cannot recover in the sale of their products.* Projecting that loss, farmers choose financial survival over soil conservation.

Production Costs Just Keep Going Up

The root of this financial judgment lies in the current economic squeeze facing farmers. Everyone faces inflation these days in virtually every aspect of life. But farmers face a constant level of inflation in the cost of the things they purchase, without an offsetting steady rise in the price of the goods they sell. And for farmers, another, less well recognized, factor that keeps levering costs higher is the shift away from natural productivity toward artificial productivity as the basis for producing farm crops.

No longer does the farmer plant a crop with the intention of harvesting the increase created by soil, sun, and rain—all factors that don't take a whole lot of cash. You could lose a crop like that—or, as the old-time farmers used to say, "barely get your seed back," and you might be able to hold on for another year and try again. Today's farmer plants a crop that may need several hundred dollars worth of seed, insecticide, herbicide, fertilizer, and, perhaps, irrigation water invested in *each acre* before a harvester ever enters the field. Lose *that* crop and your banker may want to have a serious talk.

A few examples suffice, with Figure 3.1 illustrating the general trends. Total agricultural chemical use in U.S. agriculture has grown 50 percent since 1967, with nitrogen and potash fertilizer each up almost 75 percent. The average horsepower of tractors has nearly doubled, as has the amount of diesel fuel used on farms.[22] So it is not just inflation or rising costs that is hitting farmers, it is the double whammy of using a great deal more of those inputs that must be purchased with cash while, at the same time, inflation is driving up their prices.

Figure 3.1. Use of Selected Farm Inputs

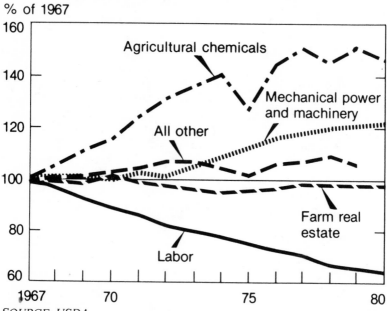

SOURCE: USDA

Figure 3.2. Energy Prices Paid by Farmers

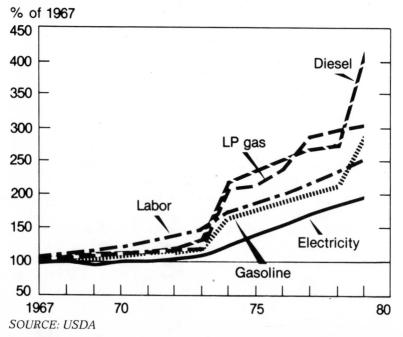

SOURCE: USDA

The impact on young farmers, or those trying to overcome cash flow deficits, is devastating. If, in the process of paying for all those inputs, he has to borrow operating capital, the interest cost was nearly six times as high in 1980 as in 1967![23]

Figure 3.2 shows what has happened as a result of recent energy price increases. There has been a pronounced shift from gasoline to diesel engines as tractors and other equipment have grown bigger and bigger. Look at the difference in how the two prices have shifted. Gasoline prices have been rising, but not as fast as diesel prices. The net result is that farmers have been shifting to a fuel whose price has been rising the fastest of all.

If the nation were trying to plan a strategy that would make American farmers most dangerously vulnerable to price inflation based on a petroleum shortage, it would be hard to plan one more effective than the recent trends in the use of agricultural chemicals and power machinery shown by Figures 3.1 and 3.2.

The irony of the whole situation is that the farmers have been the leaders in pushing these changes. Minnesota farmer Pat Benedict, featured on the cover of *Time* magazine in 1978 as a prime example of "the new American farmer," gave the main reason: "Eliminate uncertainties, that's the golden rule. You eliminate uncertainties in production through technology and careful management."[24] In other words, if you are uncertain about having time to get all your crop planted at the right time in the spring, you buy another tractor, or a bigger tractor. If you are uncertain about summer rainfall on your crop, you install a center-pivot sprinkler system, then double the fertilizer rates and turn on the electric pump whenever the plants need water. You have reduced the uncertainties associated with the weather and replaced them with the certainty that your costs are now going to be tied more than ever to two things: inflation and Arab oil decisions. *Time*'s conclusion that the process was an unqualified success, giving "growing production, rising innovation, expanding exports—and reasonable costs to customers," totally ignored the rising vulnerability facing U.S. farmers.[25]

One thing is certain: During periods of chronic inflation, farmers face the prospect of constantly escalating production costs. Between 1959 and 1977, the prices paid by farmers increased every year except one, when there was no change. In 9 of those 18 years, the increase was 5 percent or more; in 2, the increase was over 15 percent.[26] The only uncertainty involved in the farmer's cost ledger is how big the annual increase is likely to be.

On the other side of the economic ledger, there is no accompanying rise in income to help farmers pay those added costs. During the same 18 years, farm prices dropped five times, once by 6 percent and once by 4 percent. The USDA index of prices rose by 5 percent or more in 5 other years, and from 1972 to 1973 the price index jumped by over 43 percent. Emery Castle of Resources for the Future calls this tendency for farm costs to rise steadily while revenue jumps up and down erratically a "fundamental part of the social environment of U.S. agriculture."[27]

One effect of this, understandably, is to make farmers very cautious about the costs they incur and the investments they make. The general rules are "hold down costs wherever possible," and "wait on making optional investments." Too often, unfortunately, the kinds of costs that can be avoided are those of applying soil and

STAMPEDE   By Jerry Palen

"My banker's charging me 22% interest. I'm paying
$1.25 for diesel, seed prices have gone up 40%
and this jerk thinks he's gonna rob me?!"

water conservation practices; the "optional" investments are those needed to build the soil's strength. Costs for soil and water conservation seldom return an offsetting profit in the first year—and may not return one in five years, or ten. In a study of 12 Pennsylvania dairy farms, for example, researchers found that the implementation of erosion and sediment control plans would in fact *reduce* net income in 10 of the cases. Those losses would come, not only from the initial cost of installing conservation practices, but also from the revenues lost by shifting erosive soils from row crops to grass for erosion control.[28]

Other research points to similar conclusions. Several years back, studies on water pollution in Indiana showed that the installation of low-cost grass or close-drilled grain "filter strips" along the borders of the streams would do as much or more to prevent water pollution as anything else a farmer could do.[29] But seeding a 16-foot strip (the width of most farm drills) to wheat, oats, or grass on each side of the water courses could add up to a lot of acres taken out of corn or soybeans. Lost income was the real cost.

In short, *the full price of establishing soil and water conservation systems must include reduced income as well as out-of-pocket costs.* If the public is going to help carry a portion of that burden, we need new ways of determining how that is to be done. Typical methods such as cost-sharing and tax deductions for farmers, based on the amount of cost actually incurred, won't suffice.

Who Will Get the Benefits?

Although the costs of soil and water conservation are immediate and direct, the cash benefits are not. Using cross-slope tillage, for example, results in extra fuel costs because more turning and travel may be required, but there will be no additional crop grown on the field that year as a result. There may not, in fact, be any additional crop for a number of years, until the soil-saving begins to add up enough to affect yields. The benefits of soil conservation efforts start very slowly, pick up gradually each year as soil condition improves, and begin to pay back dividends some time far in the future, if at all.

To someone facing that type of investment, a slow return on benefits is a distinct disadvantage. High interest rates increase that disadvantage. If a project won't return a 15 percent profit, it is hard to justify borrowing money at a 15 percent interest rate to invest in it. From an economic point of view, a dollar spent on soil and water conservation will not be as profitable as a dollar spent for fertilizer that may return $1.10 worth of crop in five months.

That is not to say, however, that conservation investments

are not of any benefit to the farmer. He may value the approval and respect earned from his neighbors, the feeling that he has done the "right" thing with his land, or the notion that whoever is farming the land in the future will have a better, more productive soil as a result and therefore be more secure. None of those benefits are financial in nature, but they are important. Their value, however, may depend on the farmer's forecast for who will be farming his land in the future.

If the person making the decision is 35 years old, looking for a future in farming, and calculating how to make that future secure, benefits 10 or even 25 years from now are important. They will benefit him and his family directly, and taking steps today to secure them are sensible. Next, his children may be farming it. If the person is 55 years old, but has sons or daughters who will be taking over the farm, an investment today is an investment in their future. A farmer may make that investment as readily (or, maybe, even more readily) than he would invest for his own personal benefit. The pride in saying, "I left it better for my children than I got it," is a powerful social force.

The third possibility is that some stranger may be farming the land. The farmer may be 55 years old and the kids all married, moved away, and not interested in farming or able to return to the farm. In this case, an investment made today with a potential pay-off in soil productivity some 20 or more years down the line can't be justified in terms of either self-interest or parental pride. Economics provides no incentive for conservation: If you are not going to receive any of the benefits from an investment, the value of a dollar invested today is *zero*, no matter how you calculate. It takes some higher kind of ethical and moral motivation, a commitment to leave things better for humanity in general, or a feeling of responsibility for stewardship of the land in the environmental or religious sense, to make that kind of an investment.

Finally, there is the chance that *nobody* will be farming the land. Maybe the land is to be converted to a subdivision, torn up by a surface mine, or buried under asphalt in a shopping center within 5-10 years. A soil conservation investment (or any other kind of investment) today that takes 10-25 years to pay back benefits is not only uneconomic, it is senseless. The benefits will accrue to no one, not to society or to the land, in an environmental sense. Neither economics nor ethics supports this kind of investment.

The problem is that more and more farmers have shifted out of the first category — young farmers planning for the future — as the average age of U.S. farmers has slowly crept upward. Today's average farmer is between 50 and 55 years old, and his future plans may be centered more on Florida than on the farm. More and more farmers have shifted out of category two—those planning to pass the farm to their children—as the structure of agriculture has changed and economic conditions have made it more and more difficult for young people to take over the family farm.

The far more difficult situation—category three—farmers who must make conservation investments on the basis of a strong ethical or moral conviction because the benefits will only accrue to strangers—is growing, as categories one and two shrink. And the very worst situation—category four—farmers who doubt the future of agriculture on their land—has risen dramatically in recent years, due to the impacts of buckshot urbanization, farmland conversion, topsoil loss, competition for water, and economic stress.

Farmers in these last two categories become speculators, not stewards, and make management and investment decisions based on a phase-out plan. Productivity left in the soil will be of little value when the land is sold, since farmland sale values seldom hinge on the condition of the topsoil. A farm composed of soils that have been properly cared for and maintained over the years may sell for no more per acre than one with only half of the topsoil left, and it virtually exhausted of organic matter. Even when the buyer is a neighboring farmer, he may simply fail to recognize that the soil is run down or feel he can overcome that deficiency with a good fertilizer and management program—or he may simply have to pay top dollar to bid the farm away from other would-be buyers. Because there is no evidence that "mining" the quality of the topsoil results in a financial penalty when the farm is sold, the incentive is to mine it as part of the phase-out plan. Where the land is to be paved over or destroyed, any residual productivity is simply lost, so the incentive to mine the soil is even stronger.

Those circumstances identify another basic cause behind the lack of soil and water conservation today: *Too many farmers feel that conservation investments will not benefit their family because it will soon be out of farming or may not benefit either the land or society because agriculture has no future on their land.*

Farmers Must Recognize the Problems

A recent study in Nebraska revealed another common cause of farmers' reluctance to take action to stop erosion—the farmers just didn't feel that the rates of erosion that were occurring were a serious threat. In the study, farmers were shown a set of situations where soil erosion rates in the 10-15 tons-per-acre-per-year range were occurring. Overwhelmingly, they said that was no problem. Soil conservationists, shown the same examples, overwhelmingly disagreed.[30]

The farmers, questioned about their reasoning, pointed out that such erosion rates had not, in the past, resulted in declining yields. The productivity losses from erosion had been effectively masked by yield increases from fertilizers and other technological changes, so the farmers were right. While yields were no doubt below what they would have been if soil erosion had not taken place, how can you measure a loss in terms of something that might have been? In addition, the farmers pointed out, the soil damage was not so bad as to slow down normal field work or prevent normal farming operations. Had it been that bad, they would have recognized that it imposed a real cost on their operations.

The soil conservationists saw that the rates of soil loss were far faster than the rate of topsoil formation, meaning a loss of productivity that would drive future costs higher and future yields lower until, at some point, farm production on the land would be uneconomic. That was a crisis, as far as they were concerned. But it was not a risk *for this year*, which was the primary concern for the farmers. It was a threat for the distant future, of concern only to future-thinkers.

As a result, the soil conservationists would be led to identify needs that the farmer would not see, then propose solutions (with costs) that the farmer would think both unnecessary and impractical. Had they both seen the same threat, agreement on a feasible solution would have been easy. The basic cause of the dilemma is that *too many farmers may not recognize the full implications of soil erosion on their land.*

Soil conservationists are going to have to develop better ways to measure soil damage and illustrate its effects in terms that are relevant to farmers. They are going to have to learn how to make slow, imperceptible soil damages today mean something that farmers can associate with their personal planning horizons. And they may need to help farmers see how to extend those planning horizons beyond just this year to a reasonable future.

In addition, there are some challenges for the agricultural colleges and the USDA. Agricultural colleges are educating most of today's farmers, either through regular campus education or in the continuing education and extension programs. The next generation of farmers will be preponderantly college graduates, it is fair to guess. Will they know what they should about managing the land properly?

USDA, too, needs to put more emphasis on this subject. The fact that USDA, in 1980, after over 100 years in business and over 40 years of soil conservation programs, still could not provide a convincing analysis to demonstrate the full cost of soil erosion to either an individual farmer or to society, is a disgrace. It is proof that such problems have not had a very high priority over the years.

Big Tractors: Straight Rows

The big, fast tractors of today's farm are impressive to watch at work, and a delight to operate, but they don't turn on a dime. In order to be efficient, they must be working on a long, straight run, not wasting time wriggling in and out of tight corners. Since property boundaries in much of the country are straight, and usually a half-mile or more in length, the need for tractor efficiency leads farmers to reorganize small fields into large units, bounded only by roads, property lines, or absolutely impassable pieces of terrain.

But such "square" organization ignores the topography of the land. Operating on the contour, or even keeping farm operations generally across the slope, is virtually impossible. In an earlier time, small areas of very steep or rough land in the middle of an otherwise farmable field, would have been left in trees, brush, or grass, with farming operations dodging around them. Not any more. Now the only efficient way to farm is to include those "rough" areas in the field, even though cultivating them may result in severe soil erosion and, in terms of the product from those specific acres, a crop return that doesn't cover the costs involved. Those marginal acres are easier to go straight across than to go around, so they are cultivated.

Windbreaks become a real pain, so thousands of miles of them have been torn out to enlarge fields so that big tractors or center-pivot sprinkler systems can have free travel. Terraces — particularly those that wander around on the contour — can't be tolerated. Those old contour terraces, with all the "point" rows and "stub" rows that slowed down planting and cultivation, were feasible with two-row equipment. A farmer could drive a two-row rig to the end of a set of rows that met a terrace at a sharp angle, turn around, and catch the next two rows where they joined the terrace. But with six- or eight-row equipment, an operator can't even get to the ends of some of the rows without jamming the other end of the machine in the terrace bank! So the old terraces have had to go, victims of technological obsolescence. Harvesting equipment imposes the same kinds of limits.

Closely spaced windbreaks were planted in 1939 to protect these fields on the Texas-Oklahoma border. Few of these windbreaks remain today, as farmers have torn them out to make room for larger farm equipment. SCS photo by Hermann Postlethwaite.

This 1955 terracing design in Iowa worked with small tractors, but large tractors and six- or eight-row cultivating equipment made most of these systems obsolete. SCS photo by Erwin Cole.

In Idaho, potatoes used to be planted on the contour for erosion control. The rows would be carefully surveyed out before planting, and the irrigation water would follow the gentle slope of the row without causing erosion. But with the new potato harvesters, the potatoes are carried from the ground up over the machine on flat conveyor chains. The soil falls through the gaps in the chains and people stand on either side as the potatoes pass by, picking out the rocks and large dirt clods before the potatoes pass on up the conveyor and into the truck. It is a fast, efficient harvester, but it doesn't work on a hillside because the potatoes roll off the open chains unless the machine is kept fairly level. The answer: Abandon contour potato rows, switch to sprinklers for irrigation, and run the rows up and down the hills. The result: More efficient machinery operation and potato harvesting at the cost of soil-loss rates that jumped, on the steeper irrigated lands, from under 10 tons per acre per year up close to 100.

In sum, *the new farm technologies, all designed to substitute capital, petroleum, or technology for labor, have made many of the basic conservation techniques obsolete, without providing new techniques to replace the old ones.*

The Government Promotes Soil Mining

In addition to financial pressures, growing uncertainty about the future, and emerging agricultural technology, farmers don't need any other pressures to push them toward poor soil management, but they have them. Public policies, too, encourage the abuse of land.

Farm policies that push farmers to plow "fence row to fence row," as were common in the early 1970's, have not really changed despite widespread recognition of the soil damage that is occurring as a result. To meet the calls for more crops, farmers are still encouraged to use marginal land, where the productivity is lowest and the cost of conservation treatment is greatest. The government provides financial support for added production through loans, price supports, or crop insurance programs, but these are not in any way related to the use of the land. The farmer who misuses marginal land can qualify for government assistance as readily as the one who does not.

The history of disaster assistance programs shows that the government has actually encouraged the farming of unsuited lands by subsidizing the risk. On the marginal lands of the Great Plains, the promise of federal disaster relief payments has served as a form of collateral for farmers. Bankers knew that, if it rained, the farmer would get a wheat crop and be able to pay off his loan. If it didn't rain, which was just about as likely, the federal disaster program would cover the banker's loss. As one Texas farmer reported, "Hell, I never see any of that money. I just sign the back and hand the damn thing over to my banker."[31]

The disaster relief programs discouraged conservation practices, as well. In Kiowa County, Colorado, a wheat farmer planted grass strips in his wheat fields to cut down wind erosion. The result, he was informed by USDA, would be a reduction in his disaster relief check, which was based on the total acres *planted* in crop. Since the grass strips were not considered a crop, they served to reduce his total acreage and cut his payment. The next year, the farmer plowed out the grass strips and planted solid wheat again.[32]

The 1977 Farm Bill contained a provision that based a farmer's eligibility to participate in farm programs on how much crop he planted as compared to the "normal crop acreage" (NCA) for the farm. In computing the NCA for each farm, the crop acres already planted to grass as a conservation measure were not counted as part of the "normal crop acreage." Therefore, in order to get credit for diverting crop acres, the conservation farmer needed to set aside additional lands. The effect was to penalize the farmer who had already diverted his marginal cropland, while providing a relative advantage to the ones who had not. The farmer who had voluntarily planted marginal acres to grass was forced to divert good acres; the one who had been abusing marginal land could now set it aside and collect a bonus check from the government. In case after case, after reviewing the implications of the new law, and counting the financial costs, farmers did the obvious—they broke out the plows and plowed down the grass before the deadline for computing the NCA. Then, after it was labeled "cropland," they could "divert" it again to qualify for federal programs. It was like digging ditches in order to fill them up again, and the farmers knew it, but the long-term economic penalty that might accrue if the NCA ended up too low was more than they could tolerate.

Thus, *despite the public opinion that soil conservation and farmland protection should be encouraged by public policy, the net effect of many public programs is to do the exact opposite and encourage farmers to waste or destroy resources.*

The conclusion one can reach from all this list of pressures on farmers is that the business of stopping the waste of farmland is not going to be either simple or easy. It is not just a matter of knowing how to plug gullies or plant grass and trees. It is not a few dollars to encourage a farmer to build a terrace or install a tile line. It is not a matter of providing a few million federal dollars or building a new research laboratory somewhere.

Encouraging farmers to stop wasting farmland will require the development of a complex set of incentives that helps farmers see that it is in their own best interest, as well as the interest of society, to not waste farmland. That set of incentives does not exist today, and the best way to demonstrate that fact is to look at what is happening to the land itself. It is dying, and the public programs that should help farmers save it are, in too many instances, dying as well. The signs of trouble are clearly visible. One need only look at the land to see them.

4.

The Measure of the Land

The largest concentration of prime farmland anywhere in the world lies in the United States, and this tremendous land bounty has been intensively developed in our economic system which prizes individual enterprise and competition. A highly effective public and private information system makes the modern American farmer the most well advised in history. As a result of information and the social pressures that push him toward always trying to improve his operation, he uses new agricultural technology almost as soon as it becomes available.

Because of our farmer's quick adoption of new yield-increasing methods, the amount of land growing crops has stayed fairly stable for many years — somewhere around 400 million acres — and the need for new production has been met largely by growing more and more crops on each acre. But a great deal of land was being used up at the same time, either through soil erosion, conversion to non-farm uses, or through other factors that make the soil unusable for farming. Policymakers seldom saw that to be too much of a problem. There was always a large reservoir of unused land that could be converted into cropland to replace whatever was lost. As a result, wasting farmland was not considered a major threat to future security, and there were few incentives to encourage farmers not to waste.

Today, the situation looks very different and much more threatening. Part of that is due to the new data available as a result of the 1977 National Resource Inventories (NRI) done by the Soil

Conservation Service (SCS), but part is also due to the fact that each of the forces that waste farmland are gaining momentum and speed. At the same time, the reserve of good land that can be used to replace the lost acres has been steadily used up, until today it appears that the limits of the nation's farmland reserve are clearly in view.

Despite a significant investment in conservation practices by the federal government and private landowners, soil erosion from agricultural lands continues at a massive rate, though certainly less than that which would have occurred without the present soil conservation programs. Phase I of the 1977 NRI shows a national 1977 average annual loss from sheet and rill erosion in excess of 4 billion tons. Since it is spread all over the country, that quantity means little in any one place. In total, however, it indicates a serious decline in the nation's soil resource. If it were all concentrated in one area, 4 billion tons of soil loss would mean the removal of all the topsoil (six inches) from 4 million acres. With that kind of loss each year, *it would only take 100 years to wash away every single acre of cropland now being farmed in the United States.*

Much more difficult to evaluate, but perhaps equally important, are the changes occurring within the farmland soils even though they remain in farming. There are strong indications that the quality of the topsoil is being badly damaged in many ways by the intensive methods of modern agriculture. To the extent that this is the case, improved methods of crop production will be absorbed in offsetting the diminishing productivity of our nation's soils before they result in any new *net* productivity.

About one-eighth of America's cropland owes its productivity to irrigation, either as a supplement to natural rainfall or to provide virtually all the crop's water in very dry climates. Where water resources are underutilized and new lands could be irrigated, the potential for expanding agriculture still remains. Where water is being taken away from agriculture by growing competition from non-farm users, or where underground water pools are being pumped dry, irrigated agriculture is doomed to die out. Because irrigated cropland produces, on the average, twice as much food per acre as non-irrigated land, every acre gained or lost from the irrigated cropland is thus doubly important in affecting our future need for land.

Emerging technologies also affect pressures on the land. The amount of food produced can be increased as much by introducing a new higher-yielding variety of grain as by expanding the acres of grain. In the past, U.S. farmers have used many such yield-raising breakthroughs. In addition to new crop varieties, important yield-raisers have included fertilizers, pesticides, management methods, irrigation techniques, tillage machines, and a host of other scientific tools. But the rate of increase in crop yields has fallen off significantly in the past decade, and there are signs that continuing to extract a constantly rising yield from America's farmlands will become more and more difficult in the future.

It becomes apparent, then, that in order to evaluate the significance of wasting America's farmland, many complex factors must be weighed. We need to know how much land there is, and its quality; how much land is in reserve, and what it will cost to convert to cropland; the outlook for water and irrigation; and, finally, any new technologies that may reduce our need for land by making each acre more productive. We must also assess the speed with which farmland is being converted to other uses and the rate at which that process is likely to continue in the future, along with the possibility of new agricultural crops that may begin to compete with traditional food and fiber crops for the use of the available farmland.

Each of the many factors and trends that are important to the future of America's farmland is a complicated topic. But they cannot be viewed separately, because it is the combination of all and the way in which they interact that forms the total picture. Too often those who look at the soil erosion problem recognize its severity but assume that new cropland or new research developments will be available to offset any losses. Agricultural researchers may realize that few new yield-enhancing scientific developments are on the verge of being ready for commercial adoption but assume that there is no particular problem because land is available to meet all the emerging needs. The public may feel that both the soil scientist and the researcher have considered all aspects of the problem and thoroughly evaluated it, when in truth the two have never talked to each other.

The following chapters will explore more of the separate forces bearing on America's farmland situation today and demonstrate how they will affect the nation's farmland security in the future. Before starting that discussion, however, which must deal with a great many numbers and scientific terms, it may be helpful to start with a look at the ways we can measure and evaluate land as a resource. In addition, a survey of how Americans use their land today will provide a basis for evaluating the severity of current trends affecting the land.

Measuring the Land

There are about 2.25 billion acres in the 50 states. But what does a number of that size indicate? There are also about 2 billion acres in the Sahara desert, but any school child knows that the

United States is infinitely more productive, so what kind of comparison can be made from size alone? Obviously, we need to know a great deal more about the land than simply its size. If we are going to concentrate on the farmlands of the nation, we can eliminate many millions of acres at the outset, because while it may be possible to grow food on glaciers, deserts, or mountaintops, it is certainly not very likely.

We can eliminate about one-third of the nation's land from consideration as farmland, because these lands are either in federal ownership or are within urban areas. What remains is just under 1.5 billion acres of rural, non-federal land — a huge, complex, and almost infinitely variable landscape. On this land, we produce virtually all the wheat and beans, fruit and nuts, meat and milk for 215 million Americans. Crops like cotton, flax, and tobacco provide clothing, shelter, and a host of other products. Broad vistas of grasslands support not only grazing cattle, but also horses for work and pleasure, sheep for wool, and a broad array of wildlife. Timber products, ranging from wood for the family heating stove to sawlogs for lumber or wood chips for paper, are harvested from this land. In short, these are the nation's "working lands," where people can harvest the natural productivity of the soil to get the products that are vital to human existence. It is the land that lies between the "living-space lands" of towns and cities and the "wild lands" of parks, deserts, mountains, and lakes. It is the land which provides most of the basis for our national wealth and the strength of our economic system.

Still, understanding 1.5 *billion* of *anything* is virtually impossible, and when we begin to try to visualize and understand 1.5 billion *acres of land,* the problem is monumental. If you were to drive a car steadily all day on one of the modern highways that lace America, you might make 500 miles. When the weather is good, and the view not blocked by trees, hills, or 18-wheel trucks, chances are you could get a fairly good look at the land for about half a mile on either side of the car. If that were the case, you could look at, roughly, 320,000 acres of land in the course of that day-long drive. At that rate, it would take something on the order of 13 years of steady travel for you to personally inspect America's non-federal rural lands. Even then, however, you would have gained only a fleeting glimpse of the landscape and what it really means. That 55 mph view of the land's surface completely hides the third dimension of the landscape — the soil and its productivity. Land is more than space, and its full measurements include not only the flat dimensions of the surface but also the productive dimensions contained in topsoils, climates, water resources, and other factors.

America's farmland offers an endless variety of scenes, from the rolling hills of West Virginia (top) to the broad sweep of a Minnesota prairie (bottom). USDA photos.

Before trying to understand land statistics in terms of millions of acres, concentrate on what *1 acre* looks like. (The acre is the measurement of land most commonly used in the United States. The rest of the world uses a metric measurement, called a *hectare,* that equals about 2.5 acres.) An acre equals 43,560 square feet, just a little less than an American football field. A common city block covers somewhere around four acres; the ordinary suburban house lot is somewhere between one-fourth and one-third of one acre.

Out in the rural countryside, land is commonly laid out in squares a mile long on each side called *sections.* A section of land contains 640 acres. Farmers speak in terms of "half-sections" (320 acres) and "quarters" (160 acres), because these are the most commonly traded pieces of farmland. Much of this dates back to the early days of the land survey in America, when the public domain was laid off in one-mile squares for the purposes of identifying and describing land boundaries. Each square section was then divided into four equal quarters, and farmers were given the opportunity to take one up as a 160-acre homestead.

Today, those individual 160-acre farms have vanished almost as completely as horse-drawn plows, but their outlines may still be visible. Land titles are still often described in terms of the original 160-acre blocks, which may have been bought, sold, or traded many times since the original owners. Ask a Colorado wheat farmer how much land he is farming and the answer is liable to come back, "nine quarters." Because many of these "quarters" have been added to the farm at different times, and because they may not all be adjoining each other, they are often farmed as separate fields and the boundaries show clearly from an aerial view.

Another graphic evidence of the lingering influence of early land measurements can be seen from the window of a modern jetliner over the central part of the nation. Those green circles-within-the-squares that show up so clearly on the land are center-pivot irrigation systems, each operating on its own 160-acre plot. With a well in the very center of the "quarter," the sprinkler pipe pivots around and around, carried on wheels driven by the high-pressure water flowing through the pipe. Around 135 acres are sprinkled in the process, with the remainder of the "quarter" lying dry beyond the reach of the squirting water guns. This modern technology developed in response to the common pattern of land ownership. People buy, sell, and manage much of America's farmland in "quarters," so that was the logical size to build the sprinklers.

A careful look at the surface of the land shows several things that contribute to its agricultural productivity. Perhaps the most obvious is its topography, or slope. Flatter lands are better for farming, as a general rule. They are the easiest to cultivate and work with machinery, and rain falling on them will have less tendency to run off down a slope, so more of it can soak into the soil to nourish the growth of plants. Visible soil characteristics often tell much about the land's value for farming. Land littered with huge boulders or studded with rock outcrops, for example, is obviously not very desirable. The natural vegetation on the land indicates the type of growing conditions that exist. In general, the more vegetation and the larger it is, the more moisture the land receives, and the more productive it will be for agricultural purposes.

What can't be seen are the soil characteristics that lie buried from view but which have important effects on the growth of plants. The rolling hills of the Palouse region in northern Idaho and eastern Washington, for example, are made up of layers of wind-deposited soil that may run 40 or more feet deep. There are no layers of rock to stop plant roots from penetrating, and alfalfa roots have been found growing 20 feet below the soil surface. On the eastern edges of this highly productive region, however, only a few inches of soil were deposited on top of rocky outcrops. Here, the surface of the land looks similar, and the rainfall is not significantly different, but the soils are far less productive. Roots can only penetrate a few inches, there is far less water-holding capacity in the soil, and the total amount of plant nutrients available to the plant is far less.

The job of identifying soil characteristics, grouping similar soils together, and mapping their location on the land has been the task of the National Cooperative Soil Survey, an effort carried out

Center-pivot irrigation sprinklers form lush green circles on the Nebraska landscape, leaving the corners of the 160-acre square parcels brown and dry. SCS photo by Erwin Cole.

by the Department of Agriculture, the land grant colleges, and the states for nearly a century. In the process, soil scientists have identified over 20,000 different kinds of soil that are used for growing agricultural crops, and the job is still unfinished. For each kind of soil, complete descriptions are prepared on physical and chemical characteristics, as well as the maps of the location and extent of each separate soil type. These surveys, available locally in virtually every county in America through the SCS, can be very useful to the homeowner or gardener, the farmer, or the urban developer. They contain a wealth of information about the soil and its proper management, as well as providing the scientific basis for the national inventory of farmland quantity and quality contained in SCS's NRI (see Appendix B). It is when the data from all these inventories is compiled that we begin to see more clearly the limits of the American landscape. It is huge, but not endless.

The Quality of the Land for Agriculture

Not all acres are the same, of course, and so one of the first ways to gain a better understanding of the land base is to divide it up into categories based on the capability of the soil to support intensive cropping. Such a system, known as the Land Capability Classification System, has been developed by the USDA. The details on this system are explained in Appendix A. The accompanying photo shows visually how it separates soils into eight major classes of land, indicated by roman numerals I-VIII, based on the capability of the land to support intensive agriculture without suffering permanent topsoil damage in the process. In using these land classes, it is common to refer to soils that fall in Classes I through III as land that is "suitable" for cropland, land in Class IV as "marginal," and land in Class V through VIII as "unsuitable." Figure 4.1 shows how the land base of the United States is divided when we split it up into the USDA land capability classes. Although almost half (43 percent) of America's rural non-federal land is suitable for cropland, less than 3 percent is Class I land. Class I land is a rare, valuable resource; not one to be taken for granted.

It is clear that not all regions are equally endowed with high quality land, and Figure 4.2 shows how the non-federal rural land in each region divides up in terms of farmland capability. The Corn Belt, with 76 percent of its land suitable for cropland, is the best-endowed region in the nation, if not in the world.

The quality of soil resources is commonly indicated by land capability classes and subclasses. The land that will have the least conservation problems if it is used for cropland is Class I, with each class getting harder to farm safely. Classes V-VIII are normally considered unsuited for cropland. SCS photo.

Figure 4.1. Percentages of Rural Non-Federal Land in Each Capability Class, 1977

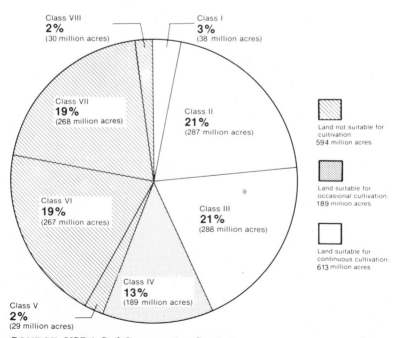

Class VIII
2%
(30 million acres)

Class I
3%
(38 million acres)

Class VII
19%
(268 million acres)

Class II
21%
(287 million acres)

Class VI
19%
(267 million acres)

Class III
21%
(288 million acres)

Class IV
13%
(189 million acres)

Class V
2%
(29 million acres)

Land not suitable for cultivation:
594 million acres

Land suitable for occasional cultivation:
189 million acres

Land suitable for continuous cultivation:
613 million acres

SOURCE: USDA Soil Conservation Service

Figure 4.2. Percentages of Soil in Capability Classes I-VIII, by USDA Farm Production Region

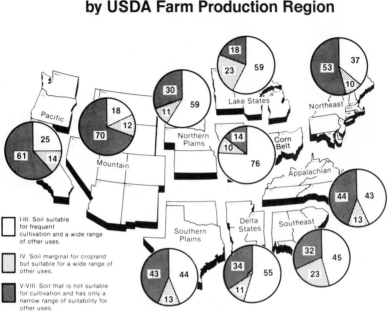

SOURCE: USDA Soil Conservation Service

While the land classes often reflect the productivity of the soil as well as its susceptibility to erosion, they do not always define the most productive land. Some Class III or Class IV soils are very susceptible to erosion when they are intensively cropped but are also very high yielding if properly managed. On the other hand, there are Class I and Class II soils that have little erosion hazard but also are infertile or unresponsive to modern agricultural methods and are, thus, not too productive despite their favorable capability. So in addition to the land classes related to erosion, we need to identify the land that is the "best" for farming when all the soil factors that affect productivity are considered.

By the mid-1970s, the terms "prime farmland," "prime cropland," and "prime agricultural land" had been used in many ways, by many people, but there was no real definition of the term, and everyone who used it seemed to have a different thing in mind. The need for a scientific definition of prime farmland was considered urgent by many in USDA so that some rational decisions could be made on land-use conflicts. If a power plant in Indiana was to be built with federal money, and take 10,000 acres of farmland out of production permanently, was there any loss involved that should concern the nation? The first question in such cases, and there were hundreds of them cropping up each year, was: "What kind of farmland is it? Is it prime, or just mediocre?"

To answer these kinds of questions, SCS developed a standard definition of prime farmland, using soil survey data and taking into account all of the soil factors associated with high soil productivity and stability in response to modern agricultural management systems. The result was a definition of prime farmland that hinges on nine specific soil factors.[1] In general, prime farmland is the land that has the best combination of physical and chemical characteristics for producing food, feed, forage, fiber, and oilseed crops. Prime farmland has an adequate and dependable water supply, a favorable temperature and growing season, acceptable acidity or alkalinity, acceptable salt and sodium content, and few or no rocks. Air and water flow readily through the soil, and the soil is not subject to excessive soil erosion. It is protected from flooding and is not saturated with water for long periods of time.[2]

Steep soils erode badly if they are not well protected. This Class VI hillside in the Palouse region of Washington suffered not only serious rills, but also lost whole sections of topsoil in miniature landslides. Soil slips such as these are fairly common when sudden spring rains fall on frozen or saturated soils. SCS photo by Verle Kaiser.

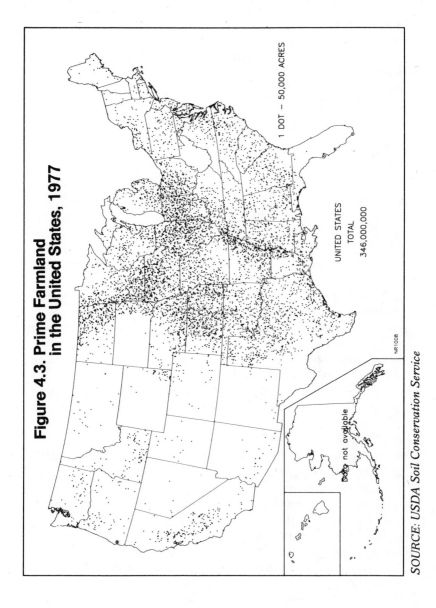

Figure 4.3. Prime Farmland in the United States, 1977

1 DOT — 50,000 ACRES

UNITED STATES TOTAL 346,000,000

Data not available

NR1008

SOURCE: USDA Soil Conservation Service

With the 1975 prime farmland definition in hand, SCS has been able to include estimates on the amount of prime farmland in its resource surveys. The 1977 NRI indicate that there are about 345 million acres of prime farmland in the United States. As Figure 4.3 shows, most of the best farmland in America lies in a broad belt reaching from the Great Lakes to the Gulf of Mexico. The Corn Belt contains 20 percent of the nation's total supply of prime farmland, with the Northern Plains close behind at 19 percent and the Southern Plains at 18 percent. Appendix E gives a state-by-state and regional breakdown of the prime farmland acreage that is useful in developing statewide evaluations of this important resource base.

The Use of the Land

Patterns of land use show a great deal about the relationships that have developed between people and land, and those patterns can also give some pretty good indications about the likely stability of the people on the land. A civilization that has developed a pattern of use in keeping with the environmental characteristics of the land has great stability; living in harmony with the land takes advantage of natural productivity without a constant battle to reshape and fight the elements. On the other hand, people who are misusing the land are only temporary residents. They may think they have the money, the energy, or the skills to resist the realities of the environment permanently, but history teaches that this has not been true, and environmental science shows why. Mother Nature is a strong and relentless adversary, with little respect for humans or their trappings of civilization. An agricultural economy built on the inappropriate use of soil is a house with a foundation of sand. Whether or not it will wash away is not a question. It is only a matter of time.

Table 4.1 gives a national overview of rural land use in 1977, showing that people in the United States make many of their land-use decisions in keeping with the land's capability. Almost all (82 percent) of the Class I rural lands are being used for cropland, with two-thirds of the Class II land being cropped. Since these are the soils that are most easily cropped, with the least problems in the process, this matching of land use with the capability of the soil is one encouraging fact that emerges from the data. On the other hand, we can see that there are 43 million acres of Class IV cropland and 20 million acres of Classes V-VIII cropland. From our brief study of Capability Classes, we would anticipate that these lands would be susceptible to damage from erosion and, indeed, might be literally washing out from under many of the farmers who are trying to use them for crops. Unfortunately, that is all too often true.

Table 4.1.**Use of Rural Non-Federal Lands in 1977 by Capability Class**

Land Use	I	II	III	Capability Class IV	V	VI	VII-VIII	Total
				(million acres)				
Cropland	32	187	131	43	3	13	4	413
Pastureland & Rangeland	4	48	87	76	8	165	153	541
Forestland	2	41	58	60	17	83	109	370
Other Land	1	10	11	9	2	6	72	111
Total	39	286	287	188	30	267	338	1,435

SOURCE: SCS National Resource Inventories, 1977.

While the capability of the land stays pretty much the same from decade to decade, people constantly change their methods of using it. Table 4.2 shows the trends that have been observed since 1958. The amount of land being used for cropland has been slowly dropping, while the acreage converted to urban, water, and other land uses has been slowly increasing.

There has been considerable shifting of use between pastureland, rangeland, and forestland as well, although Table 4.2 may overstate these changes. Between the 1967 survey and the 1977 survey, USDA scientists agreed on new definitions for forestland and rangeland. As a result, some land that was counted as forestland in 1967 was called rangeland in 1977. Despite these changes, USDA experts still feel that the trend is toward additional pastureland and rangeland, with a decreasing amount of forestland.[3]

The constant changing of land use, as land owners respond to economic situations, environmental factors, and their own personal preferences, are in fact, much greater than Table 4.2 really shows, since these are the *net* shifts observed during the decade. There were a great many more changes that took place during the ten-year period, but the decade-end numbers do not capture them.

Table 4.2. **Trends in Use of Non-Federal Land**

Land Use	1958	1967[a]	1977
		(million acres)	
Agricultural Land			
Cropland	448	431	413
Pastureland & Rangeland	485	507	541
Forestland	453	444	370
Urban and Built-Up	51	61	90
Small Water Areas	7	7	9
Other Land Uses	60	50	77
Total	1,504	1,500	1,500

SOURCE: SCS National Resource Inventories, 1977.
[a] Unadjusted 1967 Conservation Needs Inventory data.

For example, the cropland acreage declined by 18 million acres between 1967 and 1977, but that does not mean that only 18 million acres were shifted out of cropping. Instead, as Appendix C demonstrates, there may have been as many as 52 million acres taken out of crops in the past decade. This was partially offset by the conversion of 34 million acres of pastureland, rangeland, forestland, and other land into cropland use, so the net result was an 18-million-acre decline in cropland. In total, however, this net shift may have resulted from land-use changes that affected over 85 million acres, or about 1 acre for every 5 acres of cropland.

Several other land-use trends from recent decades are shown by the USDA inventory data. Table 4.3 shows, for example, that American farmers have been slowly shifting their crops to the better soils. The percentage of cropland on Class I-III lands has been rising, while the cropping of unsuited (Class V-VIII) lands is gradually dropping. Most, if not all, of this shift is probably due to economic pressures. Marginal and unsuitable lands are more expensive to farm, particularly if people try to farm them well and protect them from soil erosion. If people don't farm them right, the topsoil soon washes or blows away, and the land stops producing. In either case, the land demands greater investment for each unit of yield, so there is little mystery why people tend to farm the better lands.

Table 4.3. **Trends in Use of Non-Federal Land for Cropland by Capability Class**

Use of Cropland by Capability Class	Year			
	1950	1958	1967	1977
Classes I-III	82%	83%	84%	85%
Class IV	10%	11%	11%	11%
Classes V-VIII	8%	6%	5%	4%
Total	100%	100%	100%	100%

SOURCE: *RCA Appraisal 1980*, Part I.

The crops that farmers are growing on their croplands are also changing, and this gives some important insights into how intensively America's lands are being used. Table 4.4 indicates some of these trends, and also shows what the term *cropland* means to USDA. For most people, the word *cropland* conjures up visions of Iowa corn fields, Kansas wheat, Idaho potatoes, or Texas cotton. It is all of those; but in addition, it is Cape Cod cranberry bogs, Washington apple orchards, and California vineyards. It is hops for brewing beer, flowers for decorating a wedding, and sod for planting a lawn. It is summerfallow, where the land is clean-cultivated all summer long in order to kill weeds and store moisture in preparation for next year's wheat crop. It is hay and pasture, often planted for two-five years as a soil-building part of a crop rotation, but sometimes little more than a wild meadow that may be irrigated or fertilized for added production, but never cultivated. It may be idle cropland that is simply left alone, either through the personal choice of the owner or at the urging of government pro-grams that pay people to set aside cropland for a year or two in order to hold down supplies of certain crops. In short, cropland grows just about every kind of agricultural crop, and its appearance can be very different from our vision of a standard field of grain.

The most obvious trend over the last decade has been the move away from soil-conserving crops to row crops. While rotation hay and pasture is down 40 percent, row crops are up 27 percent in the past ten years. This shift from crop rotations containing soil-building grasses and legumes to monocultures of corn, soybeans, potatoes, and other high-cash crops means rapidly increasing problems in soil erosion, compaction, tilth, and organic matter maintenance, all of which add to the difficulty in maintaining soil productivity.

Table 4.4. **Cropland Use in 1977 Compared to the 1967 Conservation Needs Inventory**

Cropland Type	1967 Conservation Needs Inventory Data	1977 National Resource Inventory Data	Change
	(million acres)		
Row Crops	160.4	203.2	+ 27%
Close-Grown Crops	100.6	104.9	+ 4%
Rotation Hay & Pasture	50.4	30.2	− 40%
Hayland	27.2	33.1	+ 22%
Orchards, Vineyards, etc.	5.1	5.5	+ 8%
Other Cropland	61.9[a]	6.8	− 89%
Summerfallow	32.0	29.3	− 8%
Total	437.6	413.0	− 5.6%

SOURCE: Developed from data in the 1967 Conservation Needs Inventory and 1977 National Resource Inventories.
[a] Other cropland in 1967 included conservation acres and idle cropland (set aside) as well as "other" cropland.

The other major fact that is readily seen in Table 4.4 is that the nation has no "surplus" cropland. Farmers had over 50 million acres "set aside" in 1967 under government programs designed to hold down wheat and feed grain production, but virtually all those acres were back in production by 1977. This fact is important in evaluating proposals that suggest, for example, that the nation can easily produce enough corn for a gasohol industry simply by "putting extra acreage" into corn production. Such a proposal was widely circulated in late 1980 by the U.S. National Alcohol Fuels Commission, based on the findings of a report prepared by Schnittker Associates, a Washington, D.C., consulting firm. The Schnittker analysis said that 13 million acres could be added to the 84 million acres growing corn in 1980 without bringing marginal land into production, by simply using high quality cropland already in use.[4] Looking at Table 4.4 makes one wonder where the "extra" acres were, or whether such studies are suggesting that farmers can abandon soil-building and soil-conserving rotations completely without adverse effect on productivity.

There isn't any surplus land in the current cropland base. Farmers are using virtually all of the cropland they can get a plow into and using it hard. In addition, there are many forces taking land out of cropland, and these seem to be increasing in pace. Historically, farmers have always been able to increase their production by coaxing higher yields out of each acre, or by plowing new acres. Some of that is still possible, but as the amount of farmland being lost and the amount of topsoil being damaged continues to increase, the rate at which that must happen accelerates dramatically.

The stability of the people:land relationship is threatened in that process. We find ourselves essentially "using up" the principal in our soils "savings account." The total land supply available to U.S. farmers is shrinking rapidly, in both quantity and quality terms, while the pressures being applied by people is growing; that much is clear. The next important fact to understand is the speed at which those trends are taking place and the potential for them to either continue or accelerate in the future.

5.

Farmland Conversion

Every hour, about 320 acres of U.S. agricultural land are converted to non-farm uses. Not all of that land is cropland, but about 100 acres are, and another 100-120 acres have the physical capability to be used for cropland. Losing 220 acres of existing and potential cropland an hour means losing over 5,000 acres every day—the equivalent of 23 average Missouri farms. It means an area the size of the average Missouri county every 2½ months, and the equivalent of the whole state by the year 2000. It means paving over a one-mile-wide strip from New York to California every year.

The loss can be calculated in economic terms, as well. Each day, America's potential to produce corn is reduced by some 5,800 tons, worth over $620,000 when corn is $3 a bushel. Each year we lose the capacity to produce about 2 million tons of corn, worth some $220 million at the same market price.[1] And this year's losses must be added to last year's losses, and next year's losses will add to both, until the total economic impact from our wasteful land use practices over only a few years adds up to an appalling economic toll.

Cropland is not the only agricultural land that is affected, either, and there are other important agricultural values being lost each day. Nearly 1,500 acres of pastureland and rangeland are converted on an average day, which would, if kept in agricultural use, continue to provide meat, milk, wool, and leather. Conservative estimates of the amount of meat production lost from each year's land loss would add up to some 8 million pounds. The future supply of lumber and paper products is likewise diminished, by the conversion of 2,250 acres of forestland each day.[2]

Those images have led to a widespread public concern over the fate of America's farmlands. They have been the source of innumerable headlines and newspaper stories since 1975, when the farmland issue first began to come into public view. Headlines tell the story: "America's Vital Farmland Is Vanishing," trumpets the *Tulsa Tribune,* "Resource Crisis Perils State Farm Output," was the banner in the *Fresno Bee,* and "Crops Die Where Cities Sprawl," was the verdict of the *Chicago Tribune.*[3]

This prime Virginia farmland is still in crops, but its ultimate fate has been sealed as houses continue to be built closer and closer. SCS photo by Tim McCabe.

But there is far from widespread agreement about the legitimacy of headlines like this, and the search for a national policy statement that would express the appropriate degree of concern for the loss of farmlands has, thus far, been a frustrating one. Will the United States have enough agricultural land to meet the rising demands for food and fiber in the coming years?

That question, or some variation of it, has been the subject of five major conferences during late 1979 and early 1980, was a major focus of the Resources Conservation Act (RCA) study carried out by USDA, and is the target question of the National Agricultural Lands Study (NALS), a cooperative effort between USDA and the Council on Environmental Quality. But the answer remains elusive, in large part, according to Pierre Crosson of Resources for the Future (RFF), because people are really not addressing the question from the same premise.[4]

On one side are the economists, who generally tend to agree with RFF's Emery Castle, who proclaimed at the conclusion of an RFF conference on the adequacy of agricultural land in 1980 that "the loss of agricultural land is not a high priority national concern." This point of view, or one generally like it, views the question of farmland adequacy in strictly economic terms: Land is a factor of production but is only one of many factors. If land becomes more scarce, or if its productivity drops due to soil erosion or some other factor, farmers will simply substitute other factors for land. They will use more fertilizer, more irrigation, or more intensive management. They will meet production goals irregardless of the land loss. If food prices rise enough, they will bring other, more costly lands into crop production.

Even those economists who feel uncomfortable with the idea that the ongoing loss of farmland does not matter come back, generally, to the basic point of view: Land is but one of many factors in the production process, and the alleged threat to the nation's food supply from the conversion of agricultural land does not appear to be well founded.[5]

But that apparent complacency does not satisfy those who look at the farmland losses and suggest that not only rural values, but also significant food production values, are being lost.[6] The thought that more fertilizer, pesticides, or intensive production methods will be enough to overcome the current rate of land loss does not satisfy the concerns of people who feel that agriculture is fast becoming too addicted to these high-cost inputs. As evidence mounts that the costs of petroleum-based farm inputs are going to continue to rise, and rise significantly, farmers are looking for ways to use less, not more of them. This could lead to what RFF's Pierre Crosson calls "land-using technologies" instead of "energy-intensive" methods.[7] Such a trend would mean we would need more land, not less. But the USDA surveys in 1975 and 1977 have not been very reassuring on that score. What they show is that the land supply available to U.S. farmers is shrinking, particularly the supply of prime farmland—the land that is most productive for agriculture.

Developing America's Farmlands

Concern over the rate of farmland conversion began to move beyond a few concerned conservationists when the results of the 1975 Potential Cropland Study were made public in 1976. This survey showed two worrisome trends: (1) Agricultural land was being converted to other uses at the rate of about 2.9 million acres per year, a figure much higher than had earlier been thought; and, (2) a disproportionate share of the land being lost was prime farmland. As Table 5.1 shows, only about one-third of the land converted was in Capability Classes V-VIII, meaning two-thirds of the land lost to agriculture was either good cropland or had potential for crops or pastures.

Table 5.1. **Land Converted to Urban Uses and Water Areas in the United States between 1967 and 1975**

	Converted to Urban Uses	Converted to Water Areas	Total Converted	Percent
	(thousand acres)			
Prime Farmland	6,550	1,493	8,043	34
Non-Prime Land in Capability Classes II-IV	6,302	1,567	7,869	34
Land in Capability Classes V-VIII	3,783	3,649	7,432	32
Total	16,635	6,709	23,344	100

SOURCE: 1975 Potential Cropland Study.

The tendency of urban uses to take the best land is clearly shown by Table 5.1. When all the non-agricultural uses are taken together, 32 percent of the land converted in the eight-year period was prime farmland. When only the urban uses are considered, however, the proportion is about 39 percent. Since prime farmland makes up only about 25 percent of the nation's non-federal lands, it is clear that urban development tends to concentrate on the best lands.

There is little mystery why this trend should be the case. In addition to the fact that prime farmlands are the best suited for building, as well as farming, they have normally been more intensively developed for agriculture. That means more roads and more attraction to homebuilders. Another factor is the tendency of urban areas to be located near prime farmlands. Studies of the 1967 Conservation Needs Inventory data showed that, although urban counties make up only 43 percent of the nation's land, 52 percent of the Class I and Class II soils are in those counties.[8]

The 1977 National Resource Inventories (NRI) showed that the rate of land-use change affecting agricultural lands had speeded up dramatically in recent years.[9] By 1977, there were 90 million acres of urban and built-up areas larger than 10 acres, up from 61 million acres in 1967 and 51 million acres in 1958.

Nor is there much evidence to suggest that the trends will slow down. Even though housing costs are skyrocketing as inflation drives all prices upward, it is estimated that 11 million more Americans will reach the prime home-buying age of 30 in the 1980's than did so in the 1970's. The number of new families will increase 25 percent in the next decade. When added to the number of families occupying substandard housing and the number of housing units destroyed each year, the net result will be the need for 23 million new housing units in the 1980's, according to the National Association of Home Builders.[10]

Perhaps the most significant thing about the statistics on the conversion of prime farmland are the regional and local impacts they reveal. Not every region is equally endowed with prime farmland, and some of the regions with the least resource capability are where most of the farmland conversion is taking place. Table 5.2 shows the regional data for prime farmlands that emerged from the 1975 Potential Cropland Study. It was not surprising that regions such as the Northeast, which lost more prime farmland between 1967 and 1975 than it has remaining in reserve, would be the region where public concern for farmland would be highest. But the 1975 data also gave some clues about what was to come. It seemed apparent, for example, that concern for this issue would soon rise in the mountain states, the Pacific states, the Corn Belt, and the Southeast as well, even though there was little indication of concern from those regions in 1975.

In addition to the land lost to water impoundments and urban and other types of construction, there is considerable indication that the future will see mounting conflict over using farmland for energy production and transmission facilities. These kinds of competition for land are not new to many farming areas, but the scale of the new energy programs is far larger than ever before in history, and much of the development will affect both agricultural land and the agricultural industry itself.

Table 5.2. **Regional Distribution of Prime Farmland in the United States, 1975**

Farm Production Region	Total Prime Farmland	Prime Farmland Cropped (1975)	Total Lost (1967-1975)	Remaining Reserve
	(million acres)			
Appalachia	26.4	13.2	0.77	4.3
Corn Belt	76.9	61.2	1.44	3.0
Delta States	29.6	15.2	0.50	1.7
Lake States	38.4	30.4	0.59	0.9
Mountain States	18.1	15.7	0.21	0.2
Northeast States	17.9	8.9	0.95	0.4
Northern Plains	72.0	58.1	0.52	3.8
Pacific States	13.8	8.9	0.58	0.8
Southeast States	23.6	9.0	1.60	2.9
Southern Plains	67.2	29.2	0.84	6.0
Total*	384.0	250.0	8.00	24.0†

SOURCE: 1975 Potential Cropland Study.

*Totals are rounded to the nearest million.
†Prime farmland estimated in 1975 to have high potential for conversion to cropland under 1974 economic conditions.

Mining and Farmland

Surface mining, largely for coal, is taking an ever-increasing amount of agricultural land, disturbing more land since 1965 than was done by all previous generations. Before 1965, about 3.2 million acres had been surface-mined for all minerals in the United States. Since that time, disturbed land has increased to 4 million acres in 1972, 4.4 million acres in 1974, and 5.7 million acres in 1977.[11] The Soil Conservation Service estimates that the current rate of land disturbance due to surface mining now averages about 400,000 acres per year.[12]

About half of this land is cropland. In states such as Illinois, for example, over half of the acreage in the strip-mining permits issued between 1972 and 1977 was former cropland.[13] Much of this land is prime farmland as well, particularly in the west central and northern parts of Illinois.[14]

A great deal of surface-mined land has been left unreclaimed (top) or partly smoothed out and seeded to grass (bottom). Unless the topsoil is saved during the mining process and replaced carefully after the mine spoil has been smoothed, however, it is virtually impossible to restore the soil to its premining productivity. SCS photos by D. W. Kiser (top) and Charles A. Foster (bottom).

When cropland is surface-mined for coal it is, theoretically, to be restored to its former productivity upon the completion of the mining process, under the terms of the Surface Mining Control and Reclamation Act of 1977. This act contains strong language requiring the restoration of prime farmlands. The value of protecting prime farmland from destruction through mining was articulated even by James Schlesinger, former Secretary of Energy and a strong proponent of surface mining, during consideration of the federal Surface Mining Act. On February 15, 1977, while Congress was debating whether or not to include strict provisions for the restoration of soil productivity on prime farmlands, Schlesinger wrote to Representative Morris Udall, the bill's prime sponsor in the House:

> Fortunately, the great abundance of coal in this country allows us to declare certain areas off limits to strip mining because of their greater value for competing purposes. Protection of alluvial valley floors in the West and prime agricultural land should be considered on the basis of the most valuable use of those lands to the nation. It is wise planning to utilize land that is more productive for agriculture for that purpose.

Since passage of the law, however, there has been a great deal of battling between the mining companies, state reclamation agencies, and the Department of the Interior over how this is to be done, and there is still skepticism among scientists about whether mined land can be restored to its former productivity. Soil scientists in Illinois have cautioned that reclaimed land "tends to suffer from inferior soil structure and excessive compaction" which may permanently impair productivity.[15]

There is little doubt that surface-mined land can be restored economically to some valuable after-mining use. The creation of lakes, recreational areas, homesites, and similarly useful land and water areas following mining is common. Soil scientists have pointed out that some soils with hardpan layers may be actually improved in the mining and reclamation process. But to the person evaluating the impact of surface mining on prime farmland, most of this is meaningless. If surface mining cannot be done without returning those lands to prime farmland condition, the agricultural productivity of that land has been reduced *for all time*. That may be too high a price to pay for coal, particularly when there are some 300 years worth of coal that could be mined *without* stripping prime farmlands.

Coal company advertisements show cattle grazing in lush pastures and corn growing tall and green. But they don't always say how much fertilizer, lime, irrigation, or other efforts were needed to produce such a result, or how that compares to similar soils that have not been mined. Only when a farmer can handle "restored" mined land in the same manner as unmined land, and get the same results, will the productivity of the land be truly restored. There is far too little experience to know when, and under what conditions, this can be reliably done.

So the issues are far from resolved. Passage of the 1977 act has been followed by three years of intensive litigation between coal companies, trade associations, and the federal government. As one observer wrote, the results have been "enough legal paper to fill an abandoned strip mine," but little consensus on what can be done to prevent mining from destroying millions of acres of America's prime farmlands.[16]

What is obvious in all this legal hassling is that the stakes are not trivial. Although there have been other legal arguments beside the prime farmland restoration issue, it has been a prominent part of the controversy. The fate of millions of acres of prime Midwest farmland lies in the balance, and there is no way of telling right now how that balance might tip. If economic and political conditions swing toward the "mine first, worry about agriculture later" philosophy of the miners, the land is doomed. Even the few years of uncertainty associated with the lawsuits is costly. As Walter Heine, former Director of Interior's Office of Surface Mining has pointed out, 1,800 acres of prime farmland will be mined in Indiana alone during the legal hassle. By 1982, he predicted, some 2,500 acres more will be mined while the nation waits to see whether reclamation really returns the land to its previous productivity.[17]

There are additional losses to agriculture in a surface-mine area, and not all of them are confined to the land where the mining is carried out. Haul roads, storage areas, and other mining activities use half again as much land as is actually mined. Stripping may also disrupt drainage on adjoining land, and it can pollute surface water or groundwater used by livestock and farm families. In the mid-1970's, coal surface mines produced an estimated 6,000 tons of sulfuric acid daily, which found its way into about 13,000 miles of streams.[18] For many of those streams, the result was severe changes in the stream ecosystem or even biological death, as native plants, fish, and animals simply disappeared from the stream, to be replaced by the yellow and red stains of metal oxides and the algae that can live upon them.

The conversion of coal to liquid fuel can make even further inroads into agricultural production, since it will use water as well as land. The Synthetic Fuels Corporation, created in 1980 by the Energy Security Act, has the potential to stimulate the development of 2 million barrels of synthetic fuels per day by 1992. The Department of Energy (DOE) has located 41 counties in 8 states—including 10 in Montana, 8 in Illinois, 7 in North Dakota, 6 in Wyoming, and 5 in West Virginia—that might be logical locations for a syn-fuel plant.[19] Many of these areas, particularly in Montana, Wyoming, and North Dakota, are water-short; any water used for syn-fuel will be water that otherwise would have gone for irrigation. This threat has greatly concerned citizens in the Intermountain West, who have stridently opposed both coal gasification and coal slurry pipelines that would require huge quantities of the region's limited water supply.

And, as Wendell Fletcher of the American Land Forum points out, the DOE analysis of environmental impacts failed to contain any consideration of prime agricultural lands, disputes over water transfers, or the "cumulative impacts of large-scale facilities"—all crucial factors for assessing the impacts that such a program might have on agriculture and agricultural land.[20]

While there are many factors that may affect the rate of future surface mining, coal companies currently plan to mine another 312,000 acres between now and 1985. This would require an additional 120,000 acres to provide for storage areas, haul roads, etc. Of this acreage, approximately one-fourth lies in states with major prime farmland regions.[21]

In addition, surface-mining for phosphates in Florida, North Carolina, and Idaho; for oil shale in Colorado; and for uranium in several states will add to the toll. For these minerals, surface-mine reclamation standards are not yet established by federal law, so the chances for restoration of agricultural productivity after mining are less certain than for coal. Many of these mine sites are not associated with farmland, so the reclamation issues revolve more heavily around restoration of forestland and rangeland productivity, protecting water quality, and preventing the visual blight of scenic areas.

Electrical Generation and Transmission

No small amount of farmland is used in the generation and transmission of electricity, and forecasts circulated by the electric power industry suggest that future demands will be significant as well. In 1978, with existing generating capacity somewhere in the range of 579,000 megawatts, the industry projected that some 308,000 megawatts of added capacity would be in operation by 1987.[22] Power plants, whether coal or nuclear, need land: 2,000 acres or more is common. One nuclear plant in the Chicago area took 4,480 acres, most of which had once been farmland.[23]

The amount of land needed for a plant depends on its design. Some plants use lakes to provide the necessary cooling for the nuclear core. Water is pumped through the plant, heated by the nuclear "engine," then pumped through a lake or series of lakes where it cools into the atmosphere. The water not lost to evaporation is recycled back through the plant again. This reuse of water makes such a design less water-consumptive than a plant which uses cooling towers depending on high evaporation rates to provide needed cooling, but more land is needed.

In Sullivan County, Indiana, for example, the Merom Generating Station will take 11,600 acres of rural land, much of it prime farmland. The site will include a power plant, a cooling pond, waste disposal areas, and transmission line rights-of-way. In Wisconsin, nearly 3,000 acres of cropland and pastureland have been purchased or condemned for the Tyrone Energy Park.[24]

Security factors weigh as well. Proposals of only a few years ago spoke of constructing huge, multimillion-acre "nuclear parks" where several power plants and fuel processing plants could be enclosed in a very tight security net. This would minimize the logistics of protecting each plant from terrorism, but would mean the loss of millions of acres of agricultural land, thousands of farms, and many rural communities. The predictable uproar created by such a proposal seems to have stilled it, at least for now.

In estimating the amount of land likely to be required in the near future, American Land Forum's Wendell Fletcher uses the figure "well over 100,000 acres" before 1987, while Professor Dick Esseks of Northern Illinois University estimates that "new coal and nuclear power plants may result in the permanent conversion of yet another 1.5 million to 2 million acres."[25]

Although it is difficult to compare their land requirements with those of other types of power-generating facilities, hydroelectric power plants are also important land users. Reservoir sizes of 1,000-25,000 acres are common, and much of this land is valley bottomland that will qualify as prime farmland.

In a widely publicized case, the Tellico Dam, near Knoxville, Tennessee, is being built after years of controversy. Most of the publicity centered around the damage to the Snail Darter, a small, insignificant, but endangered species of fish. What was less well known, but arguably more important, was that Tellico will bury some 16,000 acres of farmland. Nationally, the Soil Conservation Service estimates that about 200,000 acres are being lost to reservoirs each year. Between 1967 and 1975, it was estimated that some 1.5 million acres of prime farmland were lost to reservoir development.[26]

Hydroelectric reservoirs also serve a variety of other purposes, such as flood control, irrigation, and recreation, so their land requirements are difficult to compare with other forms of electrical generation facilities. In addition, the hydro plant contains its own "fuel" source within the acreage taken for the reservoir. This is in contrast to a coal-fired plant such as the Antelope Valley generating station in North Dakota, which needs 600 acres of farmland for its site and will also entail the strip-mining of another 2,000 acres of North Dakota farmland to provide its fuel.

Moving all this energy to consumers creates major issues as well. About 4 million acres are dedicated to rights of way for electrical transmission, with the estimates of future needs ranging from 1.5 million to 3 million additional acres before the year 2000.[27]

The aspect of these developments that may prove most difficult for farming concern the effect of new extra-high-voltage transmission lines on surrounding land use, crops, livestock, and human health. Research on this problem in the United States has been sketchy, at best, but enough controversy has been generated to suggest that farmers next to transmission lines may need to use caution in the ways they use the land in extra-high-voltage rights-of-way.[28] Stories of people getting badly shocked, livestock refusing to graze beneath the wires, and even crop growth being stunted are sketchy yet, but pervasive enough to raise serious questions. Even without these problems, farmers face disruption and added costs in their operations where power line structures are placed in fields.[29]

In a letter to James Schlesinger, Secretary of Energy, on June 26, 1978, Lloyd E. Payne, a South Dakota farmer, explained the frustration, bitterness, and fear of farmers faced with a proposal for a power line. In Mr. Payne's case, the Basin Electric Power Cooperative was planning a 500,000-volt transmission line that would cross his farm. The original proposal would have put the power line directly through the Payne's cattle shed. Payne protested and was assured that the design was not yet final. Before long, however, Basin Electric was back, this time with an easement to be signed, but with the power line still in the same location. Only after a great deal of negotiation was the line moved 400 feet away, a distance that still is not satisfactory to the farmer. In his letter, Payne articulated the fears that many farmers have about the extra-high-voltage lines, and the frustrations they face when trying to get solid information about the potential harm these installations may bring:

> We feel that long-term hazards of 500,000-volt lines have not been adequately researched; we also feel that research which has been done has yielded results which are highly disquieting; we feel the hazard information (as well as other information provided by Basin Electric) is so unreliable as to be utterly worthless. We think it is reasonable, where there are many, many feasible alternate locations for the transmission line, that the line should be moved so that the near side of the 200-foot easement is at least 650 feet from the cattle shed.[30]

Huge 500,000-volt power lines move electricity economically, but take up a significant amount of farmland in the process. Farmers face increased costs of operation as they must work around the towers in their fields, and there are critics who question the safety of these lines, citing radiation and ozone effects on nearby people, livestock, and crops. Department of Energy (DOE) photo.

Controversy about the Data

By 1980, it was clear that there was a serious problem with farmland loss. But while the data at hand was the best ever assembled on the topic, that did not settle the debate over whether or not farmland loss should be considered a national concern. Careful study of the 1977 NRI by NALS showed that Florida, New Hampshire, and Rhode Island could lose most of their remaining farmland in the next 20 years if the current annual losses were to go unchecked. Other states with serious rates of loss compared to their remaining land base include West Virginia, which could lose 73 percent; Connecticut, 70 percent; Massachusetts, 51 percent; New Mexico, 50 percent; Maryland, 44 percent; Vermont, 43 percent; and Utah, 35 percent.[31]

Release of these estimates by NALS staff caused considerable controversy within USDA. Assistant Secretary M. Rupert Cutler called the NALS survey results "explosive," and by-lined a newspaper article that appeared first on the Op-Ed page of the *New York Times* and then was reprinted widely across the country. "America is on the brink of a crisis in the loss of agricultural land that may soon undermine our ability to produce sufficient food for ourselves and other nations of our hungry world," Cutler's opening statement said.[32]

But many analysts in USDA thought this assessment was overly dramatic, and some classified it as a gross exaggeration. Pressures began to build, both within USDA and the NALS study itself, to stem the flow of interim data from the study to the public. USDA economists, seeking to have the NALS study take a more sanguine view based on their analysis of agriculture's potential to produce ample supplies of food despite the land losses through a combination of technological improvements and other inputs such as fertilizer and energy, argued that attempts to build public concern over farmland loss through such "crisis" talk would be proven to be wrong, and would backfire on the advocates of farmland protection. "You cannot sell farmland preservation with the scarcity rationale," said one analyst.[33]

There has to be some reason, however, why public concern continues to mount, despite the fact that the "official" line from the USDA economists has scarcely changed one bit since 1975. At that time, a brochure entitled *Farmland: Will There Be Enough?* answered its title question with the reassuring statement that "fortunately, the U.S. land resource base appears ample for future domestic needs, at least beyond the year 2000. It also seems likely that U.S. farmers will have adequate land and technology to make sizable commitments to the export market."[34]

That statement leaves me (and a lot of other people) not quite satisfied that the USDA economists have found an appropriate way to explain the fact that 44 states and hundreds of counties keep searching diligently for some way to stem the flow of good lands out of agriculture. People place no value on a resource that is in ample or surplus supply, and this helps explain our cavalier attitude toward farmlands in the past. But farmlands, and particularly prime farmlands, are no longer felt to be a surplus commodity by a fair percentage of Americans. A 1979 public opinion poll conducted by Louis Harris revealed that 53 percent of Americans consider the loss of good farmland to be a "very serious" concern.[35] Have those people been misled by "overly dramatic" statements from the farmland protection advocates, or do they see realities that the USDA economists are not addressing?

Perhaps the deficiency of the data we have at hand today is that it tells us a great deal about the land, but not enough about the real toll that the farmland conversion process inflicts on American agriculture and its ability to produce. Maybe it will be necessary to look further than the number of acres covered with asphalt, or even the speed with which certain areas are losing a limited supply of prime farmlands, to see why the public concern with farmland loss continues to grow almost daily.

The Problems Associated with Farmland Conversion

Pierre Crosson of RFF thinks that much of the public concern over the loss of farmland to non-agricultural uses can be traced to the loss of rural amenity values.[36] That is partially true but also incomplete. If it were only "amenities" that are being lost, the concerns would be limited to such things as the loss of open space, visual quality, or a community's "openness." Those kinds of concerns are, indeed, a part of what people are saying. But there is more. Agriculture is more than acres; it is an industry. From individual farm businesses through the many industries that provide agricultural inputs or process agricultural products, to the thousands of small American communities that rely on agriculture as their major source of economic support, agriculture is an interdependent system that must be considered in total. In many communities, it is the entire agricultural industry that is in trouble, and many of its woes begin with the loss of agricultural land.

The loss of agricultural land does more to damage agricultural production than simply reduce the land base. When some acres are urbanized, or covered with water, or developed in any way, other acres are cut off from commercial agriculture. It may be leapfrog development on the edge of a city, or a highway that cuts diagonally through an area and messes up field after field and farm after farm, or it might be a power line right of way that prevents free and easy access with farm equipment. A conservative estimate is that, for every acre lost to development, at least one more acre is affected in such a way that intensive agriculture is no longer possible. That would mean, at the least, the impact on agricultural production is at least twice as serious as the estimates of land loss would otherwise indicate.[37]

There are social and political effects as well. When urban people begin to make up a significant portion of an agricultural community, rural residents begin to lose their political and economic status in the community; agricultural interests give way to residential concerns. These changes result in certain "spillover effects": regulation of farming activities deemed a nuisance (e.g., fertilizer use, manure disposal, slow-moving vehicles on commuter roads); increased taxation to pay for new services and facilities needed by the new residents; increased destruction of crops and equipment and harrassment of farm animals; use of eminent domain to acquire farmland and other rural land for public uses aimed at serving the growing non-rural population (e.g., roads, reservoirs); air pollution damage to crops; disruption of farm infrastructure (e.g., farmers face increasing inconvenience as traditional farm suppliers begin to accommodate suburban gardeners rather than commercial farmers); and diminishing availability of farmland frequently accompanied by increased land price.[38]

But that is still not the entire story. As a farming community begins to undergo the development process and subdivisions start springing up here and there, farmers begin to change their plans for the future. They wonder—and rightly so—whether their land will stay in agriculture much longer or if it will soon become more profitable to sell out. Some worry about the fact that their new neighbors may put pressures on them that will force them to sell out whether they want to or not. Once that feeling of insecurity begins to grow in a farming community, agriculture is in real trouble. New Grade-A dairy facilities don't get built—and old ones don't get modernized. It takes too many years to recover that kind of an investment, and the farmer who wonders how much longer he'll be in business will put that long-term investment off until he can see the future more clearly. As he does, the farm starts wearing out, productivity goes down, and his uncertainty becomes a self-fulfilling prophecy. Pretty soon he has little choice but to sell out.

In the meantime, the farm kids will have left the farm and taken jobs in the city, since nobody thinks the farm has a future anymore. So the little rural town either dries up and blows away or becomes a haven for city workers who come out to live in the country and drive to the city every day for a job. And the farms? They're likely to go into five- and ten-acre ranchettes with a few horses and gardens.

That is the kind of agricultural future that faces many rural communities in America today, and it is not particularly comforting either to agricultural leaders or to local governmental officials. It is particularly critical along the Eastern seaboard, in New York, New Jersey, Massachussetts, Pennsylvania, Maryland, Delaware, and Virginia. In every one of these states, special legislation to keep farmland in production is being considered or special programs have already begun. In county after county, county planning commissions and supervisors are trying to find alternatives for the futures their communities are facing. But slowing the conversion of agricultural land is not easy, because the problem has so many difficult political aspects, and because the trends in population movement and settlement pattern that are causing much of the problem are still poorly understood. We are facing new kinds of development in rural America, and many rural leaders still can't fully understand the extent of the changes occurring in their regions, let alone agree on whether it is good or bad and if it is bad, what to do about it.

The People are Moving

For about a decade now, there has been a major change in the migration patterns of people in America. Reversing past trends, more people now are moving into rural areas than are moving out of them. Rural counties are growing at a faster rate than metropolitan areas. When this trend was first identified in the mid-1970's, it had already been in progress for several years, but people viewed the reports with a great deal of skepticism, according to Calvin Beale, the USDA demographer who is widely credited with identifying the new movement. Beale says now, however, that "as the decade has proceeded, every source of formal data has confirmed the new trend, whether it be population estimates, sample surveys, or employment statistics."[39]

Scarcely anyone anticipated this trend. In the 1960's the prevailing view was that almost all growth would occur in expanding suburban rings around large cities.[40] One study, looking to the year 2000, forecast that the nation's people would be concentrated in a few "super-cities" labeled Chi-Pitts, Bos-Wash, and San-San that would consist of almost complete urban and suburban development between Chicago and Pittsburgh, Boston and Washington, and San Francisco and San Diego.[41] These projections were not entirely wrong, as many people are still being added to these metropolitan areas. But the proportion of growth captured by urban areas has dropped significantly. Between 1970 and 1979, metropolitan areas gained 8.3 million people, while non-metropolitan areas increased by 7.2 million.[42]

What this means, of course, is that rural areas are now experiencing a rate of growth — and a kind of growth — that is unprecedented in recent decades. For the period between 1970 and 1978, non-metropolitan areas averaged 10.2 percent growth, compared to 6.1 percent for metropolitan counties. The increase was greatest (11.2 percent) for those counties adjacent to metropolitan areas, but there was still significant (9.6 percent) growth in the counties that were entirely rural in character.[43]

This rural growth does not mean more people on farms, however, nor is it necessarily tied to agriculture at all. "Growth in smaller areas does not represent a return to the land," according to the Advisory Commission on Governmental Relations. "The new residents of these areas enjoy 'rural amenities,' but they are part of the non-farm urban economy. Thus, the emerging pattern of development could be interpreted more as a dispersal of urban America than a rural renaissance."[44] This new growth has caught local officials off guard and has raised new kinds of issues for agriculture.

In Illinois, for example, rural counties like Jo Daviess are experiencing growth from both population growth and second-home development. The county grew over 10 percent in the 1970's — the first time its population had substantially increased in 100 years. New residents either commute to Dubuque, Iowa, to work, or are drawn to the area by Galena, a restored nineteenth-century mining town or several man-made recreational lakes.[45] New second-home developments are springing up, and the Jo Daviess Soil and Water Conservation District estimated that about 8,000 acres of farmland were converted to non-agricultural uses between 1974 and 1979.[46]

In Stephenson County, Illinois, concern over the loss of farmland led to a 1976 zoning ordinance that sets a 40-acre minimum lot size on Class I and Class II soils outside incorporated areas. This ordinance, designed to direct growth away from the county's prime farmlands, was an attempt to protect a large dairy and swine industry. But county officials are not sure that it will work as well as they want. Del Schieder, local farmer and county planning board commissioner, points out that "scattered housing can ultimately destroy our farm economy, in spite of our best efforts to save it."[47] The risk, local farmers fear, is that even if housing is guided to the poorer soils, the mix of farm and non-farm residents will cause conflicts that will ultimately put farmers out of business.

What is at stake, in Illinois as elsewhere, is not just saving land from development, but saving the farming industry and its ability to function, and this is the point that is missed by the economists who count up the acres that have been converted to urban use and decide that nothing alarming is occurring. Much of the real concern is due, not to the amount of growth or development that is occurring in rural areas, but to the type of growth, its dispersed nature, and the expectations of the people who are moving to rural America.

In preparing the report on "Land Impacts of Rural Population Growth" for the Council on Environmental Quality, field researchers for the American Land Forum found that, "In every area visited, new growth presented problems for maintaining the integrity of farming and farmland."[48] In the Medford Valley in Oregon, a fruit-growing and specialty-crop industry is being threatened by new residential development. In the valleys of New Mexico, irrigated cropland provides not only the resource base for agriculture, it is also the only place for new housing. Most of the surrounding land is in public ownership and not available for purchase. This irrigated land—only 1.3 percent of the land in Santa Fe county, for instance—is going out of agriculture rapidly, and with it, the traditional way of life of the Hispanic community, which is based, in part, on agricultural self-sufficiency.

In Stephenson County, Illinois, there are serious development pressures and a great concern over farmland retention, even though the county lost population in the decade of the 1970's, according to the Census Bureau. This is a sign that rural regions can have difficulties with retaining agriculture that are not necessarily tied simply to the numbers of people moving in or out, or the number of farm acres being urbanized. A look at the type of growth pattern that is occurring and what it means to farmers may help explain some of the concerns.

Buckshot Urbanization

Farming near a city has always been accompanied by tensions, particularly if the farm was right at the city's very edge. City folks with their kids and dogs have always found it tempting to run through a farm field, regardless of the crop damage; hunt upland game or birds, sometimes with permission, sometimes without; leave gates open—or closed—as they travel (or trespass) through farm fields; and, in dozens of ways, make farming a little more difficult—and costly. The problems are not all one way, either. A farmer who starts plowing at 6:00 A.M. can be a noisy neighbor, and the odors of his barn, feedlot, or silage pit can become pretty unpleasant to those who happen to live in a downwind subdivision. If he uses a weed killer that happens to drift, the damage to ornamentals, fruit trees, and gardens can be serious.

But there has always been land-use conflict where two different types of use meet. That is the genesis of the zoning ordinance of today. Zoning was not invented so that government could tell people what to do with their land, but zoning is essentially a private device, with its main function the lessening of private battles between adjacent land users. A quick look at a modern urban zoning map shows that different, and incompatible, uses are carefully kept apart. No one wants a "skunk works" in the middle of a residential neighborhood.

Farmers have basically distrusted zoning as a tool, and it has been little used in rural areas. As a result, even though the cities carefully keep the industrial and commercial land uses from imposing on the residential areas (and vice versa) for all sorts of good reasons, there is no such protection in most of rural America. Homeowners are free, in all but a small (although growing) number of places, to buy a tract of rural land and build the house of their dreams. When they do, all too often they do not become farmers; they become rural-dwelling urbanites, and they don't like the smell of cow manure or the drift of poisonous sprays any more than they did when they lived in suburbia.

This kind of uneasiness—even conflict—has always been a feature of the farm-urban interface. The problem today is the different shape of that interface. It used to be limited to the edge of town, where the urban uses met the farm area in a fairly well defined boundary. Then America went through a period of suburban sprawl, or strip sprawl, where a subdivision leapfrogged over some open farmland and isolated it within a suburban region, or growth followed a highway far into the rural area, bringing a "strip" of commercial and residential neighbors to a number of farmers.

Development in rural areas not only takes up farmland, but also affects nearby farms. The random pattern that characterizes much of today's rural housing has been called "buckshot" urbanization (top). Dairy farmers are often affected the most, with expensive dairy barns and outbuildings that have no sale value to developers (bottom). When nearby neighbors begin complaining of odor and other farm sights and sounds, the dairy may be forced out of business. SCS photos by Tim McCabe (top) and Erwin Cole (bottom).

While the land-use planners were trying to figure out how to cope with those kinds of growth, an entirely new pattern developed. First termed *buckshot urbanization* by land-use analyst Charles E. Little of the American Land Forum, the new pattern is readily shown by the land-use map of almost any county in rural America today. In the typical pattern, small spots of growth, many only a few houses in size, grow up in a random pattern throughout the farming region, until the map looks like someone stood back and shot at it with a shotgun. There is little logic to the shape of the growth, except that it will almost always be defined by the system of farm-to-market roads.

Buckshot urbanization brings the urban-rural interface to most, if not all, the farmers in the region, and with it, the potential for land-use conflict. The stories abound. I spent an afternoon with a frustrated farmer in Idaho in the early 1970's who had developed a very sophisticated irrigation system. He had constructed concrete head ditches that were slightly higher than the field grade. The ditch could be filled with water, then siphoned over the ditch bank onto the field with siphon tubes. Once the siphon tubes were all flowing, the amount of water in the ditch could be adjusted to get just the right flow to irrigate the field properly.

The system worked beautifully—except on the weekends. On Saturday and Sunday, his neighbors had time off from their jobs in town and wanted to irrigate their one-acre yards and gardens. Ignoring the strict rules enforced on the farmers as to when each could and could not take water, these "ranchette-er's" needed to water when they were home, not when it was their turn on the ditch. Besides, each felt that the little bit of water he or she needed wouldn't make much difference. But it did.

All too often, the total withdrawal from the lawn-waterers would drop the flow in the farmer's ditch to the point where the siphon tubes would drain the water level in the ditch down too low. What would then happen, as the siphon tubes sucked air and quit taking any water at all, was a disaster. The water would continue to run into the ditch, but with no place to go. It would then begin to overflow, washing away the earth foundation under the ditch, and making huge gullies down through the farmer's fields. Thousands of dollars of damage could be done in a matter of minutes.

The farmer was at a loss as to how to protect himself and was considering whether or not commercial farming was any longer possible on his land. He reminisced with me about the time, a few years earlier, when he had been the only person on the ditch, but now he had 20 or more homeowner neighbors—several on one-acre lots that he himself had sold from the roadside fields on his farm. What had once seemed to him like an easy financial bonanza had turned into a boomerang against his business.

In the Midwest, the story was different, but the theme was familiar. In 1975, at Fort Wayne, Indiana, conservation district officials told me of the frustrations being encountered in cleaning out major drain ditches which provide the outlet for the tile drainage systems that are vital to the high yields of many of the soils in that region. Weeds, grasses, and brush grow in the ditches and, if not cleaned out periodically, begin to trap silt and build up the elevation of the ditch bottom. Because the entire area is so flat, it takes only a few inches before the outlet grade is gone, and the on-farm drains begin to back up. The results, of course, are clogged drains and wet fields, so the farmers and the drainage districts they have formed operate a constant clean-up campaign to keep the main outlet ditches flowing free.

Conflicts begin to crop up as non-farm rural residents begin to protest the ditch maintenance program and, in places, stop it. These people liked the natural beauty of the brushy drains, they valued the wildlife habitat and visual relief offered, and they did not want to see the brush cleaned out. At times, they either owned property adjacent to the drain or could swing public opinion in their favor, and were able to halt the drain maintenance program. The long-term result, the conservation district officials feared, would be the removal of many prime farmland acres from agricultural production as the soils returned to a wet condition. The short-term result was serious land-use conflict, with increasingly bitter feelings between the farmers and their rural neighbors.

It is not just farmers who are caught up in these conflicts, because the basic dichotomy lies between urban-oriented homeowners and industrial land users of all types. In Southern Oregon, researchers for the American Land Forum found that timber industry officials were worried about the future of timber management because new rural residents objected to the use of brush-killing agents on forestland. The problems of spray drift, with its attendant vegetative damage and possible health effects, was leading the rural residents to seek restrictions on the foresters.[49] Since that time, the removal of 2,4,5-T from most approved uses by the Environmental Protection Agency has borne out the fears of the lumbermen.

In that same area of Oregon, the scattering of homes and other development into remote forest areas has been accompanied by an increase in forest fires. In addition, the influx of new residents, and the views they bring as to how the natural resources of the region should be used, are affecting the management of the public lands. Forest Service Rangers must begin to plan timber sales and other resource uses with the desires—and political power—of the new residents, as well as those of the cattlemen and lumber industry, in mind.[50] In every instance, no matter who appears the aggrieved party, industry ends up the loser.

The irony of this is that the industrial infrastructure built to make farming, ranching, or forestry more productive is what makes buckshot urbanization possible. Most important, of course, is the rural road system. In many farming areas today—particularly the highly intensive farming areas such as the Corn Belt or the irrigated areas of the West—a paved road runs through at every mile interval. These farm-to-market roads were built to improve farm productivity, and they have. But they also provide free access to millions of homesites. Often, the road is accompanied by a rural power line, a rural telephone line, an irrigation or drainage ditch, and sometimes even a natural gas line. It becomes a great place to build a house!

When a person buys an acre next to one of these farm roads, access and services are readily available—and cheap. There will be no extra charges, as there would have to be in a conventional subdivision, for extending access and services to each lot. The result, unfortunately, is that when too many people take advantage of those rural systems, the system begins to be overloaded. Farm-to-market roads become commuter highways, and harried drivers have little patience with slow-moving farm vehicles or loose farm animals. Roads designed for limited traffic break down under heavy use, or become inadequate. The farm support systems, built at great public expense to facilitate commercial agriculture, have attracted the kind of growth that now, in many regions, threatens the continued existence of the farm economy.

And, finally, there is the effect on productivity due to land speculation. There is little incentive to maintain soils, soil conservation practices and structures, barns, drainage, and fencing if a farmer is going to sell his land to a land speculator or developer in the near future. Studies have shown that farmers tend to exploit or mine land that they do not own or on which they don't have long-term leases to provide secure tenure.[51] With a disproportionate amount of the nation's highest quality cropland lying in proximity to urban areas, the impact of diminished productivity due to a lower level of soil care in these regions is increased.

One attempt to quantify the relationship between buckshot urbanization and agricultural productivity found measurable effects, however, only in relation to dairying.[52] The probable explanation is that the proportion of farmland idled or whose productivity decreases as the result of these indirect effects is generally incremental and diverse. As such, these effects appear as little more than "noise" in the data currently being collected. Although it is very difficult to estimate the lost productivity resulting from population spread into rural areas, the damage is there and we must learn how to mitigate it somehow.

Farming, ranching, and forestry are resource-based industries that have as much difficulty co-existing with residential properties in a rural area as a foundry does in the city. Farmers in some parts of the country are now looking to land-use regulations as a way to protect their interests. "Right to farm" ordinances are being widely discussed in the East. These would establish the types of farming activities that were protected within the agricultural regions of the community, and protect farmers from any nuisance suits brought by non-farm people who might move into the farming area at a later date.[53]

That may not keep the roads from getting overcrowded, or the farmers from losing livestock to roaming dogs, but it will at least assure them that they will not be subject to a nuisance lawsuit because of tractor noise, animal odors, or other aspects of farming. Next, farmers are going to have to look for a way to prevent the buckshot urbanization pattern itself. If they don't, a thin overlay of urban development may jeopardize agriculture in many of America's most productive farm regions.

That danger to agricultural productivity won't be spotted in land conversion statistics or population surveys. It will be picked up, first, in attitude surveys among the farmers themselves, then in close monitoring of their investment patterns. When farmers begin to doubt that farming has a strong economic future in their area, it may not have any future at all. The land conversion statistics are but a single-dimensional indicator of a serious multidimensioned problem. They reflect what has happened on the land long after the real changes in agricultural productivity have taken place. The problem, of course, is that we do not have adequate ways to measure what is really happening to U.S. agriculture as a result of the complex set of changes that are affecting farm productivity. What is happening today is clearly not an extension of what has happened in the past, so a simple review of history as a preview of where farmland trends are leading is inadequate. At the same time, the data we have contains no magical prophecy either.

It also contains no answers to what, in the final analysis, is largely a political question. The relationship of people to the land is at question, but even more important, so is the relationship of people to each other. Those are the kinds of conflicts that the political process is supposed to solve, but modern-day politicians, at all levels of government, have found the political thorns of land-use issues almost impenetrable.

Farmland as a Land-Use Issue

Aldo Leopold admonished us that we were "remodeling the Alhambra with a steam shovel, and we are proud of our yardage." As urbanization and growth continue to wipe out the best croplands of the United States, it is clear that we need a new way to decide the best long-term use of land. "We shall hardly relinquish the shovel, which after all has many good points," Leopold said, "but we are in need of gentler and more objective criteria for its successful use."[54]

Land-use issues, including those that involve agricultural land, are more than issues of how we should use land. They are outward manifestations of deeper questions troubling society. How shall we grow? Which technologies are beneficial and which harmful? What kind of an environment do we want for our children, and their children? How can we afford it? These abstract questions become very real when a community is drawn into a land-use battle.

A new subdivision, highway, or power plant may take agricultural land out of production. In addition, it may attract other kinds of growth onto nearby lands and take more farmland, or encourage a land-use pattern that increases competition for land and conflict between farmers and adjoining neighbors. Local citizens, sensing conflict, get concerned and question the proposal. The argument hardens and emotions rise. Private citizens and public officials are drawn into a battle.

The argument may be about a highway, but the questions are about growth and change in the community, and most communities do not know how they want to grow. There are too many unknowns and too little agreement about the most desirable future. People are no longer comfortable with the notion that all growth is good or that all construction is progress. But they know that no growth is not a real option, either.

The questions that rise strike to the heart of economic and political activity. How do you balance public needs, while protecting private rights? How, and how far, should government intervene in private land-use decisions? These are questions of deep concern and far-reaching significance. But they are not, for the most part, being debated in the high halls or supreme courts of the land. They are being argued in county courthouses or city halls by ordinary citizens locked in a battle over a subdivision or rezoning proposal.

The fact that farmland may be lost as a result of a development proposal often serves simply to enter more unknowns into the debate. Many times the land involved is prime farmland, so the potential loss of agricultural productivity is, on a per-acre basis, fairly high. By its definition, prime farmland is level, free of stones, and protected from flooding damage. Because of its obvious attraction for intensive farming, often it has been cleared of trees; is served by roads, electricity, and other support services; and is assembled in large contiguous tracts. It is no riddle why it is highly prized as development land. In the economic market, farming can never outcompete more intensive uses for such lands. But the growing public concerns about the continued conversion of the nation's best farmland may make the debate even more strident.

Many characteristics make the farmland issue one of the most complex of all land-use issues for local politicians to handle. Among them:

1. Retaining prime farmland in production not only has economic implications; it also raises ethical and moral questions about producing food, preventing hunger, and distributing the benefits of natural resource wealth equitably.
2. There are hundreds of millions of acres of farmland in America, and most developments involve only a few acres or a few hundred. The developer can always argue that the 20 or 40 acres in question is not important in either the local or national agricultural picture. Where and when in this "nibbling process" is the public interest threatened, and how should government establish a way to discourage it?
3. Retaining farmland in production requires the protection of the agricultural industry as well as the protection of the land. It is not enough to retain open space if the community really wants to retain agriculture as a producing, healthy part of the local economy.
4. In many parts of America, competition for limited water supplies, depletion of underground aquifers, and increased energy costs threaten to reduce agricultural production despite the existence of available land. In much of the West, where irrigation is essential to prime farmland, land-use decisions must not only protect the land, but also the water that makes it prime.
5. Public works, such as highways, sewers, reservoirs, power plants, or airports provide major attractants for new growth. These governmental actions heavily influence subsequent private land-use decisions. A developer may decide to buy a certain farm because the sewer is nearby. If government has built the sewer through

prime farmland, then tries to use zoning to prevent normal economic land conversions and development from taking place, it is fighting against itself, and the land will be the ultimate loser.

6. Prime farmlands are privately owned lands, and efforts to retain them in production affect their value as well as their use. Farmers often have most of their accumulated wealth and their major source of retirement security locked up in their land. Public actions to limit the use of these lands must recognize these private values.

The farmland question is not an absolute issue. Local government can't prohibit the use of prime farmland anywhere for any type of development activity. Such a policy would be impossible to enforce. In much of the country, needed development must be placed on farmland if it is going to be built at all.

Often there are alternatives that can be considered. A few years back, the Iowa highway department studied some projects that had been approved for construction but where construction had not yet started. They found that several thousand acres of prime farmland could be saved by simple design alterations, without compromising transportation or safety goals. When those highway designs had originally been done, people were not concerned about impacts on farmland. Once that concern was expressed, a second look by the planners revealed that alternatives did, in fact, exist.

But changing a road design here or a subdivision plan there is not going to eliminate the pressure on agricultural productivity being created by farmland conversion. An entirely new approach to land-use planning, growth management, and protection of resource-based industries like farming, ranching, and forestry will be required. The old approaches simply won't work, because the circumstances are completely different from ever before.

What is happening is nothing less than the complete redevelopment of the rural landscape, without any real planning or guidance by any level of government, and without the singularity of purpose which united the first developers. Those pioneers, with the common goal of making the new land agriculturally productive, could usually agree on what should and should not be done to make their communities best for the type of farming, ranching, or forestry that was adapted to the region. They established local government, and dictated what it did and did not do. But that unity of purpose is gone from much of rural America today, and the new people, with their new patterns of growth, and their new political strength, are redeveloping those farming regions in ways that farmers often cannot tolerate but, so far, cannot prevent.

A basic change has occurred — one from which America will never retreat: Agricultural land and its fate are now a public concern. On February 7, 1980, the U.S. House of Representatives, for the first time in history, seriously considered the notion that America's land base for farming is limited and that the federal government ought to do something about it.[55] They weren't ready to enact legislation that day, and H.R. 2551, the Agricultural Land Protection Act of 1980, sponsored by Congressman James M. Jeffords of Vermont, was defeated by a vote of 177 to 210.

It is a fair guess that the issue is far from dead, however, and that Jefford's bill, or something similar, will be back. Back in 1977, when the farmland issue was first raised on Capitol Hill, "all but a few congressmen either went glassy-eyed or thought the environmentalists were at it again, trying to zone all the farmland for scenery," says land-use analyst Charles E. Little, who was deeply involved in the issue at that time.[56] That is no longer true and probably will never be again. Protecting America's farmlands is liable to be the most important land-use issue of the 1980's, from county courthouses all over America to the halls of Congress.

The National Agricultural Lands Study

One of the reasons farmland loss will stay on the public policy agenda is the National Agricultural Lands Study (NALS). This 18-month, $2 million study was a cooperative venture between USDA and the Council on Environmental Quality (CEQ), growing out of the defeat of the Jeffords bill.

That bill would have: required all federal agencies to adopt procedures to minimize the adverse effects of their activities on farmlands; established a presidentially appointed committee to study farmland protection issues and determine how serious the losses are and what should be done about them; and provided funding and technical assistance to local and state governments willing to try some pilot projects for farmland protection.[57]

USDA, under severe pressure from the Office of Management and Budget, opposed the legislation on the grounds that the study proposed could as easily be done without the bill, and the other actions were not warranted until such a study was completed. Largely because of that position and because of the concern for farmland loss held personally by Secretary of Agriculture Bob Bergland and CEQ Chairman Charles Warren, the administration-sponsored study was initiated.[58]

The NALS study did no original data-gathering in its short life span but assembled a great deal of information on farmland issues and presented a strong argument for federal action to shore up assistance to state and local government in an attempt to aid their efforts in slowing farmland loss. In addition, the report noted many ways that the federal government could "clean up its own house" so that federal programs and projects were not the major driving forces behind the unnecessary conversion of farmland.

It is, of course, much too early to assess the impact of the study on federal policy or public opinion relating to farmland loss. It was finalized in January, 1981, during the final days of the Carter administration. Its findings were quickly embraced by incoming Secretary of Agriculture John Block, but the details of how the study will be incorporated into future USDA policies and programs are yet to be developed.

One positive indication of the most likely fate of all the information generated by the NALS* occurred in the new administration's 1981 Farm Bill proposals, where a title on farmland resources constituted the only new natural resource conservation initiative to be proposed. Block's proposal would have Congress recognize that agricultural land is an important strategic resource that ensures the nation's self-sufficiency for food and fiber. USDA would be encouraged to cooperate with and provide assistance to state and local programs for farmland protection and produce additional informational materials.

The legislative proposal is not important for what it contains, because it would authorize USDA to do no more than they already are doing, but because it exists at all. It is the first time the Office of Management and Budget (OMB) has allowed any Secretary of Agriculture to support *any* proposal to aid in slowing the conversion of farmland. That is a good omen, reflecting either a gradual shift in opinion at OMB, a resurgence of influence for the Secretary of Agriculture, or both. It could be a hopeful sign for future conservation policy initiatives.

*In addition to the final report containing the study's findings and recommendations,[59] NALS published several interim reports on such topics as farmland and energy,[60] and the effects of soil degradation[61] that address issues which affect agricultural land, and a set of technical reports, on the various aspects of the agricultural land issues studied. One of the reports contains an in-depth study of the types of state and local programs operating in 1980, along with an assessment of their effectiveness. A complete list of the NALS reports can be obtained by writing the Director of Information, Soil Conservation Service, Washington, D.C. 20250.

6.

Soil Erosion

Soil, it is commonly taught, is a good example of a renewable resource. It produces a crop this year—whether of wheat or corn, grass for grazing, or trees for wood. That crop can be removed and converted to human use; then, with management techniques that have been widely known for generations, the land can be used to produce another crop. The wheat is consumed, but the soil lives on, renewed by nature and capable of producing another crop, and another, and another.

Or does it? The truth is, soil is a renewable resource *only* as long as it is managed for sustainable production. When it is managed for maximum economic production in the short term, soil is a non-renewable resource, undergoing slow but inevitable destruction. The most common, and rapid, way for soils to be destroyed is through the removal of topsoil by either wind or water, through the process of erosion.

Soil erosion is a natural part of the environment. It has been occurring since the first wind or water struck the surface of dry earth and will continue as long as that process continues. Human actions have, however, drastically accelerated the erosion process. And the historical record has been replete with civilizations who did not properly care for their soils. Despite those warnings, soil erosion in the United States is still occurring at a rapid rate, far more rapid than the soil can be naturally regenerated.

Phase I of the 1977 National Resource Inventories (NRI)

estimates shows a national 1977 average loss from sheet and rill erosion in excess of 4 billion tons. If it were all concentrated in one area, 4 billion tons of soil loss would mean the removal of all the topsoil (six inches) from 4 million acres. With that kind of loss each year, it would take only 100 years to wash away *every single acre of cropland in the United States.*

How Erosion Works

The erosion process is complex, and too little is known about the extent of the soil damage currently affecting America's farmlands, what that really means in terms of its effect on future productivity, and how to develop appropriate solutions to the problem. In the 1977 NRI, however, the Soil Conservation Service (SCS) gathered a great deal of new information regarding the type and amount of land damage currently occurring in the United States.

Like every other profession, soil scientists have developed a "language" that is commonly used whenever soils or soil erosion are discussed. In this "language," soils are the natural, living systems where the root systems of most higher plants live, along with millions of microorganisms that carry out essential functions of converting minerals, dead plant and animal remains, water, and air into compounds that nourish the plants in a continuous cycle of life.

The size of the mineral particles determines the texture of the soil, which, in turn, affects both the productivity and erosivity of the soil. Clay is the smallest mineral, composed of microscopically fine particles, while silts and sands are names given to the larger particles. Organic matter in the soil can be recognizable bits of dead plant or animal material, or it can be in a concentrated, digested form called humus. The size of the mineral particles and quantity of organic matter determine the pore space or pockets within the soil and affect its ability to absorb, hold, and freely circulate air and water. Pore space, and soil-air-water relationships are also affected by soil structure, which is the manner in which the soil particles are held together in larger groups called soil aggregates.[1] Gardeners know the value of crumb structure, which is the tendency for high-organic soils to form small to medium-size crumb-like bits that make the topsoil loose and open to air and water. Other types of structure such as large blocks may form within the soil, where they affect the penetration of roots, air, and water.

Erosion, the wearing away or movement of soil or its component parts, may be caused solely by the force of gravity acting on soils that rest on steep slopes, and there is some soil movement from the effects of agricultural tillage. But wind and water are the dominant erosive force.

The erosive force of water starts with the raindrop. The larger the drop, the more force it has when it strikes the soil, and the more soil particles are displaced and moved. Larger raindrops are usually associated with higher-intensity rains. The more rapidly the rain falls, the less able the soil is to absorb the water, and the more water must run off the surface of the land. The total amount of rainfall, both in any given storm and over a period of time, is important as well.

The pounding of raindrops on exposed soil loosens soil particles to begin the erosion process. Where the soil surface is protected by grass, little to no erosion occurs, but where there were no grass or stones to break the force of the rain, 2 inches of topsoil were lost from this site. SCS photo by Sharpe.

As water gathers on the surface and begins to flow downhill, its force is affected by the steepness of the slope. The steeper the hill, the more rapidly the water flows, and the more soil-carrying energy it can generate. The length of the slope is important, too. As water begins to move down a slope, it forms small rivulets, then collects into larger channels as the rivulets come together. This produces the familiar candelabra pattern of erosion on a hillside, with the rills — or gullies — becoming larger and larger as they come further down the hill. Long slopes allow large water volumes to collect and greatly increase the total amount of soil per acre that will be moved in a given storm.

The smaller and lighter components of the soil, all else being equal, will be the most easily dislodged and carried the furthest before being deposited. With the lightest particles gone, the physical or chemical characteristics or composition of the soil are altered. In general, an eroded soil will be more coarse-textured, have lower organic matter content, and be less fertile as a result.[2]

The physical characteristics of the soil, including texture, structure, organic matter, and pore space are critical in determining the degree of erosion that is likely to occur. Soils range from very resistant to erosion to very susceptible.[3] In general, soils made up of a large percentage of silt (the medium-size soil particles) will be highly erodible. When these soils are cultivated on steep slopes, in regions where frequent high-intensity rains are common, soil conservation measures are essential to prevent severe erosion.

Erosion from moving water is called sheet, rill, gully, or streambank erosion, according to the type of soil movement that occurs. Both sheet and rill erosion are caused by water moving across the surface of the land. Sheet erosion removes soil fairly uniformly in a thin layer, or sheet, while rills are small channels formed as running water concentrates and flows in small rivulets down a slope. Gully erosion, in contrast, involves the formation or enlargement of small to medium-size ravines or channels that are too large to be obliterated by normal tillage operations. Streambank erosion refers to the soil moved from the banks of established streams, creeks, or rivers.[4]

Most of the soil erosion on agricultural land — particularly cropland — is sheet and rill erosion. Tremendous amounts of soil are transported in this manner. Up to 10-20 tons per acre can be lost from farm fields during winter and spring runoff in rills so small that they are obliterated by the first spring cultivation. Thus, the major type of erosion loss on cropland may go largely unnoticed.

The classic rill erosion pattern shows on this Class IV and Class VI hillside in eastern Washington's Palouse region. This hillside lost an average of 145 tons of topsoil per acre in the 1971 spring runoff season. SCS photo by Verle Kaiser.

Even relatively gentle slopes suffer severe erosion if left exposed to the rain. These rills will be erased by the first cultivation, but hundreds of tons of precious topsoil are gone forever. USDA photo.

In the Midwest and Southeast, pictures of huge gullies scarring the land were instrumental in sparking public concern about soil erosion in the 1920's and 1930's. Today gully control methods are fairly well known and widely used, but severely gullied land still exists in many parts of the nation. Once eroded, it is almost impossible to bring the gullied land back into productive crop use.

Streambank erosion, although serious in some local situations where streamside lands are being chewed away by the meandering actions of adjacent streams, moves only about one-eighth as much soil as sheet and rill erosion, according to SCS data.

Wind can equal or exceed water in its destructive force on the land, particularly on soils that are devoid of plant cover, have a fairly smooth surface, and are composed of loose soil particles that can be lifted and carried by the wind. Wet soils erode very little, if at all, but wind erosion increases as soil moisture decreases. Completely dry soil erodes one-third faster than a soil that has enough moisture for plant growth.[5]

Although dust clouds often rise thousands of feet in the air, most of the actual soil movement takes place within a few inches or feet of the ground. Soil particles are "bounced" along by the wind when they are too large to be carried aloft, and their energy in striking and dislodging other soil particles at the soil's surface adds to the effectiveness of the wind as a soil mover.

Wind speed and direction are important factors. A wind blowing 50 km/hr will cause three times as much soil erosion as a 30 km/hr wind.[6] The amount of soil that any field will lose in a wind storm is affected by the distance across the field in the direction the wind is blowing, as well as the speed of the wind. Knolls, ridges, sand pockets, or other unprotected spots start to erode at lower wind velocities than the other parts of a field, but once the process is started, it often expands rapidly in a fan-shaped pattern downwind. A blowout that begins on an exposed knoll may be one-half mile wide by the time it has reached one mile downwind. This is due mainly to the abrasive, erosive action of the wind-blown sand, which can destroy vegetation and open up soils that would have otherwise remained protected. Once unprotected, these soils are helpless against the drying action of the wind and the abrasive action of the wind-blown sand. Stopping wind erosion patterns as quickly as possible when they begin is essential to keep wind damage from rapidly escalating both in seriousness and scale.

The ability of the wind to move soil is tremendous. The winds blowing over the Mississippi River basin, for instance, are estimated to have 1,000 times the soil-carrying capacity of the river itself. Despite the fact that around half the soil moved in a wind-storm remains within 2 inches of the ground, and over 90 percent may be transported at less than 12 inches off the ground, one dust storm in 1895 was estimated to have carried 1,600 tons of dust per cubic mile over Rock Island, Illinois.[7]

These dust clouds near Lamar, Colorado, were the result of high spring winds in the mid-1950's. Similar "dust bowls" occur periodically in this dry region. USDA photo.

Soil Erosion Today: What the Surveys Show

Despite a significant investment in conservation practices by the federal government and by private landowners, erosion from agricultural lands continues at a massive rate. Nearly 4 billion tons of topsoil are moved each year by sheet and rill erosion, with half the loss occurring on cropland, as Table 6.1 shows. In 1977, erosion rates were above 10 tons per acre per year (t/a/y) on 32 percent of the lands used for row crops in the South, 9 percent in the Northeast, and 19 percent in the Corn Belt.

To make these figures more meaningful, a ton of soil is

roughly equivalent to a cubic yard. An inch of topsoil covering an acre would weigh in the neighborhood of 165 tons. Six inches of topsoil, the depth normally cultivated in modern agriculture, weighs about 1,000 tons. Thus, an estimate that a given field is losing 10 t/a/y means that it will lose an inch of topsoil every 15-20 years, and the whole plow layer in roughly 100 years.

Table 6.1. **Sheet and Rill Erosion for All Capability Classes**

Land Use	Average Annual Erosion Rate	Acres	Total Erosion
	(tons/acre)	**(millions)**	**(million tons)**
Cropland	4.7	413	1,926
Rangeland	2.8	408	1,154
Forestland	1.2	367	445
Pastureland*	2.6	133	346
Total		1,321	3,871

SOURCE: *RCA Appraisal 1980*, Part II.
*Includes native pasture.

The croplands that are suffering the most soil erosion are the steeper, more marginal lands, as would be expected. The 1977 NRI found that the average annual rate of sheet and rill erosion ranges from 2.8 t/a/y on Class I cropland to 23.1 t/a/y on Class VIIe land used for crops. (The *e* in the VIIe means that the land is placed in Class VII largely due to an erosion problem. See Appendix A for a full discussion of capability classes and subclasses.)

Table 6.2, which gives erosion figures by land capability class, shows that 25 percent of the sheet and rill erosion on cropland occurs on only 15 percent of the land; the land that is marginal or unsuited for cropland. Continuing to use these lands for growing crops is certain to continue high soil erosion rates. The only cure is to take them out of crops now rather than after the topsoil is exhausted and the land laid waste. Farmers face the choice of converting these lands to grass or trees now, while some topsoil is left, or taking off a few more crops and then abandoning the land.

Table 6.2. **Estimated Average Annual Sheet and Rill Erosion for Cropland, 1977, by Capability Class**

Land Capability Class	Land Area		Annual Sheet and Rill Erosion	
	Acres	Percent	Tons	Percent
	(millions)		(millions)	
Classes I-III	351	85	1,446	75
Class IV	44	11	287	15
Classes V-VIII	18	4	193	10
Total	413	100	1,926	100

SOURCE: SCS National Resource Inventories, 1977.

Surprisingly, the 1977 NRI also indicated serious erosion on some of the nearly level soils of the Mississippi Delta, caused by the fine-textured, highly erodible nature of the soils, coupled with very long slopes.[8] Class I cropland in Louisiana, for example, is eroding at the rate of 9.1 t/a/y, a rate that will lead to total topsoil depletion in little more than a century.

The prime farmlands of the nation suffer less soil erosion loss than other land used for crops, but they still suffer.[9] The estimated average soil loss from sheet and rill erosion is 3.8 t/a/y on prime farmlands, compared to 5.8 t/a/y for the other croplands. This difference, though significant, is not as great as might be expected and suggests that even the prime farmlands, which are the very easiest for farmers to farm without soil damage, are not receiving the type of conservation care they need.

Depending on how the land is managed, erosion on pastureland and forestland is generally less than that on cropland. Well-managed forests represent a minimal risk of soil losses from sheet, rill, or wind erosion. Likewise, pastures do not lose much soil unless they are overgrazed. While all land uses can be managed to limit excessive erosion, current estimates indicate that cropland and rangeland produce the bulk of the sheet and rill erosion today (Table 6.1).

On forestlands, those that are grazed lose by far the most soil, eroding at a rate of 3.9 t/a/y, six times the rate experienced on non-grazed forestland. As a result, some 55 percent (239 million tons) of all sheet and rill erosion from non-federal forestland comes from the 17 percent that is grazed.[10]

Grazing management is the key to erosion control on range and pasturelands. The field on the left is well protected from pounding rains, but the overgrazed one on the right is vulnerable and could be seriously damaged by a storm. SCS photo by Carl Holzman.

Non-federal pastureland and rangeland account for an average of about 1.5 billion tons of yearly soil loss to sheet and rill erosion. Pastureland erodes at rates of over 5 t/a/y occur in states like Hawaii, Illinois, Kentucky, and West Virginia, where farmers graze cattle on steep slopes. About two-thirds of all soil lost from pasturelands occurs on the Class VI-VII lands, although they make up only about one-fourth of the acreage.[11]

The estimated average annual soil loss on rangelands goes from a low of less than 1 t/a/y to 10 t/a/y, with a national average of 3.4 t/a/y. Most rangelands are in arid and semi-arid regions where rainfall is infrequent but often intense. Because native plant cover in arid regions is often sparse, much of the soil erosion is a natural phenomenon, but the kind of livestock management practiced also directly affects the amount of land damage that occurs.

To get the most complete picture of the soil erosion problem available from Phase I of the NRI, Table 6.3 adds the wind erosion estimates from the ten Great Plains states, along with the sheet and rill erosion estimates on pastureland, rangeland and forestland.

Table 6.3. **Sheet, Rill, and Wind Erosion on Non-Federal Cropland, Pastureland, and Forestland in 1977.**

	Non-Depleting	Depleting	Rapidly Depleting
	(1,000 acres)		
	(less than 5 t/a/y)	(5-13.9 t/a/y)	(14+ t/a/y)
Cropland	272,224	93,053	48,000
Pastureland	119,021	9,485	5,062
Forestland	353,047	11,721	4,895
	(less than 2 t/a/y)	(2-4.9 t/a/y)	(5+t/a/y)
Rangeland	283,478	55,501	68,882

SOURCE: *RCA Appraisal 1980*, Draft Summary and Program Report.

These totals give an idea of the amount of land that is eroding at rates that will rapidly deplete the topsoil. With almost 12 percent of the cropland losing more than 14 t/a/y and 17 percent of the rangeland losing over 5 t/a/y, we face the task of either stopping this loss or writing those acres out of the productive inventory within a few short decades. At a time when more productivity is being demanded, a significant percentage of our productive lands are rapidly approaching ruin.

Erosion in Urbanizing Areas

New information developed in the National Agricultural Lands Study (NALS) indicates that soil erosion may be more serious in areas undergoing urbanization than in the rural counties nearby. A study of the 1977 NRI data shows that, for six metropolitan areas, the average rate of sheet and rill erosion on cultivated cropland was

17.5 t/a/y. The average rate in adjoining counties within the same Major Land Resource Area (a geographical area established by SCS that has similar soil and climatic conditions) was 11 t/a/y. Thus, at least for these six metropolitan areas, significantly higher rates of soil erosion seem to be occurring.[12]

Table 6.4. **Erosion in Urbanizing Areas, 1977**

Metropolitan Area	Crop	County Group	RKLS	Erosion in t/a/y
Milwaukee, Wisconsin	Corn	Urban	23.35	6.14
		Rural	15.25	3.94
Memphis, Tennessee	Cotton	Urban	48.04	23.50
		Rural	45.47	20.67
	Soybeans	Urban	95.44	38.22
		Rural	46.30	15.92
Sioux City, Iowa	Corn	Urban	69.59	20.21
		Rural	32.08	10.42
Kansas City, Missouri	Soybeans	Urban	66.03	28.19
		Rural	49.98	19.26
Chicago, Illinois	Corn	Urban	12.88	4.50
		Rural	12.07	4.72
	Soybeans	Urban	17.80	6.27
		Rural	9.61	3.95
Harrisburg, Pennsylvania	Corn	Urban	58.28	12.90
		Rural	30.79	9.30

SOURCE: Compiled by Allen Hildebaugh and Charles Benbrook, NALS Technical Paper No. 14, *The Economic and Environmental Consequences of Agricultural Land Conversion.*

One of the reasons for this phenomenon could be tied to the "impermanence syndrome" that causes farmers to slow down investments on the land. Farmers in urbanizing areas often feel that the land will soon be lost, and it is pointless to be spending money on soil conservation practices that will soon be rendered useless by a change in land use. Another factor, however, may be that farmers are farming poorer lands in the urbanizing areas. In Table 6.4, the column *RKLS* refers to the average product of the rainfall (R), soil erodibility (K), length of slope (L), and slope steepness (S), factors in the Universal Soil Loss Equation. The higher the number, the more erosion-prone the soil. (See Appendix D for details on the use of this equation.) As can be seen, the RKLS number for the urban areas is higher than for the adjoining rural areas. Whether this is due to the fact that urban growth has taken the better soils and pushed farmers onto more marginal lands, or simply that the lands in these urbanizing counties was more marginal at the outset, is not known. It does appear, however, that there is a surprising difference in the soil quality of the cultivated cropland between the urban and rural counties.

Where Does Eroded Soil Go, and What Does It Do There?

The damage from the soil erosion process doesn't end with the land that has lost topsoil. Much of the soil that is eroded ends up in waterways, lakes, and reservoirs. Although the relationship between the amount of soil eroded from a field slope and the amount of sediment ultimately reaching the waterways is poorly understood, studies have shown that from 10 to 50 percent of the soil eroded from the hillsides may eventually end up in the water.[13] Large, gently sloping watersheds deliver the least sediment to the stream, since there are more opportunities for the eroded soil to drop out of the moving water prior to reaching watercourses. Small, steep watersheds can deliver more than half the eroded soil directly to the stream. Thus, the relationship between the soil erosion occurring on the land and the amount of soil reaching the stream can only be determined by a complex calculation on each individual watershed.

Sediment from soil erosion is the largest single pollutant by volume in the nation's waters, with some 4 billion tons reaching streams annually. Of that amount, about 1 billion tons reach the oceans, while the other three-fourths of the sediment remains in streams, rivers, reservoirs, and lakes.[14]

When soil reaches waterways, it does many kinds of damage. Silt and sediment restrict water flow, increasing flood damages, reducing the quality of the water for other recreational or consumptive uses, and damaging fish and wildlife habitats. Spawning beds for fish may be destroyed as silt smothers the gravel beds needed for egg laying and incubation.

In a January, 1975, report, the Council for Agricultural Science and Technology (CAST) estimated that soil erosion cost $83 million for dredging channels and harbors, $50 million for floodplain overwash and reservoir sedimentation, and $25 million for added water treatment costs and system maintenance in munic-

ipal and industrial water systems.[15] The CAST report also estimated that, based on 1974 prices, it would cost some $1.2 billion to
replace the plant nutrients lost through soil erosion that year.

*Topsoil that has left the farmer's field becomes a nuisance somewhere else.
Here, silt deposits have virtually clogged a stream channel in Mississippi.
Without expensive clean-out, this stream will become more and more flood-
prone as the channel capacity diminishes. SCS photo by W. L. Watts.*

Soil particles, along with the nutrient elements (particularly
phosphorus) and pesticides they carry, have an important ecological
impact on streams as well. They affect temperature, light, dissolved
oxygen, and food production in the water; all of these will have
impact on the biological life in a stream. The added nutrients increase
the growth of algae, which, as it dies and decomposes, lowers the
oxygen levels in the water too low for fish. The result can be fish
kills and a shifting of dominant fish species toward those tolerant of
lower quality water. Other effects of sediments include the reduction of light penetration into the water, which can reduce aquatic
plant growth or alter species composition, and increased water
temperatures, which can harm fish species like trout.[16]

The soil that does not reach streams also causes important damages. If sand and coarse soil particles come to rest on fertile bottomland soils, the result can be a lower productivity. The opposite situation is also possible, however, if reasonably fertile soil material is deposited on top of sand or gravel bars. For that reason, it is difficult to generalize about the net effect of sedimentation on the productivity of bottomland soils. My intuitive estimate is that it is negative, but there is little data to support such a conclusion, let alone calculate its magnitude.

In addition, eroded soil material fills roadside ditches, gutters, and culverts, causing significant maintenance costs each year. Wind-blown soil can come to rest in dunes and drifts up to several feet in depth, burying topsoil, roads, railroads, fences, and even buildings. Airborne dust adds to the discomfort and costs suffered by people with respiratory ailments and adds to the total problem of air pollution. In short, soil out of place is a costly nuisance, a prime example of a valuable resource that has been converted into a liability through human neglect.

How Much Soil Loss Is Too Much?

In order to assess the impacts of soil erosion on the future of agriculture, we must know what a "tolerable" soil loss limit might be. Soil erosion cannot be totally eliminated under any feasible management plan, but it can be reduced to levels that allow the soil to retain productivity. But what is that level? What is "tolerable"? Tolerable soil loss might be defined as the maximum rate of annual soil erosion that will permit a high level of crop productivity to be sustained economically and indefinitely. It will have to vary for each soil, depending on how fast new topsoil can be formed to replace the soil lost to erosion.

Soil formation is a slow, continuous process, with new soil material being gradually formed as minerals break down due to chemical and biological processes. Scientists estimate that, under agricultural management, around one inch of new topsoil will be formed every 100 to 1,000 years.[17] At the most rapid rate, this is about 1.5 t/a/y. The rate varies widely, and is influenced by climate, vegetation and other living organisms, soil disturbances, the nature of the soil's parent material, and topography.[18]

If there is to be no loss of the long-term productive capacity of the soil, the thickness of the topsoil and a sufficiently favorable rooting depth must be maintained, along with favorable physical, chemical, and biological conditions for plant growth. No single tolerance rate is applicable to all types of soil, but the multiple judgments of soil scientists, agronomists, and geologists is that 5 t/a/y is the maximum rate of loss for which indefinite and economic productivity can be maintained. Why the "tolerable" level accepted by SCS is roughly three times higher than the rate of soil formation is a good question, one that needs attention. As now accepted, however, erosion tolerance levels used by SCS range from a low of around 1 t/a/y to a high of 5 t/a/y on some soils, with a national average close to 4 t/a/y.

The estimates of tolerable erosion levels fail to consider the impacts that eroded soil may have on other lands where it may come to rest. Soil erosion estimates only count the acres from which soil has been removed, and since an average of nearly 75 percent of the eroded soil is eventually deposited on another site, much of the soil that has been eroded has not been truly lost. But qualitative changes occur on both the eroded soil and the soil where sediments are deposited. The smaller, lighter, nutrient-rich organic and clay particles are most likely to leave the original site and, being the most transportable of all soil materials, these often end up in a stream or lake. So their removal depletes the eroded soil without enriching intermediate soils during the soil-transporting process. It is this loss of organic matter that is the primary reason for the lower productivity of eroded soils.[19]

That leads to a criticism of the basic concept of soil tolerance levels. If there is a qualitative change in the soil as the result of erosion, and that change is generally negative in its effect on the soil's productivity, how can it be said that the soil losses below 5 t/a/y (or whatever level is established as the tolerance level) are of no harm? If a crop field is losing 10 t/a/y and we can, through conservation methods, reduce that rate to 5 t/a/y, has more been accomplished—in terms of enhanced or prolonged productivity—than if we were to take a field losing 6 t/a/y and reduce it to 1? Unfortunately, current knowledge doesn't answer this question. My feeling is that it may make little difference—that the amount of improvement achieved on both fields (a 5-t/a/y reduction in the erosion rate) is more important than the meeting of an arbitrary target established on what may be fairly shaky technical ground.

Finally, official estimates of tolerance levels give little consideration to the effect of erosion on rooting depth. The weathering of parent rock into a favorable root zone is a distinctly different phenomenon from the formation of topsoil. In most soils it proceeds more slowly; thus, while limiting erosion to 5 t/a/y might maintain the topsoil, the total root zone would become thinner and the soil poorer as a result.[20]

One example of this occurs in soils that are shallow over rock. Several years back, I assisted farmers in eastern Idaho who were working soils that had, in their native state, been 12-15 inches deep over bedrock. By 1968, after less than 50 years under the plow, the rock was only 6-7 inches below the soil surface. Cultivation was becoming difficult, but there was nothing that could be done to speed soil formation. If those soils were seeded to grass and left undisturbed, it would be thousands of years before they would again have 12 inches of soil depth.

There are equally difficult problems on range and forest soils. Many of these soils have very thin topsoils under natural conditions, often occurring in areas where high altitude, short growing season, or limited moisture slows the soil formation processes dramatically. On some of these soils, it may take 10,000 years to produce an inch of topsoil. What is a "tolerable" soil loss under these conditions?

In short, there are serious questions about the concept of "tolerable" soil loss. Work is underway today to address those questions, but what is important to the public is the concept that should undergird that work. The objective of humans in using soils should be that those soils will maintain their productive potential despite that use. Any goal short of that is simply transferring the cost of today's excesses on to our children or grandchildren.

The Effect of Erosion on Soil Productivity

Research on soil loss and its effects on productivity demonstrate that yields of crops other than legumes go down on eroded soils unless liberal amounts of plant nutrients are added to replace the lost soil fertility. In addition, losses may result from lower water-holding capacity in the soil or restrictions in root penetration and growth. Whatever the form, soil erosion always increases the cost of production.[21]

Converting these facts into quantitative estimates of the amount of loss is somewhat more difficult. Field studies show that losing six inches of topsoil will severely compromise the productivity of most cropland, if not make it unprofitable to farm. In the southern Piedmont, for example, a six-inch reduction in topsoil was found to reduce average crop yields by 41 percent.[22] In western Tennessee, similar soil erosion reduced corn yields 42 percent and vetch and fescue forage yields 25 percent, and wheat and oat grain yields dropped nearly 30 percent.[23]

Hugh Bennett, in his later life, told a story about one of his early encounters with soil erosion. The place was Louisa County, Virginia; the year 1905, but Bennett's story could be repeated today in virtually any part of the United States.

> Bill McLendon of Bishopville, South Carolina, and I were stirring through the woods down there in middle Virginia when we noticed two pieces of land, side by side but sharply different in their soil quality. The slope of both areas was the same. The underlying rock was the same. There was indisputable evidence the two pieces had been identical in soil makeup.
>
> But the soil of one piece was mellow, loamy, and moist enough even in dry weather to dig into with our bare hands. We noticed this area was wooded, well covered with forest litter, and had never been cultivated.
>
> The other area, right beside it, was clay, hard, and almost like rock in dry weather. It had been cropped a long time.
>
> We figured both areas had been the same originally and that the clay of the cultivated area could have reached the surface *only through the process of rainwash*—that is, the gradual removal, with every heavy rain, of a thin sheet of topsoil. It was just so much muddy water running off the land after rains. And, by contrast, we noted the almost perfect protection nature provided against erosion with her dense cover of forest.[24]

The qualitative change in soils as a result of the erosion process is also significant, although little scientific work has been done to measure this. A field experiment, done in Dodge County, Wisconsin, clearly illustrates how soil deposited by erosion may appear to be of high quality, but in fact be inferior to the topsoil it covers up.[25]

Soil profiles were studied on a hillside, both in a forest and in an adjacent field, and at the bottom of the slope in both cases. The topsoil was 14 inches deep in the woodland, but only 5 inches remained in the cultivated field. At the bottom of the slope in the field, 7 inches of the sandier portions of the topsoil from above had been deposited. About 2 inches of nutrient-rich organic and clay material from the hillside soil had been washed beyond the field into the natural waterways.

As a result of being buried under the sand, the soils at the bottom of the hill required more fertilizer and lime than the eroded soils at the top. Plant roots, instead of having normal topsoil near the surface, were forced to feed in soil from which the natural nutrients had been largely removed.

Wind erosion degrades soil in much the same way as water erosion, but often in a more dramatic and visible manner. The wind action on the soil is just like that of a fanning mill that separates wheat from chaff. Organic matter, clay, and fine silt are removed first and carried furthest away. The lighter the particle, the more likely it is to be carried into the high air currents, which can often transport the soil for hundreds or even thousands of miles. The coarser, less-fertile sand and coarse silt particles hop, slide, or roll over the surface of the soil and often are not moved far. These tend to pile up in drifts wherever some change in topography or surface features cause the wind to slacken and lose its carrying energy.

Each time a soil is shifted by the wind, it loses more plant nutrients and becomes coarser textured. After a soil has been moved several times, it will be largely sand, regardless of the original texture. One study in Oklahoma in the 1930's showed that, as a result of cropping and wind erosion, the organic matter in the cultivated soils had dropped 18 percent and the nitrogen was decreased by 15 percent.[26]

Additionally, the damage to growing crops from wind erosion can be significant. National data is lacking except in the ten Great Plains states, but in that region, from November 1977 to June 1978, SCS estimated that 2.2 million acres of crops or cover were destroyed on land where the erosion rate was not enough to damage the soil permanently.[27]

The abrasive action of wind-driven soil particles on growing plants also damages crops by uncovering plant seed and by changing plant metabolism. Crop damage may occur even though the soil erosion tolerance is not exceeded. Estimated crop tolerances to wind erosion range from nearly zero (onion, cucumbers, lettuce) to above the soil tolerance limit (buckwheat, barley), with most crops showing some damage at a soil erosion level of 1 t/a/y or less.

Soil deposited by water (top) or wind (bottom) is more sandy and less fertile than in its original state, due to the separating action of wind and water during the soil-moving process. In both fields, covered crops were destroyed by the silt. SCS photos by Darrell Vig (top) and Elvin W. Jenkins (bottom).

In the past, many adverse effects of soil erosion on the productivity of land have been masked by other factors. New and more productive crop varieties coupled with the heavy use of fertilizers, better control of pests and crop diseases, and improved tillage and planting methods have resulted in yield increases despite topsoil loss. While these technological increases have masked the permanent effects of soil erosion, they have not eliminated them. We are now dependent on such technology (much of which is growing more expensive as petroleum prices rise), and must, it appears, continue to have similar technological breakthroughs in the coming years to maintain or improve crop yields. How seriously this affects our future ability to produce will depend on our success in controlling soil erosion. If we continue to let topsoil slip away at the rates found in 1977, we are doomed to a future of spending more and more trying to coax less and less from a dying land.

The Effect
of Soil Erosion
on National Productivity

One of the really difficult tasks in a study of the land is estimating the total effect of soil erosion on the nation's productive capacity. Even if we are able to show the impact of erosion on one soil, in one place, at one time, what does that mean to the nation as a whole? How can we quantify the national problem? Two methods have been proposed, but neither provide entirely satisfactory estimates. One was used by USDA as part of its appraisal of soil and water resources under the Resources Conservation Act (RCA).[28]

Using crop yield statistics, USDA estimated that if the current level of erosion were allowed to continue for the next 50 years on the 290 million acres of cropland contained in its model, erosion would cause a reduction of productive capacity equivalent to the loss of 23 million acres of cropland, or 8 percent of the total land base included in the calculation.[29]

The implications of this analysis can be seen by an example from the Corn Belt, one of the most productive farming areas in the world. This region, despite its reputation for outstanding soil and climatic characteristics, has serious erosion. About 43 percent of the land used for row crops in the Corn Belt is composed of highly erodible soils. If erosion in this area is allowed to continue at the 1977 rate, USDA estimates that potential corn and soybean yields would probably be reduced by 15-30 percent on some soils by 2030.

Another approach to estimating productivity loss due to erosion is to examine the loss in terms of acre-equivalents.[30] This method, which I first proposed in 1979, would offer a way to evaluate the total amount of soil erosion in terms that can be related to other estimates of productivity. There is little way to equate a ton of topsoil directly in terms of bushels of grain or dollars of value. What we can do, however, is approximate how many acre-equivalents of productivity this loss of topsoil might represent. That will be a rough estimate of the magnitude of the loss suffered, in terms that can be compared to the acreages lost to other land uses, or added by the conversion of other lands into cropland.

In 1977, the total excess erosion (sheet and rill erosion greater than the soil-loss tolerance) shown for cropland was 1.01 billion tons. If a six-inch layer of topsoil weighs approximately 1,000 tons per acre, and the loss of six inches of topsoil will destroy the productivity of most cropland, then it would appear reasonable to equate 1,000 tons of erosion with the loss from productivity of one acre (acre-equivalent).

The fact that the loss is the sum of many smaller losses, spread over many acres, makes it no less a loss. These acre-equivalents are not individual acres that are being washed away in one dramatic event, but they represent the slow, imperceptible losses being cumulatively suffered over millions of acres of American farmland.

On this basis, cropland is losing—to sheet and rill erosion alone— about 1.01 million acre-equivalents annually. Wind erosion in the ten Great Plains states would add another 0.24 million acre-equivalents. This would add up to about 62 million acres of lost productivity over the next 50 years, assuming a continued uniform rate of loss. That is roughly triple the estimate that was developed by USDA, but it applies to the entire 413 million acres of cropland in the United States, where the USDA estimate includes only the 290 million acres of major crops that are covered by their computer model.

The two approaches suggest that over the next 50 years the loss of productivity due to erosion on cropland will be equivalent to the loss of between 25 and 62 million acres. To get some idea of the magnitude of that loss, 25 million acres could produce 50-75 million metric tons of grain, or half of the total exported from the United States in 1980. If we lose 62 million acres, it could represent the loss of virtually all of 1980's exportable surplus. So these estimates are significant to America's productive future. At the same time, they are both incomplete.

None of the gully or streambank erosion estimates are included in either estimate. Wind erosion data outside the Great Plains was not available. No erosion calculation was made for the effects of applied water on irrigated land, and the estimates tend to understate erosion west of the 100th meridian. Thus, both estimates are probably low.

Yet that staggering soil loss is not inevitable. Table 6.2 shows that, by converting 18 million acres of Class V-VIII cropland to grass or trees, American farmers would cut sheet and rill erosion by 10 percent, but affect productivity by less than half that amount. Installing effective conservation measures on Class III and Class IV cropland could cut soil losses by 20-30 percent. Better grazing management on rangeland and forestland would significantly reduce erosion there, particularly if efforts were concentrated on the steeper lands.

In short, *preventable* soil erosion is depleting the nation's productive soil resource base at a rate that is a serious cause for concern. The main point in these calculations is not one of the scientific accuracy in tons or acres. The main point is that the current rate of productivity loss, however one chooses to convert it to numerical values, is *far too high*. It can, and should, be drastically lowered. To do less is to accept the fact that our farmlands, and our society, are destined to die.

7.

Losing Soil Quality

Topsoil doesn't need to be bulldozed, paved over, or washed away to lose its productivity. If it is mismanaged, or polluted, it can lose many of the natural soil properties that directly contribute to plant growth or the soil's capability to respond to agricultural management. In farmer language, it can be "worn out," and most farmers know that when they decide a piece of land has reached that stage, whoever tries to grow crops on it will have a difficult and expensive time of it.

Unfortunately, soil science doesn't give us a very good definition of what "worn out" means, at least in scientific terms. One way to approach the subject is to define what makes soil good for farming, so that as those qualities are lost we can measure the effect on soil productivity. Soil quality can be defined as the sum of those factors that makes the soil productive and/or responsive under modern agricultural management. It cannot be completely separated from soil erosion, since eroded soils are nearly always lower in quality than they had been in the natural state, and many times soils that have been buried by sediments carried from another site may be reduced in quality as well. But soil quality can also be lost through the depletion of soil organic matter or an increase in soil compaction, salinity and alkalinity, acid buildup, or even toxic substances. The topsoil may still be there, but it is worn out.

At the outset, we should recognize that there are many ways that good agricultural management can improve soil quality. The natural soil state was not the most advantageous for crop production in the desert soils of the West, for example. Under natural conditions, rainfall and vegetation were both sparse, so there were few plant roots in the soil. As a result, organic matter was low, soil structure was often nearly non-existent, and plant nutrients such as nitrogen and phosphorus were often lacking. When farmers first began to irrigate these soils, water intake rates were often very slow, since the soil tended to puddle and seal at the surface because of the lack of organic matter and structure.

This exposed bank shows the distinctive layers of soil formed under normal conditions. The dark topsoil contains most of the organic matter and nutrients to support crop growth. If the topsoil is lost, the soil's productivity will drop dramatically in most cases, since the soil material below is less fertile and less able to absorb and hold moisture. SCS photo by B. C. McLean.

A few years of good management, however, has often resulted in a marked improvement in these soil qualities. Organic matter levels in the topsoil can rise from less than one-half of 1 percent to as high as 2-3 percent. Water intake rates improve, water-holding capacity goes up, and total fertility is higher. This is the result of management techniques such as crop rotations containing grass and alfalfa, applications of barnyard manure, and proper management of irrigation water, crop residues, and fertilizers.

Similar examples could be used in circumstances where the natural soil condition was too sandy, had a hardpan that limited root growth, or was very deficient in one or more plant nutrients. Good agricultural management can help remedy these deficiencies.

This topsoil has a high percentage of organic matter and a good granular structure and should be easy to cultivate and manage. In this highly magnified photo it is easy to see the many grass roots that are instrumental in building up the soil's favorable structure. USDA photo.

Unfortunately, poor agricultural management can also create problems. In many of the grassland soils, where high organic matter levels, good soil structure, and high fertility greeted the first settlers, today we find far lower levels of organic matter, along with reduced nitrogen and phosphorus levels in the soil, lower water-holding capacity, and poor soil structure. Because soil quality is reduced, more artificial inputs are needed to replace the natural values that have been lost. That is a built-in increase in the cost of agricultural production that will continue to extract its toll each and every year.

Although there is far too little data to support any firm conclusion about the current trends in average soil quality, there are many indications that it may be going down. Crop rotations that include grasses or legumes are being abandoned in favor of either monoculture or two-crop rotations such as corn/soybeans or wheat/potatoes. Soil compaction is a concern, and there are growing problems with both salts and acids. No one knows just how fast this trend is proceeding, but it is possible to look at some of the individual parts of the picture and get some general ideas about its seriousness.

Organic Matter

Organic matter is the single most important indicator of soil quality, and its reduction over a period of years is a sure sign that the productivity of the soil is being lost. The source of organic matter in the soil is partially decayed plant and animal remains that remain in an active state of decay through the action of soil microorganisms. As a result, organic matter is a transitory soil constituent that must be constantly renewed by the addition of plant residues.

Most mineral soils do not contain much organic matter— usually from 3 to 5 percent. In desert soils, organic matter content may run less than one-half of 1 percent. Under wetter, cooler conditions, the natural organic matter content may run much higher, up to 15 percent or more in some cases.[1] Organic matter content increases as the soil gets wetter, since the added moisture encourages more plant growth without increasing the rate of decomposition. Cooler temperatures slow down the decomposition process in the soil so that higher organic matter levels can build up.[2] Most of the organic matter is concentrated in the top few inches of the soil, as Figure 7.1 shows.

Figure 7.1. The Average Organic Matter Content of Three North Dakota Soils before and after 43 Years of Cropping

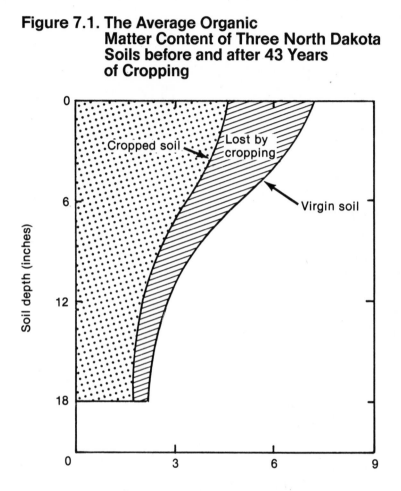

Organic matter content (%) (N x 20)

SOURCE: H. J. Haas et al, Nitrogen and Carbon Changes in Great Plains Soils as Influenced by Cropping and Soil Treatments, *Tech. Bul. No. 1167* (Washington, D.C.: USDA, 1957)

The influence of organic matter in the soil is much greater than its small amounts might indicate. It serves as the "glue" that binds soil particles together into the granules and crumbs that make good soil structure, keeping the soil open for the passage of air and water. Organic matter increases the amount of water a soil can hold and also increases the proportion of soil water that will be available for plant growth. It is the major soil source of phosphorus and sulfur, as well as essentially the sole source of soil nitrogen. It darkens the color of the soil, causing the soil to absorb heat more readily and warm up more rapidly in the spring. Finally, it is the major energy source for soil microorganisms.[3]

The microbial population of the soil is concentrated in the top few inches, as is most of the organic matter. The topsoil of a heavily manured field may contain up to 30 million bacteria per gram of soil, along with many other organisms such as actinomycetes, fungi, molds, yeasts, protozoa, algae, and others.[4] These microorganisms perform essential functions in converting raw plant and animal materials, soil minerals, and atmospheric elements into usable plant food in the soil, and their numbers, as well as their benefits, must necessarily drop as organic matter content goes down.

Natural levels of organic matter usually drop rapidly as soon as a soil is opened up for cultivation. The reductions, on an average of about 35 percent, seem to be virtually inevitable unless extra organic matter is consistently added to the soil, as the additional aeration created by cultivation speeds up the "burning" process that oxidizes organic matter into carbon dioxide and other compounds. Yet it is also clear that organic matter levels should be stabilized at some point and not allowed to drop further. Figure 7.1 seems to indicate that the cropped soil in the North Dakota tests has begun to stabilize at somewhere around 2.5 percent. It would be helpful if we could forecast that retaining the organic level at 2.5 percent — or any other percentage figure — would hold soil productivity at satisfactory levels in the future. I don't know of any research that would establish that kind of threshold level, but it is certain that letting organic matter levels slowly slip further and further downward is not an acceptable management system.

Farmers can keep organic matter levels at stable levels in the soil through proper crop rotations and residue management. Test results at the Ohio Experiment Station at Wooster, where crops have been grown continuously since 1894 without manure or fertilizer, are shown in Table 7.1.[5] Here, nearly two-thirds of the original organic matter has been lost from the continuous corn plots, while only about one-sixth has been lost from plots with a three-year rotation of corn-wheat-clover.

Organic matter can be added to the soil in several ways. Green, immature crops such as rye, oats, sweet clover, and vetch can be plowed into the soil, a practice called green manuring. The legumes such as sweet clover and vetch add nitrogen from the air, so are extremely valuable sources of soil-building plant residue. Barnyard manure is a valuable source of organic matter, as are composts made up of various plant residues, municipal sewage sludge, garbage, or other organic materials.

Table 7.1.**Effect of Crop Rotation on the Organic Matter Content of Soils**

Cropping System	Carbon	Nitrogen
	(pounds in top $6^2/_3$ inches of soil)	
Initially (1894)	20,400	2,180
After 30 Years Continuous Corn	7,400	840
After 30 Years Continuous Wheat	12,800	1,300
Five-Year Rotation (corn-oats-wheat-clover-timothy)	15,500	1,550
Three-Year Rotation (corn-wheat-clover)	17,100	1,780

SOURCE: R. M. Salter and T. C. Green, *Journal of the American Society of Agronomy* 25 (1933): 622.

The most important source of organic matter in cropland soils is the residue of the crops that are grown each year. Stubble, stover, and other aftermath, as well as the roots of the crop plants, make up the major contributions. Farmers who maintain high crop yields can, under many cropping systems, grow enough organic matter to retain soil levels, provided the residues are returned to the soil following harvest.

Table 7.1 shows that, after 30 years of a corn-oats-wheat-clover-timothy rotation with no barnyard manure added, the organic matter of the soil had dropped about 25 percent. With 8 and 16 tons of manure per acre per year in the rotation, the organic matter was 93 and 102 percent of the original soil levels, respectively.[6] Thus, it seems that intensive organic management of cropland soils can keep organic matter levels at or near the original soil levels, particularly in cool, moist climates.

At the garden level, applications of organic compost at rates of ten tons per acre per year over a period of six years was shown by Hugh Ward, a New York banker and avid gardener, to be highly effective in raising the levels of organic matter and plant nutrients on a sandy soil. When he began, the soil had 120 pounds of phosphorus, 150 pounds of potash, and 10 pounds of nitrogen per acre. At the end of six years, phosphorus was up to 800 pounds per acre, potash was at 450 pounds per acre, and nitrogen was up to 40 pounds. Organic matter levels had reached almost 6 percent — in a sandy soil — and crop yields were reported to be "highly successful." [7]

So the question is not whether or not farmers can maintain organic matter levels in the soil, it is whether or not they *are*. The general consensus is that soil organic matter levels are dropping in the humid cropping zones such as the Corn Belt, but there is little empirical evidence upon which to base that or any other conclusion. The 1977 National Resource Inventories (NRI) done by the Soil Conservation Service (SCS) did not measure organic matter levels, and there is little benchmark data against which to compare such measurements if they were done. So it is up to farmers to test their own soils and keep their own records on this most important indicator of soil quality. Soils where organic matter levels are holding up may not be totally protected against degradation — soil erosion may be occurring or there may be a slow buildup of pesticides or toxic materials — but they are certainly better than soils where the organic matter levels are continuously declining. If that is happening, it is only a matter of time until the productivity of the soil is marginal.

Soil Compaction

In its natural state, soil is about one-half solid material and one-half open pore space. This pore space provides channels for air, water, plant roots, and soil microorganisms, and is critical to proper functioning of the life in the soil. When soils are compacted, or crushed, the amount of pore space is reduced and the quality of the soil is changed accordingly. The major factors in compacting farm soils are the number of times the soil is compressed by the wheels of farm equipment, the size and weight of that equipment, the condition of the soil at the time the equipment passes over it, and the measures taken to relieve the problem or "open up" the soil once more.

The trend toward larger and more capital-intensive farming, as evidenced by the statistics showing that row cropping increased by 27 percent and rotation hay and pasture dropped 40 percent between 1967 and 1977, is evidence of a similar increase in soil compaction.[8] In addition, farm equipment has gotten larger and heavier, and the pressures of getting crops planted, cultivated, or harvested on larger farming units, coupled with the added capability of the new machines, may encourage farmers to work the soil when it is wet and subject to additional compaction.

The bigger machinery is heavier, and so it compacts the soil worse than the older machines. In many cases, the tires are far bigger, so the pounds per square inch of pressure exerted on the soil may be no greater, but the total percentage of the soil affected is increased. Tractor wheels cover about 20 percent of the soil surface with six-row machines and 30 percent with eight-row, so the increase in machinery size significantly increases soil compaction.[9]

It is not uncommon for field operations to compact soil to a depth of at least 12 inches. Compaction varies with the kind of soil, amount of organic matter, surface texture, and soil moisture. Wet soils and soils lacking significant levels of organic matter are especially susceptible, but compaction is not easily measured, and provides few visible signs for the farmer to recognize. He can often feel the effect of soil compaction, as his tractor demands a lower gear to pull the plow than was previously required, but the degree of soil damage and added cost to his operation is difficult to quantify.

There have been some indications that "lenses" of compacted soil may also be building up below the level of cultivation in some farm soils, perhaps caused by shock waves created in the soil by the passage of heavy, vibrating tires over the surface or simply by the pressure transmitted downward. If these layers impede the flow of roots, moisture, air, or nutrients, they may reduce the effective root zone of the plants and cut yields accordingly. Since they occur below the level of normal cultivation, breaking up these newly created "hardpans" may require deep chiseling or some other equally expensive practice.

The full effects of compaction on soil productivity are not known with certainty. It affects the transmission of water, air, and heat through the soil and reduces water infiltration, which can result in more surface runoff and soil erosion.[10] Roots may be physically restricted from growing freely, or the reduced air and water available to them may interfere with their normal uptake of water and plant nutrients.[11] Both seedling emergence and plant growth of sugar beets and peas were shown to be affected in recent experiments in Britain, with pea yields reduced as much as 50 percent under compacted soil conditions.[12]

The amount of compaction in farm soils may build up slowly, as a result of commonly accepted tillage practices, and its effects may be difficult, if not impossible, to measure accurately. It may affect crops only under certain weather conditions, and then it will be hard to tell how much of the yield reduction was due to the soil compaction and how much was due to lack of moisture, low temperatures, or some other factor.

Freezing and thawing can have the effect of breaking up compacted soil, but it can only relieve part of the problem. The most pronounced effect of freezing is in the upper two-three inches, where up to 90 percent of the compaction can be relieved. Much less benefit is realized at greater depths, so freezing and thawing probably does little to aid in relieving subsoil compaction. In addition, about one-third of the nation's cropland normally will not freeze enough to realize any reduction of compaction at all.[13]

Estimates of yield reductions due to compaction are at best localized in nature; there is no regional or national data collection. A study done on two Minnesota soils in the early 1960's indicated that compaction caused slower seed germination, lower plant populations, and delayed maturity. Where only the surface soil was compacted, yields dropped 7.5 percent. When both topsoil and subsoil were compacted, yields were 14 percent lower than on an uncompacted soil.[14] More recently, a survey of corn and soybean farmers in south central Minnesota identified soil compaction as the worst problem facing the modern cash crop farmer in that region.[15] Farmers in that survey said that decreased yields during wet years when compaction was worst averaged 7.5 percent for corn, 10 percent for wheat, 13 percent for sugar beets, and 54 percent for potatoes.

The loss of crop value in the United States as a result of soil compaction was estimated to be $1.18 billion in 1971 and $3 billion in 1980, but this estimate was admittedly based on very sketchy data.[16] In 1971, for example, compaction was estimated to be serious enough on 2 million acres of California cropland to reduce crop yields. In addition, another 2-5 million acres were estimated to be rapidly approaching the same condition.[17] In sum, soil compaction is an increasingly important element in the soil damage being vested on American farmlands. A great deal of additional knowledge is needed to know how to identify all of the aspects of the problem, as well as the remedies needed to halt its spread.

Desertification

Desertification is a word that has only recently crept into the language of land-use analysts, and then usually in reference to Africa. There, the continuing outward creep of the Sahara is a threat both to Algeria and Tunisia on the north and to the entire band of countries that rim the desert's southern edge (called the Sahel). The 1968-1973 drought in the Sahelian countries brought world attention to the plight of the people there and to the need to be concerned with the rate at which the desert's advance was destroying vitally needed cropland.

When rangelands have been overgrazed, they can be easily con-
verted into desert by a prolonged drought. If the abuse continues
or the drought cycle goes too far, the land can be permanently
ruined. SCS photo by C. A. Rechenthin.

In the United States, desertification has attracted far less attention, but it still holds some serious implications for the future productivity of the nation's agricultural lands. Desertification is defined by Australian geographer John Mabbutt as a "change in the character of the land to a more desertic condition" involving the "impoverishment of ecosystems as evidenced in reduced biological productivity and accelerated deterioration of soils."[18] In other words, desertification is a process that saps the ability of the land to produce vegetation and support life.

Desertification is often thought of as an invasion from a nearby desert, with sand dunes rolling over previously arable lands. While that may, in fact, occur, the more common form of desertification comes when the land itself begins to be unable to respond to the stresses of drought. This may occur when the vegetation has been overgrazed and seriously weakened. A prolonged drought may kill the vegetation, opening the soil to severe erosion and reducing its ability to absorb rains when they do fall. Perennial vegetation, killed by the combination of overgrazing and drought, is replaced by invader species such as tamarisk or Russian thistle, and the soil's productivity is dramatically reduced.

The overall land area affected by desertification in the United States has been estimated at somewhere in the range of 225 million acres.[19] On a map, the areas affected reach from the southern tip of California east through southern Nevada, Arizona, and New Mexico, then take almost all the western half of Texas and extend northward through the Oklahoma panhandle into southwestern Kansas.

The principal cause of desertification in the United States, as in the rest of the world, is overgrazing by livestock. In 1879, when John Wesley Powell wrote his prophetic *Report on the Lands of the Arid Region of the United States*, there were only about 14 million sheep and 5 million cattle in the arid West. But Powell could see the incentives to enlarge the livestock numbers greatly, and he warned that the grasses of the arid region, although nutritious, were not abundant. They would be "easily destroyed by improvident pasturage, and. . .replaced by noxious weeds." The only way the grass could be safely utilized, he warned, was for it to be "grazed only in proper seasons and within prescribed limits."[20]

But the mood of the West was one of expansion and growth, not protection and conservation. Within just ten years, the number of sheep on the western ranges was up 20 percent and the number of cattle up 60 percent.[21] The results can be seen today on millions of acres of land that is, to the eye, largely worthless. It grows mainly brush and weeds, plus some early spring annual grasses that provide a little grazing. Range conservationists often calculate that it takes 40 or more of these desert-range acres to provide feed for one cow for one month. Yet the pioneer stories tell of grass "as high as the cow's backs" on much of it. The land's productive capacity has been virtually destroyed. What little topsoil is left is usually trapped in the roots of sagebrush plants, while between the plants is nothing but the pebbles left when the wind and water carried the topsoil away.

One such area is the Challis Planning Unit near Challis, Idaho, which has been the subject of an intensive study by the Bureau of Land Management (BLM). Of the 358,000 acres on the Challis unit, some 30 percent is classed as being in "poor" range condition and 64 percent is called "fair."[22] Over half of the area is undergoing moderate to severe soil erosion, according to the BLM. The area has been grazed heavily, by domestic livestock, wildlife, and feral horses and burros, since the late 1800's. BLM reduced the number of domestic livestock allowed to graze on the land in the mid-1970's, but these reductions have come too late to prevent the soil damage, and the chances of the land ever building back up to its former productivity are very slim now that the topsoil has been lost.

In another area, the San Joaquin Basin of California, an interagency study showed that of the 4 million acres of private rangeland, 3.2 million have soil erosion and plant condition problems. On some 338,000 acres of rangeland, the study found that the forage was so badly overgrazed that it could not revegetate, and that woody or noxious plants were replacing the forage vegetation.[23] All told, the study estimated that soil erosion causes an annual loss of forage amounting to $1.2 million in the San Joaquin Basin.[24]

The process of soil erosion and rangeland deterioration are closely interrelated. In the San Joaquin, a fierce wind storm in 1977 stripped as much as 23 inches of soil from some of the denuded rangelands, carrying soil away at the astounding rate of 167 tons of soil per acre *in just one storm.* The huge losses were caused by the combination of very high winds — up to 186 miles per hour — and the poor condition of the soil.[25] Erosion was most serious where the vegetation had been overgrazed. Obviously, once this much soil is removed, the chances of reestablishing good plant cover are very low. Thus, the cycle of land destruction, or desertification, continues.

Cropland is not immune from desertification, either. In Kiowa County, Colorado, some fields lost an estimated 150 tons of soil per acre in one huge dust storm in February, 1977.[26] In this dry area, the average annual rainfall is only 10 to 17 inches, and drought is a common occurence. The native grasses that would, if properly utilized, protect the soil, have been plowed up. In the western part of the county, some 150,000 acres of Class VI lands are being cropped, with erosion rates averaging 20 tons per acre per year. These lands, if they keep losing topsoil at that pace, will soon become totally unusable for crops and may, in fact, be too marginal even to establish a good grass cover. They will be "desertified" in the sense that the soil can no longer produce the vegetation it was once capable of producing, and the fault will lie totally with their mismanagement. Drought, rather than being an aberration of nature in that region, is a fact of life. Farmers and ranchers who have ignored that reality have sealed the land's fate.

High grain prices in 1973-1974 brought outside investors to marginal areas such as southeastern Colorado. They bought up cheap rangeland, plowed it and planted it to wheat, pocketed a year or two of good profits, then sold out when the price of wheat dropped again in the face of overproduction. Because the land was then considered "cropland," they made good profits on their land sales, often to investors looking for tax shelters.[27] The land? It is, in too many instances, growing Russian thistles or marginal wheat crops. The soil erosion continues and will until a permanent grass cover is reestablished. But since there is virtually no economic incentive to convert such land to grass, the best forecast is that it will remain marginal cropland until the topsoil is so depleted that neither cropping nor productive rangeland is a possibility. Then, probably, it will be abandoned, and the process of desertification will be complete.

The threat to the arid lands of the United States is serious: billions of tons of soil being lost; perennial forage grasses being lost or seriously weakened on millions of acres; over 200 million acres undergoing severe desertification. These consequences, so accurately foreseen by John Wesley Powell a century ago, are the result of people overestimating the productive capacity of arid lands. By attempting to extract more from these lands than they are capable of producing, people open up a vicious cycle of deterioration that reduces the land's capacity even more. As the land itself is weakened, the people that depend on it for a living become more desperate and abuse the land even worse in an attempt to survive.

Salinity and Alkalinity

In arid regions, many soils contain salts that have built up over centuries because the amount of rain is not enough to flush excess salts down through the soil profile beyond the reach of plant roots or into the water table. In addition, the salt content in irrigation water is a major threat to irrigated agriculture the world over. The principal salts that cause soil salinity are the cations of calcium, magnesium, and sodium and the anions sulfate, chloride, and biocarbonate. The compounds formed in the soil range from common salt (sodium chloride) to caustic soda (calcium bicarbonate).

Increased salinity in soils restricts the kinds of plants that can be grown, sometimes severely. When dissolved salts form soda, or other highly alkaline compounds, the soil moisture turns highly caustic and particularly difficult soil conditions are created. In alkali soils the organic matter may be dissolved and carried to the surface where it is deposited as the soil moisture evaporates. It then becomes cemented on the surface and produces an impermeable, sterile soil condition that allows virtually no water penetration and upon which few plants can grow.

There are many sources of excess salt to pollute the water and soil in the arid and semi-arid portions of the 11 western states; over 20 percent of the western soils have significant salinity in the soil profile. Such areas are frequently associated with soils derived from materials deposited in the beds of ancient saltwater seas.[28]

Saline seeps have recently developed in many dry cropland areas of Montana and North Dakota because of natural geologic conditions aggravated by agricultural practices. The soils of the region are underlain by salty materials laid down in prehistoric seas. Farming systems that keep half of the land in summerfallow each year allow excess moisture to seep down through the topsoil into the salty layers below. This water dissolves the salts, then moves on down the slope to seep out at the foot of the hill. When it does come to the surface, laden with the salts it has carried from the substrata further up the hill, it can result in significant areas of topsoil being converted to a saline condition where no crops will grow.

Salts have also built up over time in the Yakima Valley in Washington, the Imperial, Sacramento, and Tulare Basins in California, and the Closed Basin of the Rio Grande in Colorado because of inadequate leaching (flushing out the salts by running water through the soil) or leaching with "low quality" water under arid climatic conditions. Irrigated crops are large consumers of water, and all irrigation water contains some salts. The growing plants extract the water from the soil and leave the salts behind, resulting in a further concentration of the dissolved mineral salts already present.

It is difficult to assess exact losses in productivity due to the increasing saline content of soils. Salinity rarely results in overt symptoms. The most common effects are seedling kill and a general stunting of plant growth. Not all plant parts, however, are affected equally. Top growth is often suppressed more than root growth; grain yields for rice and corn may be reduced appreciably without affecting straw yield. With some crops (e.g., barley, wheat, cotton) seed or fiber production are decreased much less than vegetative growth. Varieties of the same crop may react differently. Some plants are susceptible to toxicities from specific minerals; in others, salinity induces nutritional imbalances or deficiencies.[29]

In the San Joaquin Valley of California, an estimated 400,000 acres are suffering reduced yields of 10 percent or more, with a loss of farm income estimated at $31.2 million. In 20 years the affected area could reach 700,000 acres and the damages could reach $321 million, a federal-state report says.[30]

Increasing salinity will definitely impair the future productivity of cropland in the 11 western states. While management techniques could be successfully employed to minimize this loss, those techniques will increase costs and may result in much irrigated land becoming less economically competitive, particularly for lower-value crops.[31] Some of the necessary gains could be achieved by better soil management. Erosion control could, it is estimated, reduce salt by some 2 percent in the rivers of the West.[32] Such a gain would not be a total solution, but it would have the advantage of protecting topsoil productivity while simultaneously improving water quality.

Acid Rain

As a result of the combustion of tremendous quantities of coal and oil, the United States annually discharges approximately 50 million metric tons of sulfur oxides and nitrogen oxides into the atmosphere. Through a series of complex chemical reactions these pollutants can be converted into photochemical smog and acids, the latter of which may return to earth as dry particles, or as components of either rain or snow. This acid rain may have severe impacts on soil and water quality in widespread areas.

Thousands of lakes in North America and Scandinavia have become so acidic that they can no longer support fish life. More than 90 lakes in the Adirondack mountains in New York State are fishless because acidic conditions have inhibited reproduction. Recent data indicate that other areas of the United States, such as northern Minnesota and Wisconsin, may be vulnerable to similar adverse impacts.[33]

While many of the aquatic effects of acid precipitation have been well documented, data related to the impacts on soil quality are just beginning to be developed. Preliminary research indicates that the yield from agricultural crops can be reduced both from the direct effects of acids on foliage and the indirect effects of minerals leaching from soils. The productivity of forests may be affected in a similar manner. Acid rains may destabilize clay minerals in the soil, reducing cation-exchange capacity and accelerating the leaching of soil nutrients.[34]

Additionally, the components of smog, oxides of nitrogen, ozone, ketones, aldehydes, and dozens of other chemicals, many formed in the atmosphere from the compounds resulting from the burning of coal and oil, may adversely affect vegetation through direct toxic action or by interfering with important plant biological processes. In 1975 the National Academy of Sciences noted that air pollution is "an unwitting constraint on agricultural production efficiency."[35]

Crops in many parts of the nation are affected by pollution. A classic example is the abandonment of the Zinfandel grape in part of California because of damage from air pollutants. Similarly, air pollution has forced the abandonment of varieties of cigar tobacco in the Connecticut Valley. Spinach is disappearing from vegetable farms near cities, presumably for the same reason.

The yields of potatoes per acre in the Connecticut Valley, where pollution is high, have been slowly falling since 1960. Legumes are particularly vulnerable to injury from pollutants, especially sulfur dioxide.[36] Smog in the Los Angeles basin contributed to the slow decline of citrus groves south of the city and damages trees in the San Bernadino National Forest 50 miles away. Fluoride and sulfur oxides, released into the air by phosphate fertilizer processing in Florida, have blighted large numbers of pines and of citrus orchards. In New Jersey, pollution injury to vegetation has been observed in every county and damage reported to at least 36 commercial crops.[37]

Studies in Southern California measured filtered clean-air crop yields versus ambient polluted-air yields and found sizable reductions in crops grown in polluted air. Alfalfa yields declined 38 percent; blackeyed peas, 32 percent; lettuce, 42 percent; sweet corn, 72 percent; and radish, 38 percent. In Massachusetts, similar experiments showed yield reductions from ambient air pollution of 15 percent for beans and 33 percent for tomatoes.[38] Crop losses from air pollution in California in 1974 amounted to over $55 million, with loss trends steadily increasing.[39]

Data assembled in 1974 showed that acid rain covered part or all the land in the United States east of the 100th meridian (roughly the line running between the Dakotas and Minnesota, then south) and showed up in large areas of the West—around Los Angeles, Oregon's Willamette Valley, Tucson, and Grand Forks.[40] The damage potential, especially in humid areas, is extensive and massive. The problem is spreading, and rapidly. Figure 7.2 shows how the area affected by acid rain grew between the mid-1950's and the early 1970's. That even more area is affected today is virtually certain.

Both the United States and Canada are intensively researching the effects of acid rain, but there is little evidence yet that could be used to convert productivity losses from acid rain into "acre-equivalents" of lost productivity as was done for soil erosion. In some respects, the deposition of elements such as nitrogen or sulfer may aid in plant growth. On the other hand, experiments using simulated acid rain show that it washes protective waxes off plant leaves, which may increase susceptibility to disease and insect damage; decreases nodulation and fixation of nitrogen in legumes; and may interfere with fertilizer uptake or the availability of toxic elements to plants.[41] These losses, to the extent that they can be documented, may be fundamentally different from those due to such things as soil erosion, salinity, or water mining. If air pollution were to cease, further damage would slow dramatically and natural recycling or human intervention through soil amendments (such as lime) might bring back most, if not all, of the productivity previously lost. Thus, it may be that, at least to some point, soil-quality damages caused by acid rain are reversible.

Figure 7.2. Eastern U.S. Acid Rain Increase

(Shading indicates pH less than 4.5)

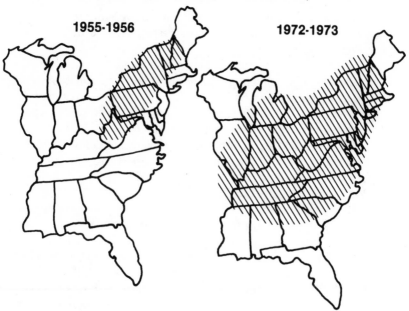

1955-1956   1972-1973

SOURCE: National Academy of Sciences

Water Problems

Many aspects of soil-quality destruction defy quantification, and that is certainly true of the damage to soil quality caused by worsening soil-water relationships. These can include waterlogging, where the subsurface drainage has been restricted, the surface of the soil itself is gradually dropping, or the water table is building up because of the sediment-caused rise of a nearby riverbed. In these cases, soils that were formerly productive are beginning to have wetness or flooding at certain times of the year, and their productivity has gone down as a result.

Talking with soil conservationists around the nation reveals many local instances of these disasters, but getting any kind of national estimate on their extent and severity is not possible. In the 1977 NRI, SCS identified 175 million acres of floodplains, of which 48 million acres were croplands. These lands are subject to periodic flooding from flood events that are likely to occur once every 100 years or oftener. What the 1977 NRI did not show, however, is how many new acres are annually being added to that "susceptible" category.

As buildings, roads, railroads, and other structures are constructed on the nation's floodplains, the ability of the floodplain to hold, absorb, and transmit flood flows is diminished. Every structure that diverts floodwaters away from their natural flood path, or prevents water from flowing freely across a historic floodplain, creates a new floodplain. The water must go somewhere. So a few more acres of nearby land, formerly above the normal flooding level, are now vulnerable to damage. The extent to which that phenomenon is occurring, and the amount of cropland involved, has not been measured.

What we do know, however, is that about 12 percent of the nation's cropland and 16 percent of its prime farmland lie in floodplains. The Resources Conservation Act (RCA) appraisal indicates that, on the upstream areas, crop and pasture damages from flooding were about $1.1 billion in 1975.[42] Their forecast was that, unless significant efforts were made, the flooding damages would almost double by 2030. Just how much of this would result from new lands being added to the flood-prone areas and how much would come from accelerated flooding due to mismanagement of upstream lands was not established.

Summing Up

There are strong indications that many soils are suffering from declining quality, although there remains little way to quantify the losses adequately. Nationally, even without comprehensive data, it seems hard to avoid the conclusion that declining soil quality is a significant factor in reducing productivity. This conclusion is supported, as well, by the absence of any significant trends toward the *improvement* of soil quality in any of the measurable quality indicators or in any region of the nation.

The examples reviewed here show, without exception, declining soil organic matter; growing areas affected adversely by soil compaction, desertification, salinity, and alkalinity; a spreading mantle of acid rain, and increasing areas where water imbalances reduce soil productivity. Not covered are the increasing number of very disturbing incidents where toxic industrial wastes such as PCBs, zinc, lead, cadmium, or mercury have polluted soils or waters, ruining milk supplies, killing plants and livestock, and threatening human health. Those situations, increasingly common as industry continues to be far more effective at creating toxic wastes than in learning how to dispose of them safely, will continue to make headlines.

There are some significant local efforts to improve soil quality. Intensive projects to reduce salinity in the Colorado River Basin and efforts to control flooding, waterlogging, and water pollution in many localities are examples. But these efforts, while important, do not begin to touch even a small portion of the growing soil-quality problem indicated by the evidence available today.

Improvement of some soil-quality factors is within the grasp of the individual farmer. He can, for example, do much to keep soil organic matter levels high. This will have a positive effect on soil compaction, which can also be helped by proper tillage techniques such as reduced cultivation. Proper soil and water management on the farm can often reduce salinity problems.

But there is little the individual farmer can do about acid rain. It may be created by a power plant many miles from his farm, and he has no protection whatsoever from it. He may have equally little opportunity to prevent flooding or soil waterlogging on his farm, if it stems from a total watershed that is out of balance.

Soil-quality exhaustion adds an important dimension to the problem of maintaining soil productivity in U.S. agriculture. Whether or not these losses will be more serious in the future lies both within the hands of individual farmers and the public as a whole, for the maladies affecting soils today stem both from on-farm management deficiencies and large-scale environmental abuse.

8.

Water for Agriculture

Classical economics teaches us that there are three kinds of agricultural inputs: land, labor, and capital. As noted economist Don Paarlberg has pointed out, that may be because the science developed in England where ample water fell regularly from the sky.[1] Budding students of agricultural economics were taught that water was a classic example of a "free" good. If those economics textbooks were to be rewritten today, at least in the 17 western states, that would not be the case. Water, while it may still be greatly undervalued, is certainly not free. Without the development and utilization of supplemental water—and lots of it—the American West would still be a land of wide open spaces, supporting a few cowboys and sheep men, but little intensive agriculture. As it is today, due solely to the use of irrigation water, the West boasts some of the most productive acres in America.

Without irrigation, there would be far less production—or, perhaps, no production—on millions of the most productive acres in the United States. Since an irrigated acre often yields two to three times what that same acre would yield under natural rainfall, the loss of *either* the land *or* the water in irrigated areas can result in a loss of agricultural production that will take two or three acres of dry cropland to replace. The future of irrigated agriculture is, therefore, an important part of any projection about future land scarcity.

When economists talk about water as a resource, they call it a "flow" resource. By this they mean that nature gives it to us on a time-installment basis.[2] You can use today's water now, and you can store some of it for use later, but you can't use tomorrow's until it comes. What people can do, however, in their insatiable appetite for resources, is overestimate what tomorrow's supply is likely to be. Thus, we see many rivers in the West such as the Colorado, where people have "appropriated" more than the normal flow of the river. In a year when streamflow is short and storage facilities can't cover the deficiency, that gets to be a serious problem, as there is far too little water to satisfy everyone's expectations and needs.

In the last three decades or so, the United States has been using another supply of water—water from underground pools— that is not a "flow" resource, but a "fund" resource just like petroleum. Many groundwater reservoirs were filled slowly, over geological time, and now are being replenished at a rate far slower than the speed at which people are pumping them dry. When they run dry, or the water level drops too low for economical pumping, that resource base is *gone* in terms of economic value. Intensive agriculture, and the communities that have grown up around it, will be lost when the water is gone.

Each of these major water-supply problems—the uncertainty, inadequacy, or wrong location of surface waters and the depletion of groundwater—has important implications for the future of agriculture, because a significant percentage of today's agriculture depends as much on dependable water supplies as it does on the availability of fertile land. Future water shortages will cut into our ability to produce food and fiber.

Water quality is also an important issue that affects agriculture. When water is used for irrigation, a significant part of it may be returned to streams, rivers, and lakes. But often this water is higher in sediment, salts, nutrients, or pesticides than the original irrigation water, and this reduced quality may harm a downstream user—another farmer, a city, or an industry. When these conflicts of interest occur, agriculture is forced to justify not only its right to use water, but also its right to make that water less useful to others in the process.

Irrigation in U.S. Agriculture

Agriculture is by far the nation's largest single consumer of water, accounting for 83 percent of total water use, some 89 billion gallons of water each day.[3] Irrigated agriculture produced 27 per-

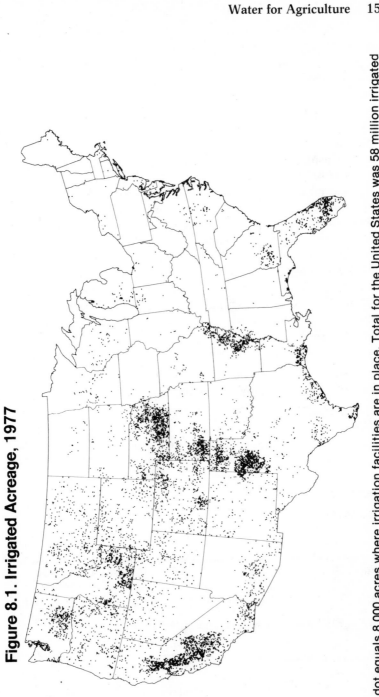

Figure 8.1. Irrigated Acreage, 1977

One dot equals 8,000 acres where irrigation facilities are in place. Total for the United States was 58 million irrigated acres, according to NRI definitions.

SOURCE: USDA Soil Conservation Service

cent of the value of farm crops harvested in 1980, on only 12 percent of the harvested acres. As Table 8.1 shows, most of that irrigation takes place in the 17 western states, which harbor 5 out of every 6 irrigated acres. There is increasing irrigation throughout the rest of the nation, however, with irrigated acres nearly doubling from 37 million acres in 1958 to 61 million acres in 1977.[4] In the 1977 National Resource Inventories (NRI), every state except New Hampshire and Rhode Island reported some irrigated acreage (see Figure 8.1).

The rapid rise of irrigation has not, however, been the result of pressures on the land base. To the contrary, irrigation has flourished *despite* crop surpluses and government programs designed to reduce the acreage of many crops. The reasons for the rapid growth of irrigation include both immediate economic benefits and the reduction of risk for the farmer. Irrigation increases crop yields significantly, making it possible to grow the same size crop on less acres. It also creates new cropland, allowing tillage on millions of acres of arid land that would otherwise be of little or no value for growing crops.

Table 8.1. **Irrigated Land in the United States, 1977**

Region	Cropland		Pastureland		Total Irrigated	
	Acres	Percent	Acres	Percent	Acres	Percent
(acres are shown in millions)						
17 Western States	46.4	23	3.9	8	50.2	20
31 Eastern States	9.3	4	1.1	1	10.4	3.5
National Total	55.7	13*	5.0	4*	60.7	11*

SOURCE: USDA Soil Conservation Service, "Basic Statistics: 1977 National Resource Inventories," February 1980, tables 3a, 4a. (Minor differences in totals are due to rounding.)

*These are percentages of the nation, so are not the sums of the percentages in the different regions. Nearly one-fourth of the cropland in the West is irrigated, but the total acreage is fairly small. In the East, with most of the crop acres, only a small percent get supplemental water.

Table 8.2 shows the enormous difference in productivity that results from irrigation. Corn and cotton yields are more than doubled in the West, while irrigated western crops yield substantially higher than the same crop grown under non-irrigated conditions in the humid East.

Table 8.2. **Irrigated and Dryland Yields of Selected Crops, 1977**

Crop	Location	Yield
		(bushels per acre)
Corn	Irrigated West	115.2
	Dryland West	48.3
	East	88.6
Sorghum	Irrigated West	77.4
	Dryland West	45.5
	East	61.4
Wheat	Irrigated West	39.4
	Dryland West	27.1
	East	38.1
		(bales/acre)
Cotton	Irrigated West	1.41
	Dryland West	0.60
	East	1.03

SOURCE: Kenneth D. Frederick, "Irrigation and the Adequacy of Agricultural Land" (Paper prepared for Resources for the Future Conference on the Adequacy of Agricultural Land, June 19-20, 1980. To be published by RFF).

That dramatic jump in crop yields is not entirely due to the effect of additional water, however. An irrigation system removes many of a farmer's weather-related uncertainties, so he is willing to invest more in fertilizers, high-yielding varieties, and intensive crop management. The resulting yield increase comes from the total management package applied to the land, not just the water alone.

This fact helps explain the rapid growth of supplemental irrigation systems throughout much of the humid East. Even when natural moisture may be adequate in two out of three years, or four out of five, a farmer may feel it advantageous to install a portable sprinkler system to be used "just in case." The system is good insurance against a periodic drought and, thus protected, he can apply intensive management for maximum yields each year without fearing a potentially disastrous crop failure.

Getting the Water to the Crop

As valuable as irrigation water can be in raising crop yields, it seldom costs much for farmers to use, and its low price is reflected in low irrigation efficiency rates. Less than half of the water diverted from streams is actually used by crops, according to the "plumbing" diagram in Figure 8.2.[5] During the year "1975" some 158.7 billion gallons per day (bgd) were diverted from streams, rivers, and reservoirs for irrigation.[6] Over 93 bgd were either spilled or lost, either in the canal systems leading to the farm or after the water reached the farmer's irrigation system.

Figure 8.2. Irrigation Water Used in the United States in 1975

Percent of diversions—billion gallons per day

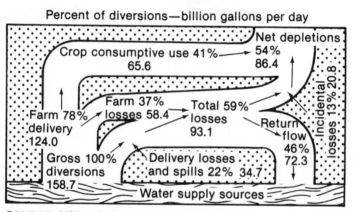

SOURCE: USDA Soil Conservation Service

About three-fourths of this "lost" water was returned to the source (often to be reused downstream), while 20.8 bgd simply disappeared, neither being used by the crop nor flowing back to surface waters. This water either ended up in underground pools by percolating down through the soil, or escaped directly back into the atmosphere by direct evaporation. Such water is not truly "lost" from the water cycle, but it was once available—often after considerable effort and money was spent to capture, store, and transport it—only to be lost before it could be put to economic use.

As water supplies become more expensive or scarce, and as the need for agricultural products goes up, the incentives for improving the efficiency of water use will also rise. Since the water delivery systems in use today only get about three-fourths of the water from the source to the farm, they are a logical place to begin.

Improvements to plug the leaks in those systems will not come without significant investment and effort, however. The Soil

Conservation Service (SCS) estimates that it will cost about $600 million a year just to repair, and replace as necessary, existing irrigation systems in the country. This would not improve their efficiency, but simply maintain the levels that now exist.

Concrete structures "step" the water down a steep slope without causing erosion in this Oregon canal. Without the drop structures, the water would carve a huge gully in a short time. This is an example of the irrigation investments that must be made—and maintained—to get water to thirsty croplands. SCS photo by Frank Roadman.

Another $150 million a year would be needed to maintain irrigation water storage reservoirs. In addition, SCS predicts that it would take another $260 million per year in the 17 western states to get a 4 percent improvement in delivery systems and a 5 percent increase in on-farm efficiency.[7] That total billion-dollar-per-year maintenance bill would amount to about $25 per acre per year, so farmers would no doubt still find it economical if they were faced with loss of their irrigation system.

Improving irrigation efficiency on the farm under current conditions does not save money and time for the farmer, however, and often the reverse is true. Many farmers could improve their water use by simply moving the water from one portion of a field to another when the optimum amount of water had entered the soil. But that may require either an automated irrigation system to turn the water on and off at various places when needed or a skilled irrigator, working virtually 24 hours a day to monitor soil and water conditions. Either way, efficient irrigation takes some effort. In the past, irrigation water has always been cheap, so it was more sensible to waste water than to spend time or money assuring that it was utilized to peak efficiency.

It is easier for a farmer to arrange his work around an irrigation schedule that runs the water in one set of furrows for 12 hours, then is changed to another part of the field. In that way, he can make one irrigation "set" in the morning, work all day at other farm tasks, check the progress of the irrigation water in passing, if at all, then return in the evening to move the water to a new location.

The stream of water entering the farm from the irrigation canal usually runs 24 hours a day and can't be shut off while he is doing other farm chores or sleeping, so the two 12-hour sets are about the only way a farmer without extra farm hands can manage. The fact that a 9-hour set would be best on some of his soils is of little consequence. Wasting the water for 3 hours may be more attractive than going through the effort (and expense) of applying it properly.

Water rights, particularly in much of the West, are another strong deterrent to water conservation. In most state water laws, the doctrines of "prior rights" and "beneficial use" determine who can use surface water and how much of it they can use. The prior rights doctrine gives an irrigator a water-use priority based on the date of the original water diversion application for the land in question. One farmer, who holds a water right dating back to 1896, may share the water from a small stream with another farmer, whose water right was established in 1902. In a normal year, each has adequate water for his crop. In a dry year, however, the farmer with the senior water right may meet all his needs while the other gets little or no water. The law allows the "prior right" to be satisfied before the later, and lesser, right is granted. Under these conditions, an early water right (which, under most western water laws, is recorded as part of the land deed) is a valuable possession, zealously guarded by the owner.

Irrigation water is directed into furrows, allowed to run for several hours until the desired amount of moisture has entered the soil, then moved to a new set of furrows by the irrigator. In this Texas field being prepared for cotton, the water enters the furrows from both ends and flows to the middle. SCS photo by J. H. Barksdale.

The second part of the water law puzzle is the "beneficial use" doctrine which is generally translated into "use it or lose it." If a farmer has an early right to a 10-cubic-foot-per-second (cfs) flow from a stream, but only needs 6 cfs to water his crops properly and efficiently, he has a dilemma. If he diverts only the 6 cfs he really needs, and lets the other 4 cfs continue to flow on downstream, another person can divert that extra water, apply it to *his* land for a certain number of years of "beneficial use," then lay legal claim to a permanent water right on that 4 cfs flow. The original water user may have, through his water conservation efforts, lost the right to 40 percent of his original water supply.

To the uninitiated, that may seem of little consequence, provided that the remaining flow of 6 cfs is ample for optimum crop production. But what happens if the farmer wants to switch over to a more profitable crop that might need more water? He could have lost that opportunity. Or there may be a dry year—or a series of dry years—when streamflow falls to one-half or less of normal. The "conservation" irrigator may now be reduced to only half of his remaining "water right," or 3 cfs, which will be very damaging to his farm operation. Had he retained his original water right, he would still be entitled to a 5 cfs flow and be more able to weather the drought.

The response is predictable and virtually universal. Farmers with a 10 cfs water right use 10 cfs of water, whether they need it or not. They run it over the fields, often semi-drowning their crops, leaching away valuable plant nutrients and causing serious soil erosion, because they must do so in order to retain their full water right. The water is not being used to grow crops, but to keep somebody else from getting it.

Under many of the state laws (Colorado and New Mexico, for example) the farmer can sell his water right separately from the land. This makes the water right itself very valuable and increases the incentive to use all the water possible, all the time, to retain the water right. In the face of these strong economic and legal incentives to waste the "free" supply of water, there is little mystery in the difficulty facing anyone who wants to promote water conservation. The economic incentives created by the laws are geared toward overuse and misuse of water. Without basic changes in the state water laws of the West, little real conservation will be accomplished; yet the political resistance to changing those laws—and particularly, to any federal intervention that would force such changes—is awesome.

Groundwater use provides more incentives for conservation. Even though the water itself is "free" in the sense that farmers do not pay a per-gallon fee for pumping it from the underground pool, the costs of pumping are often significant enough to force water conservation measures. As the price of electricity and natural gas rises, or if the water level in the well keeps going down because more water is being pumped out of the underground pool than normal recharge processes can replace, the rising costs will encourage the farmer to pump only what is needed. This can save both water and energy.

Closely spaced nozzles under the pipe allow this sprinkler system to spray the water gently on the wheat crop in this Oregon field (top). In contrast, a South Dakota system squirts the water high in the air (bottom). On a hot, dry, windy day, less than half of this water may reach the soil. SCS photos by Don Schuhart (top) and G. W. Stroup (bottom).

One of the newest water-saving and cost-cutting techniques is to convert to "low-head" sprinkler systems. The secret to these new systems is a new sprinkler nozzle design that allows the stream of water to be "broken up" into small droplets without the need for high pressures. On center pivot sprinkler systems the sprinkler heads are mounted upside down under the boom rather than on top of it. As a result, the pressures in the system can be reduced from the 50-60 pounds per square inch (psi) to as low as 20, and the horsepower for pumping can be cut back to effect an energy savings of 25 percent or more.

In addition, the water is no longer sprayed in high-pressure-propelled droplets high into the air. On a hot, dry, windy summer day in the West, it was not uncommon for over half of the water sprayed into the air to evaporate or blow away before it hit the ground. With the new "low-head" sprinklers, the water falls almost directly from the nozzle to the soil, cutting the evaporation losses dramatically. Since less water is lost to the air, less needs to be pumped to meet crop needs. As a result, "low-head" sprinklers harness water conservation to an immediate energy-cost savings for the farmer, and the new systems are likely to replace most of the older sprinklers within a few years.

Another water saver is drip irrigation. In this kind of system, the water is fed through a system of plastic pipes and tubes that have small, spaced holes to allow the water to drip out at each plant. A small, non-erosive stream of water is slowly applied directly to the crop. The first drip systems were adapted to orchards, where it is highly advantageous to assure that the available water goes directly to the tree roots and not to the weeds and grass in the intervening rows. Drip systems have now been adapted to a wide variety of high-value crops and, where they are adapted, achieve a significant increase in the amount of produce for each gallon of water.

As a result of the new irrigation technology, USDA predicts that most new irrigation systems in the future will be sprinkler systems for field crops and drip systems for high-value crops such as citrus fruit or grapes. Many of the older irrigation systems will be converted to one or the other. While the pumping systems for both of these kinds of systems require more energy than a surface irrigation system, the higher efficiencies will often offset those costs. Where water conservation is associated with either cost saving or production increases, farmers quickly adopt it for economic reasons. Where water conservation penalizes the farmer, as in the case of the western water rights situation, it simply does not, and will not, happen.

Water Quality and Agriculture

Having plenty of water is not always the whole story—not if the water is polluted and cannot be used. Agriculture finds itself in the middle of many water quality battles. In some cases farmers use the water, pollute what they don't consume, then release polluted water back into streams or rivers. Other users downstream don't appreciate that. In other cases, farmers get water onto their land that some other user has polluted. Sometimes it is not suitable for their crops, and at the very worst, it may contain poisons that sicken or kill their livestock or families.

Water quality—or water pollution—is nearly impossible to define in a general way, because it is only relevant to the ways in which people hope to use the water. To a chemist, any impurities are unacceptable, so he uses distilled water in his experiments. Everything else is "impure" or "polluted." In public drinking water supplies, water is far from "pure." Most communities add a great deal of chlorine to kill bacteria. That may make the water all right for drinking, but it may also make the water unfit for fish or for irrigating sensitive crops. (Unfortunately, in some towns, it's still unfit for drinking, too, but that is another story.)

Even "natural" waters, in many areas, are not free of salts, sediment, acids, or other impurities, and would not be "pure" if all human activities were totally removed from the scene. Such situations make the establishment of overall national water quality standards difficult, if not impossible. Water quality problems must be handled on a watershed basis.

The major water quality problem created by agriculture is caused by sediment and the nutrients and pesticides that can be carried "piggy-back" on soil particles as they are washed off farm fields by soil erosion and carried into the public waters. About 40 percent of the sediment reaching America's waters comes from cropland, according to USDA estimates.[8] This is compatible with the proportion of soil erosion that takes place under cropping conditions. Some 26 percent of the water-polluting sediment comes from streambanks, 12 percent from pasturelands and rangelands, and 7 percent from forests. Although sediment may not be as serious in terms of water use as, for example, toxic chemicals or pathogenic agents that can make waters virtually unusable, it still imposes high costs on other water users.

This Class VI and Class VII Colorado land is suffering severe soil erosion from a sprinkler system, but the farmer is still having problems keeping the sandy soils moist enough to grow a profitable crop. Both the farmer and the land are paying a high price for this improper land use. SCS photo by Jerry Schwien.

In a recent summary of state reports to Congress, the Environmental Protection Agency (EPA) reported that in 1977, water pollution affected 95 percent of the 246 watersheds in the United States.[9] That may oversimplify the situation, however. A watershed identified as being "polluted" may not be affected throughout, but only in one or two short segments. On the other hand, the monitoring data on water quality is sketchy enough to suspect that many water-quality problems were not even reported by the states.

In the same report, EPA estimated that non-point sources of pollution were contributing to excess bacteria in 61 percent of the basins, to oxygen depletion in 51 percent, to excess nutrients in 56 percent, to pesticides in 22 percent, and to other toxics (metals from urban runoff and mining) in 32 percent of the basins. The primary sources of non-point pollution were agriculture, which affects 68 percent of the basins; urban runoff, which affects 52 percent; and individual waste disposal systems, which affect 43 percent.[10] The vast bulk of this pollution, while it reduces the quality of the water, is not really dangerous to human health, but unfortunately some of the chemicals that "ride" the soil particles into the water are very dangerous.

In the United States, more than 1,800 biologically active compounds are sold in more than 32,000 different formulations.[11] For the first time in human history, we are manufacturing chemical compounds that can severely affect the natural and human environment. Some of them are quickly rendered harmless by natural processes; others are deadly poison and are virtually indestructible. Which is which, and how serious a problem do they present?

Farmers used about 450,000 tons of pesticides in 1977, and USDA projects that they will use 1.25 million tons in 1985.[12] Most research to date indicates that very little of this material enters the water, unless there is a heavy rain directly after treatment. The total amount of pesticides in the water may be a secondary consideration, however, since some of them are so toxic that *any quantity* in the water becomes a serious problem, particularly in the case of compounds that persist for a long time.

USDA estimates that between 750,000 tons and 7.5 million tons of nitrogen and between 60,000 and 600,000 tons of phosphorus enter the nation's waters from agricultural sources each year.[13] The most important pollutant, in terms of releasing algae growth in most waters, is the phosphorus. Experiments in the Maumee Basin of Ohio and Indiana indicate that reducing the levels of phosphorus entering Lake Erie was the best way to limit the growth of algae in the lake. Most of the phosphorus that moved into the lake's waters was attached to the fine soil particles being carried into the lake because of soil erosion, so reducing soil erosion is essential to protect water quality.

Salinity is a serious water-quality problem for irrigators. Most soils in the 17 western states have relatively high natural salt levels, since they were formed under arid conditions without enough natural rainfall to flush the soluble salts out of the soil profile. Much of the irrigation water contains salts as well. When irrigation is applied to a field, the water either is evaporated into the air, is taken up into the plants and then transpired into the air, runs back to the stream as drainage water, or seeps down through the soil into the groundwater. In the first two situations, the salt that was carried onto the land by the irrigation water is left there, where it can cause high—or even toxic—levels of soil salinity. In the latter two situations, excess salts can be carried into streams or groundwaters, increasing pollution levels.

To rid the soil of salts, irrigators often overirrigate to "flush" excess salts from the soils. This can result in even higher salt loadings in surface or groundwaters. In the U.S. portion of the basin, salinity caused $53 million in damage in the Colorado River system in 1973, and by the year 2000 this damage is expected to reach $124 million if no additional control measures are applied.[14] The salt content at the headwaters of the Colorado system is less than 50 milligrams per liter. By the time the water reaches Mexico, the salinity is up to 900 milligrams per liter, and experts expect it to reach 1,160 milligrams per liter by the year 2000.[15] At these levels, the use of the water for irrigation, drinking, and even many manufacturing uses is severely limited. This situation has naturally caused international tensions. In a treaty with Mexico, the United States has agreed to reduce the tremendous salt load entering Mexico in the Colorado, but much work remains to be done to fulfill that commitment.

Competing for Water

It is hard to think about water supplies without thinking of Coleridge's famous line about "water, water everywhere, but not a drop to drink." Water covers nearly three-fourths of the earth's surface, but 99.35 percent of it is contained in oceans, ice caps, and glaciers. From the remaining two-thirds of 1 percent, people must find all the usable fresh water to satisfy their needs. Still, that ought to be enough. The Yangtze River alone has enough water to supply every person on earth with 150 gallons per day, roughly 15 times more than most people in the underdeveloped parts of the world use. The dilemma, of course, is that much of the water we have is not distributed where the people and the arable land are located.

In the United States, the Water Resources Council has estimated that water consumption will increase almost 27 percent between 1975 and 2000.[16] Most of this added demand will come from growth in manufacturing and mineral industries, steam electric generation, and agriculture. To regions where water supplies are limited, that means increasingly intense competition for the available surface and groundwater. In the West, where irrigation has historically accounted for 90 percent of the water that is consumed, agriculture may not be the successful competitor for water in the future that it has been in the past.[17]

For one thing, there are far fewer opportunities to increase the water supply than there were in the past. Agricultural economist Don Paarlberg predicts that most water shortages in the future will be solved by conserving existing supplies of water, rather than by the historic method of building dams, reservoirs, and aqueducts to increase the supply.[18] That is a reasonable assumption, given the fact that most of the better dam sites have been used, and most of the readily available water supplies have already been tapped by enterprising engineers, farmers, and city water planners. If Paarlberg is right, and non-farm water demands rise as rapidly as they seem likely to do, then agriculture stands to lose some of its water. It is as simple as that.

Plain economics holds the answer. California farmers raised about $252 million dollar's worth of alfalfa in 1972, using 9,080 gallons of water for every dollar's worth of alfalfa. Grapes used the water more efficiently, but still needed 1,758 gallons of water for each dollar of crop value.[19] Farmers and non-farmers are on a collision course over the use of water, particularly in the West, with agriculture the likely loser in that battle.[20] There are too many other ways where a dollar's worth of value can be created with far less than 2,000 or 9,000 gallons of good water. When the water struggle gets to the point where tradition and precedent must give way to economic realities, farmers will lose.

Outlook for the Future

Increasing the U.S. water supply by grandiose proposals to move water great distances—from Canada or Alaska to the Great Plains, from the Columbia River to California, from the Mississippi to the High Plains of Texas—may fire the imagination of engineers and politicians, but most of these schemes were economically infeasible when they were presented and get more illogical every day.

Not only have the costs of such mammoth projects been going up like an Atlas missile, but public opposition has been rising almost as fast. People in the water-supplying areas are getting more reluctant to share "their" water, while environmentalists point out that supporting too high a human population in an arid region through the import of water may simply overload the carrying capacity of the region's other natural systems. Water from the Colorado River built the Los Angeles Basin, but in spite of its huge population and thriving economy, Los Angeles is still a desert. Whether there is enough money to maintain its artificially watered ambience indefinitely is not certain. The creation of more areas like that is extremely unlikely.

Desalination, or reclaiming salt water for drinking or irrigation, is another water supply strategy that has received an inordinate amount of attention in the United States. This technology, which was highly touted as a solution to the whole water problem during the 1950's and 1960's is no closer—and perhaps is further—from practical realization than before. It can be done, but only with high energy inputs and high costs. There are over 300 desalination plants in the United States, producing about 100 million gallons of water a day, but the costs run from 2 to 30 times more than the price of the irrigation water now being used in the West.[21]

A great deal of effort and research has gone into weather modification, and there are some who feel it holds great promise. It also holds great hazard, however, since the storms generated in one area may diminish needed precipitation elsewhere, or else be of such intensity that the damage done more than offsets the benefits. I, for one, am skeptical about this approach, and feel that it holds little practical promise, at least in the foreseeable future.

A new notion that has been attracting attention is that of hooking onto Antarctic icebergs and towing them to water-short areas for melting and piping inland. With over three-fourths of all the world's water locked up in ice caps and glaciers, and 90 percent of them at the South Pole, iceberg harvesting is an interesting technical "fix" to the water shortages of many areas. In 1977, Prince Mohammed al-Faisal of Saudi Arabia sponsored the First International Conference on Iceberg Utilization at Iowa State University, an event that gained a great deal of news coverage but resulted in little actual iceberg-towing. The challenges involved in locating icebergs, encasing them in plastic, moving them thousands of miles, and melting them down under controlled conditions will provide exciting dreams for engineers for many decades, but the realistic odds of seeing it actually happen are low.

In sum, the evidence seems to point to the fact that our ability to increase water supplies to meet all needs has about run its course. The Water Resources Council estimated that, by the year 2000, there will be severely inadequate surface water supplies in 17 subregions, located mainly in the Midwest and Southwest (see Figure 8.3). During dry weather, more subregions, including some in the East, will also be faced with a water shortage.[22]

In recent years, groundwater has been available to fill in the gaps where surface supplies were overtaxed, but this source is dwindling as well. Starting in the 1930's, and spurred by improvements in pumps and reductions in energy costs, groundwater use increased to the point where it was providing 39 percent of all

Figure 8.3. Water Resource Problems

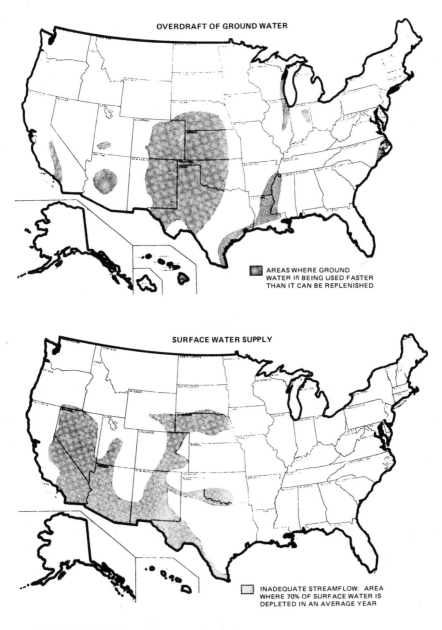

SOURCE: U.S. Water Resources Council

irrigation withdrawals by 1975.[23] But much of it is being pumped from pools where the water supply is replenished more slowly than the current rate of withdrawal. This "mining" of groundwater is now estimated to be depleting the nation's water supplies at the rate of 21 billion gallons a day.[24] There will be few new supplies available underground if that trend continues.

The most critical depletion is in the High Plains area that runs from Texas to South Dakota, but there are also serious concerns within Arizona and California, as Figure 8.3 demonstrates. SCS has estimated that at least 5 million acres can be expected to revert from irrigation to dryland farming by the year 2010, as groundwater supplies drop too low for practical pumping. Farmers in the Trans-Pecos area of Texas are already cutting back severely on irrigation. In Colorado, a state water plan has been developed that will result in one-half of the water in the Ogallala aquifer being used up in the next 40 years. In this plan, the depletion of the water resource is being carefully planned, but it will eventually be depleted just the same.

The combination of surface water shortages and groundwater depletions shown in Figure 8.3 demonstrate, not too surprisingly, that the same general regions are afflicted by shortages of both types. It appears certain that irrigated agriculture in those regions is going to decline.[25] Those losses, both SCS and the Water Resources Council predict, will be offset by the addition of more irrigated acreage, most of it in the eastern part of the 17 western states.[26] Irrigation experts in Nebraska, for example, feel that the state can support a long-term sustainable irrigated agriculture of at least 15 million acres, twice the amount that currently exists.[27] Therefore the loss of agriculture as a result of groundwater depletion is likely to be more of a regional issue than a national issue for the next few decades, at least.

Higher farm prices could make irrigation more feasible in many areas. On the other hand, there are no foreseeable levels of crop prices that would allow farmers to outbid mining, energy development, or municipal water users for limited water supplies. These users will buy whatever they need, and this may be significant in some locations in the West. Most experts feel that because there is so much water used by agriculture today, a small percentage of the national total can probably satisfy all foreseeable non-agricultural needs.[28] The fact that this is an issue of only modest national impact will not, however, keep it from causing extreme stress in many communities and regions.

Expanded use of water in the East for supplementary irrigation, multiple cropping, and high-value crops is expected to continue over the next two decades, adding as many as 8.5 million acres of irrigated crops in the East.

Despite the potential growth that may occur, it is doubtful if irrigation will make the contribution to the expansion of agricultural yields that it has in the past 30 or 40 years. Rising energy prices may double the costs of irrigating with a standard center pivot system between 1980 and 2000.[29] Kenneth Frederick of Resources for the Future has calculated, for example, that lifting an acre-foot of water from a 300-foot well with an electric pump cost the farmer $22.55 at 1970 prices, $26.64 at 1980 prices, and could cost $53.28 at 2000 prices. Since it takes 2-3 acre-feet to irrigate an acre of crop, those kinds of escalating costs will be difficult for farmers to handle and still stay competitive. Efforts to improve efficiency may drop that cost somewhat, but it is doubtful that it will wipe out the economic disadvantage that appears to be facing irrigated agriculture in the future.

The importance of these forecasts lies in their implications for the future demands on the farmland of America. In the past, expanding irrigation has contributed significantly to the rise of the average per-acre yield on U.S. farms. If there were millions of acres of irrigable land still available, and the water to irrigate it, future yields would continue to rise much more rapidly. But the land isn't there, and neither is the water, in all too many cases. Irrigation simply won't be the strong positive factor it has been in the past.

Proper management of land and water still offers opportunities for improvement and expansion of irrigated agriculture. These potentials will only be reached, however, if the attention of water engineers and researchers can be shifted away from trying to invent new methods of "manufacturing" water toward the development of new ways to get more benefit from each gallon used. Water has been too cheap in the past, and too easy to waste. As it gets more expensive, farmers will use it better, providing they are not penalized by state water laws that continue to encourage waste. Reforming those state water laws will be a challenge for all people interested in improving agricultural productivity and proper resource use.

9.

Technology and Productivity

The power to produce is the most vital and underrated asset of American agriculture. It is the measure of the system's strength. Like the horsepower in the family car, productive capacity isn't all needed all the time, but when the hills get steep and the going tough, nothing else can replace it.

There are only two ways to produce more food—use more acres of cropland or get more from each acre. It appears that we will need to double the amount of food and fiber we grow in the United States by the year 2030 if we are to meet the needs of our own people plus maintain a significant export business. To do that, we must either double crop yields per acre, double the acres of cropland, or achieve some combination of the two. Farming twice as many crop acres would mean using all the 413 million acres of current cropland, plus the 127 million acres of high and medium potential cropland, *plus* some 260 million acres that now rate as having low to zero potential for growing crops. The chances of doing that are slim, and even if we could, the costs of making marginal lands productive would be translated into food costs that the American people could find unacceptable. The enormous difficulties involved in doubling the land base leaves the other option—raising the yields on each acre so that fewer acres are needed.

The past has seen dramatic yield increases in U.S. crop yields. Will those trends continue? The question is critically important, but the evidence is not clear, and there are equally logical projections that portray far different visions of the future. The most optimistic view suggests that average crop yields could double in the next 50 years. If that happens, there is no land shortage, at least not in terms of meeting domestic needs and a significant level of crop exports. A "worst case" view foresees little, if any, yield increases from those we now achieve. That turn of events, should it occur, would almost surely lead to skyrocketing costs of food, severe cutbacks in the amount of agricultural produce available for export abroad, and serious damage to the American land.

The range of reasonable scenarios is illustrated by Pierre Crosson of Resources for the Future, who has calculated that if grain and soybean yields continue to rise at 2 percent per year, it will take 224 million acres to meet the predicted demands for those crops in 2005. If, however, yields rise at only 1 percent per year, it will take an additional 69 million acres to meet the same demand.[1] Choosing to predict the higher growth rate leads to the conclusion that farmland will be adequate in the future; choosing the lower leads to a prediction of land constraints.

If we can use research to achieve that higher growth rate, the pressures on the land base and food costs would be dramatically reduced, so technology for increased productivity is critically important to the future of U.S. agriculture. Unlike land and water, technology is man-made, created by research and development. If the nation is concerned about lagging agricultural productivity, we can increase the public funds for research and development programs. That doesn't guarantee the instant emergence of new technologies, but history suggests it pays off handsomely over the long run.

The *type* of technology that might emerge in the future is also subject to some guidance. If research priorities are aimed at increasing the immediate profitability of farm production, the continued reduction of labor in agriculture, and increased reliance on artificial inputs as the basis for productivity, the result could be continued soil and water destruction.

If, on the other hand, public research tries to develop new technologies that conserve soil and water, enhance the sustainability of agricultural production, and favor the development of profitable farm enterprises for family farms, the future could be one of added productivity coupled with improved resource conservation. Whether we choose now to design a future based on exploitive agriculture or one based on sustainable agriculture will critically affect the long-term "staying power" of U.S. agriculture.

There is an intense public debate over the priority—and money—to be allotted to agricultural research in the coming years.

It is not only the amount, but also the type of research and development that is important to the future of America's land and water resources, as well as to the future of agriculture. If technology is a man-made resource, then the American people need to understand the stakes involved and insist on development of technologies that will lead toward long-term, sustainable increases in productivity. One thing that might aid that type of public guidance would be a clearer understanding of the difference between *total production* and *productivity.*

Productivity vs. Total Production

Farm productivity is measured not by the total output, but by the units of output grown for each unit of input. The farmer who, by some change in his system, grows three bushels of grain where only two grew before, with no change in the amount of inputs (seed, fertilizer, labor, cultivation, land, irrigation, drainage, or similar factor), has realized a 50 percent increase in productivity. By the same token, a farmer who grows two bushels of wheat where two grew before, but does it with less input, has also improved productivity. But the farmer who must double his inputs to grow four bushels where only two grew earlier has increased his *total production* without changing his *productivity* at all.

Failure to understand this basic distinction has led many people to look at U.S. agriculture and marvel at its productivity. It is, indeed, productive, and that productivity has been steadily rising for many years, a trend that we need to understand if we want to guess about its chances for continuation. But the major achievement of the American farmer has been his fantastic increase in total production. He grows much more *per farmer* than ever before in history, but farmers in many other nations get higher per-acre yields of major crops than the American farmer, and often do so with lower inputs.

The statistic most often quoted to demonstrate the productivity of the U.S. *farmer* is the fact that he produces enough food for himself and 60 other people, up from only 15 just three short decades ago.[2] But that statistic measures only the effectiveness of the human labor involved and is not an indicator of total productivity. U.S. farmers have replaced labor with other inputs: fertilizer, irrigation, machinery, land. In 1939, labor made up 54 percent of the total inputs into agriculture; today that level is 13 percent.[3]

That means we don't need nearly as many farmers as most countries to produce tremendous quantities of food for our people and our foreign trade. But the number of people fed per farmer is, at best, a partial measure of productivity, because it doesn't measure the efficiency with which either labor or other resources such as fuel or fertilizer are converted into food and fiber. That today's farmer produces more food than his father is no proof that he works any harder or any more efficiently, but only that he uses more fertilizers, herbicides, and pesticides, along with better (and far more expensive) machines.

The past increases in U.S. farm productivity have been spurred by a variety of factors, one of which has been the quality of the land. The more inherent fertility in a soil, or the more responsive it is to agricultural management, the more yield will be obtained for each unit of input. American farmers have been shifting toward better, more productive lands for the past several decades, and this is part of the total productivity picture. Abandoning less productive lands in regions such as the Northeast and Appalachia in favor of bottomland soils in the Mississippi Delta has been a strong factor in raising the national average yield per acre over the last three decades. Obviously, that kind of opportunity is limited by the amount of good soil that remains to be brought under the plow. The trends of the past are not repeatable.

Another factor has been investment in land and water management systems such as irrigation, drainage, and land forming. Farmers have greatly expanded their irrigation systems, with resulting increases in yields. Because of their high installation costs, as well as the rising price of energy to pump water, these systems may have raised total production more than productivity, but they have no doubt contributed to productivity as well. Drainage of wet soils is a dramatic yield-raiser that often contributes to reduced costs as well. So it has been an important factor in increasing productivity. Land forming, the physical alteration of the land's slope by grading the soil surface, allows water to soak into the soil as needed and drain off when the soil is saturated. These soil manipulations often result in a permanent improvement in the soil's productivity, provided the topsoil is not buried or damaged in the process.

A third major factor in increasing productivity is science and technology. The classic example of research's contribution is hybrid corn, which revolutionized corn yields in a few short years. New varieties of other crops such as the short, stiff-strawed varieties of wheat and barley have given similar, if not as dramatic, yield boosts. New agricultural chemicals or fertilizers, new management techniques, new strains of livestock that convert feed to meat more efficiently, and new plant varieties that can produce under adverse climatic, moisture, soil, or other conditions have all contributed to U.S. farm productivity. Examples abound, for it is safe to say that virtually every crop field and barnyard in America today contains a plant or animal that has been developed or improved by agricultural research within the past few decades.

The larger, faster machinery available today has contributed to labor efficiency, but its effect on total productivity is far less certain. One of the positive contributions of the new machinery results from more accurate timing of farm operations. The new machinery allows farmers to work their soil faster, be ready when the time is right, and get the whole crop in more rapidly. But that doesn't mean that all farmers can take full advantage of that opportunity. Many simply add more land to help pay for the machinery, so they may still have trouble getting everything done at the right time.

In addition, it is not always certain when the time is "right," and it is still possible to guess wrong—or be proven wrong by subsequent weather events—and lose the advantage that the expensive machinery created.

In the Palouse country of northern Idaho where I put in my farm years, it was best to plant peas as early as possible. That would normally let them have a longer time for blooming before the really hot weather came. If a couple of hot days caught the peas in the midst of blooming, the blooms quit forming, and the number of pods that could mature were limited. As a result, we always tried to work our fields as soon as the soil was dry enough in the spring. With the machinery we had, it could take a week or two to get the ground ready to plant. During that time, a hard rain might force us to start all over again after the soil dried out. It was always a concern, and we often worked our tractors 24 hours a day to get the crop in as quickly as possible.

Today's machinery makes that work far easier. Soil preparation that took a week in the 1950's can now be done in a day or so, and farmers have a much better chance of getting their crop seeded at the right time. But does that always work? Unfortunately, not so. In 1980, Mount St. Helens blew her top on May 18 and buried many of the early-seeded pea fields in northern Idaho. Then the summer stayed moist and cool, so late-seeded crops had the chance to do well. When you farm, the weather can fool you no matter how fancy a tractor you drive. On the average, however, the ability to be timely with farming operations is an important advantage, so the new machinery helps the farmer who manages it properly.

Breaking through Production Limits

For two centuries, agriculture in the United States has been marked by rising yields and productivity. Each time a new barrier that might limit productivity was reached, it was breached by new agricultural technology. A review of that history provides a good basis for speculating about future yield increases. Limits are appearing on the horizon again. Will American farmers once more breach them, as they have in the past? If so, how will they do it?

In Thomas Jefferson's time, farming methods were little different from those used in the time of the Roman Empire. Farmers began to invent labor-saving machines soon after the American Revolution, but the real effects were not seen until the 1850's. In the early 1800's, farmers used broadcast seeding, walking plows, and brush harrows to cultivate their land, and it took about 250-300 hours of labor to grow 100 bushels of wheat on five acres.[4] At that rate, it appeared that the limited labor available to farm families would hold down crop production.

But by the middle of the nineteenth century, a mechanical revolution was underway that transformed agriculture dramatically. New machines were invented and, as fast as they could be produced, farmers bought them. Steel plow shares that could cut through the tough grass sod of the Midwest and the Great Plains were a good example. In 1849, John Deere made 1,000 new steel plows; by the mid-1850's, he was producing 10,000 a year.[5] Horses and mules still provided most of the power, but with the new plows, planters, cultivators, reapers, and threshers, farmers could cover a great deal more land with the labor available. Farm labor would not again pose a limit to U.S. agriculture.

In the 1920's, tractors took over from the horses and mules of the 1800's. In 1918, there were only 85,000 tractors; by 1920 the number had jumped to 246,000.[6] Mechanization accelerated the trend toward larger farms, as farmers could work a great deal more land with tractors than they could with horses and mules. A wheat farmer, for example, could now produce 100 bushels of wheat on five acres with only 15-20 hours of labor, so he could farm 15 times as much land as his nineteenth-century ancestor.[7]

But these trends, in the absence of yield increases, began to raise concerns that the limited supply of good farmland would limit the growth of U.S. farm production. The western frontier was closed, and most of the good land was being farmed, or so it seemed. The *1923 Yearbook of Agriculture* estimated that another 40 million acres of cropland would be needed, in addition to the reduction of farm exports, if the United States was to feed an additional 44 million people.[8] But irrigation and drainage, coupled with new varieties of plants and animals, fertilizers, and management methods, raised the yields steadily, while making many new lands feasible for farming, so the land constraint was avoided.

After World War II, American farmers dramatically accelerated the trend toward higher farm production. In addition to newer, more powerful tractors that would cover more acres, the continued development of new fertilizers and pesticides, improved seeds and livestock, antibiotics, and new management methods allowed farmers to harvest more from each acre and, at the same time, specialize in certain crops. Agricultural chemicals made up 3 percent of total farm inputs in 1950; by 1975 that had increased to 16 percent, largely due to a fivefold increase in the use of chemical fertilizers.[9] As a result, corn production per acre in 1980 was double what it had been in 1950.[10] With this intensification of agricultural methods, the average dryland farmer can now produce 100 bushels of wheat on only three acres, with only three to four hours of labor.[11]

Table 9.1 shows that the yield of wheat per inch of precipitation in the Great Plains has nearly tripled since 1930. Since soil moisture is the limiting factor in wheat production on the Plains and fertilizers are rarely used, these improvements represent real gains in productivity due largely to new crop varieties and the adoption of machinery and management techniques that coax maximum production from the limited moisture supply.

Table 9.1.**Wheat Production Trends in the Great Plains**

Year	Bushels per Acre per Inch of Precipitation	Bushels per Acre
1930	0.46	15.9
1970	1.05	32.2
1980	1.23	40.0

SOURCE: Pat Jordan, Director, Colorado Agricultural Experiment Station.

U.S. farm productivity has not, however, been rising as fast in recent years as it was in its heyday. From 1939 to 1960, total productivity increased 2 percent each year, but the rate slowed to

0.9 percent per year between 1960 and 1970.[12] In the 1970's, although productivity continued to climb, there were serious interruptions, as can be seen in Figure 9.1. Both 1970 and 1974 saw total farm output drop significantly due largely to an outbreak of corn blight in 1970 and bad weather in 1974.

Figure 9.1. Farm Productivity in the United States, 1967-1980

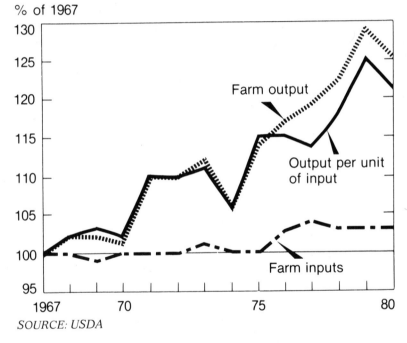

% of 1967

SOURCE: USDA

Agricultural scientists and economists have been holding a lively debate on the meaning of the recent productivity trends. The slow-down of the 1960's and early 1970's led many to predict that the era of scientific agriculture had peaked, and that yields were never again going to rise at historic rates. Then came five years of bumper crops, and by 1980, some scientists were saying that there was no evidence of a long-term slowing of yield increases.[13]

One difficulty in analyzing Figure 9.1, or any other historical record of agricultural productivity, lies in trying to assess the effect of weather on crop yields. We can't yet predict weather very effectively, let alone control it, and it can have overwhelming effects on crop yields. A recent study estimated that corn yields have histori-cally moved 12.5 percent above and below the average trend line on the basis of weather alone.[14] What this means is that the past 50 years of U.S. history shows that the corn crop is likely to vary within a 25 percent range due to factors that farmers cannot con-trol. That is a useful statistic to remember when USDA predicts next year's corn crop on the basis of the planting intentions of farmers. History suggests that those estimates could be high or low by as much as 10-15 percent, because there is no way other than sheer luck to predict any more accurately.

As Figure 9.1 shows, it was easy in 1974 to postulate that agricultural productivity was in real trouble. By 1979, it was equally easy to argue that all the earlier predictions were wrong and that yields were rising as fast or faster than ever. The events of 1980 didn't really support either contention, but simply added to the confusion. The best bet is to admit that the short-term data can be as misleading as it is helpful. When so much of our yield depends on the weather, and when climatologists cannot tell us whether the climate is getting warmer, colder, more variable, or less variable, it seems best to keep any predictions couched in fairly modest terms.[15]

Another problem in interpreting USDA's productivity data lies in trying to understand what really caused the national yield averages to rise. Were all farmers experiencing rising yields? If that were the case, one might predict a continuing rate of increase. Or were the top farmers finding that their yields were holding steady, while the poor farmers either dropped out of the picture or became more modern? This would mean that the average would rise, but only until the majority of farmers caught up with the leaders. At that point, further increases in productivity would come slowly, if at all. There is some evidence that the latter is what is happening. Many farmers are telling me that yields are not rising for the top producers and have not been for several years. Instead, they say, it is a tough struggle to hold their yields up to former levels. I have no hard evidence on this, but it is a question that needs immediate attention, because it will dramatically affect the expectations we can hold for the future.

Looking to the Future

Past trends give us a context for looking toward the future, but they do not necessarily predict it. We also need to look at current trends and efforts that might raise productivity and see how much potential seems to lie in each of them. We have already seen that there is not much new farmland to be added to the nation's cropland supply and that much of what is available is not going to be highly productive without significant investments. Much farmland going to urban and industrial uses is above average in productive potential, and much of the cropland in use today is losing topsoil faster than natural factors can replace it. These facts lead to the conclusion that the average productivity of the land itself may drop in the future, so crop yields will decline unless soil quality losses are offset by other factors.

Irrigated acreage has tripled since 1940 and doubled since 1950, even though there was virtually no change in the total amount of cropland in use during that time.[16] That rising percentage of irrigated land has contributed substantially to the rise in average yields and productivity, but there seems little way that irrigation can continue to expand as rapidly as it has in the past. With limited ability to expand water supplies in those areas where good land exists, the logical conclusion is that irrigation has about run its course as a major contributor to sustained growth in U.S. agricultural productivity.

Draining Wet Soils

On the positive side, it seems likely that drainage will once more become an important contributor to growth in crop yields. Where farmers have cropland that is too wet for maximum yields, drainage is a quick, economical, fairly permanent way to raise productivity.

Drainage has gained a bad reputation as pictures of wetlands and swamps going under the dragline gained national attention in the last decade. But millions of acres of land in the United States are not wetlands, but are simply too wet at some times of the year for optimum crop growth. These soils, labeled "wet soils" in USDA's soil terminology, make up 25 percent of all cropland, and are some of the most productive soils in the world. They are more fertile and contain more organic matter on the average than soils that have been formed under drier conditions.[17]

In all, there are about 270 million acres of wet soils in the United States that are not controlled by the federal government. While there is no sure information on how much of this land has been drained, it is probable that most of the 105 million acres of wet soil currently being cropped have been drained at one time or another. Many of the most fertile and productive acres in the Corn Belt depend on drainage systems to keep seasonal high water from swamping out the crops or delaying tillage and planting operations. Additionally, there are still around 34 million acres of cropland on which drainage systems have not been installed or where improved drainage will greatly enhance yields. USDA estimates that installing proper drainage on these wet soils may raise crop yields by 50-100 percent.[18] Draining those wet cropland soils would be the equivalent of adding 20 million new acres to the cropland supply, without putting a single new acre under the plow.[19]

Wet soils take more horsepower to till and are subject to far greater soil compaction problems, so an additional advantage to draining the 34 million acres of wet cropland would be to save as much as 137 million gallons of fuel per year, according to USDA studies.[20]

Drainage is not without its problems, however. Many of the 34 million cropland acres with inadequate drainage have some wildlife and wetland values as a result of being wet for parts of the year, and these values would be lost. In addition, many of these soils lie next to wetlands, and the drainage outlets may discharge directly into wetlands, streams, or rivers where they may cause some problems. Many wet soils serve as groundwater recharge areas, so improving the drainage can reduce recharge, which may result in lower water tables.

As a result of these problems, drainage has been actively discouraged during the 1970's. USDA policy has prohibited cost sharing for most drainage practices and, in some areas, farmers who wished to install drainage systems were denied technical assistance from the SCS. On lands where the growth of cattails or other water-loving plants indicate a wet soil condition, permits for earth moving or drainage have been routinely denied by the Corps of Engineers under provisions of Section 404 of the Clean Water Act. This section, originally designed to prevent the dredging and filling of wetlands, has been expanded and extended greatly by administrative and court actions, and now is used to prevent the drainage of wet soils and the construction of farm ponds in many instances.[21]

The future may hold a different outlook, however. Drainage clearly involves trade-offs. While it can damage wetland and wildlife values, it can also significantly raise agricultural productivity, as well as save energy. In the past, when energy was too cheap and farm productivity too high, it was easy to argue that the environmental damages created by drainage outweighed the agricultural benefits. In the future, when energy is going to be far more expensive and agricultural productivity badly needed, the balance seems likely to shift. Farmers with wet soils know that productivity is raised by proper drainage and, with a little encouragement from public opinion or programs, they will quickly expand the practice in the future.

New Machines

When you climb up on the seat of today's modern tractor, kick a roaring 200-horsepower engine into life, and move out to do a day's work on the farm, you wonder how much further the current trend toward mechanization can go. With the radio tuned to the latest music and news, the cab air-conditioned or heated to comfort, and an array of dials monitoring all the machine's vital signs, you can touch a hydraulic lever and drop a 15-foot-wide plow into the earth. At speeds faster than a man can walk, you roll 14 million pounds of soil over in an hour. By sundown, you may have plowed 80 acres, work that would have taken your father, on a 1950-vintage tractor, several days and your grandfather, with horses, a matter of weeks. And you will be ready to go to town that night, rather than being worn out from a rattling, bouncy, noisy tractor or dead tired from walking behind a team from dawn till dark.

So will we develop bigger machines? Faster machines? More sophisticated and precise machines? My prediction is "no" to the first two questions, "yes" to the last. The future costs of land, fertilizer, fuel, and machinery seem likely to continue rising relative to the cost of labor. If that happens, past trends in machinery development will reverse themselves. Up until now, new machinery has been designed to replace labor, since the long-term rise in wage rates relative to the prices of land and machinery encouraged the substitution of land and power for labor.[22] But we are looking toward a future where land will be more limited and higher priced, where energy, fertilizer, and fuel seem sure to become more expensive in comparison to other inputs, and where there are millions of people out of work. For that kind of situation, it will make sense to search for machines that cost less and use less fuel, even if the operating hours are increased.

This is not a prediction of a return to earlier, less sophisticated machines. To the contrary, it seems likely to me that future machines will feature precise placement of seeds, fertilizers, and pesticides as well as delicately guided cultivation, thinning, and harvesting capabilities. It may be true, as today's trends would indicate, that such machines have to be huge and cost 50-100 thousand dollars, but I doubt that. If American farmers don't create the necessary demand for the smaller, more sophisticated machines, Japanese farmers will. I think we will see the repetition of the trend in automobiles, where rising energy prices have made the behe-moths of the late 1960's obsolete and replaced them with more efficient, sophisticated, smaller machines. Whether U.S. farm machinery manufacturers follow the same blind rush toward giant-ism that put the automobile industry a decade behind foreign manu-facturers (and on the verge of bankruptcy) remains to be seen.

The net effect on productivity from the farm machines of the future is a toss-up. Some of the new machines will no doubt offer significant improvements. Computer-controlled functions will be commonplace, even in small, inexpensive machines, and this could mean operation at or near peak efficiency most of the time. That could result in significant fuel and cost savings, as well as maxi-mum output. For the small to medium-size farm, incorporating sophisticated mini-computers in less expensive machines could bring the advantages of modern technology into the farmer's reach.

What is not likely to happen, in my view, is the continued reduction in labor brought about by the past trends toward larger, faster machines. It appears to me that we have gone just about as far as is feasible in that direction. Many new machines may take more labor, not less. There will no doubt be some areas, such as fruit and vegetable harvesting, where machines will still replace people on the farm, but I am dubious about the Utopian vision of a computer-controlled farm where all the work is done automatically while the farmer sits in his office and watches TV screens. Too many biological processes are involved in agriculture, and the skill and judgment of an experienced farmer is too vital to be replaced by machines programmed to respond to standard conditions. That an almost fully automated farm *can* be developed is probable; that it can be a successful model for agriculture of the future is not.

Research to Spur Productivity

In contrast with new mechanical inventions, which generally replace labor, the primary effect of new advances in biological or chemical technology is to raise the per-acre yield of crops. Some inputs, like fertilizers, are associated with new costs, but as long as those costs were less than the returns received by farmers, they brought added productivity. Other new advances, such as the stiff-strawed dwarf wheat varieties, brought no new costs at all. By concentrating their energy on producing grain instead of stems or leaves, they produced more grain with the same inputs of sun, water, and soil.

Technology-developing research in plant breeding, agronomy, animal production, engineering, and farm management during the period 1927-1950 yielded a 95 percent annual rate of return, one study found. That is like getting 95 percent interest on your savings account—a fantastic rate of return. During the same period, research into basic science yielded an even higher rate, 110 percent.[23] From 1948 to 1971, technology research continued to yield over 90 percent, but research in basic science was less profitable than during earlier years. Just why that was true is not certain, but it may be that the lack of basic scientific insights was not the handicap in recent years that it was in the earlier decades.[24]

Agricultural scientists have predicted up to 12 new technologies that could raise productivity in the future. Of those, four could have really dramatic effects. These are the enhancement of photosynthetic efficiency, development of nitrogen-fixing grains, new bioregulators, and twinning in beef cattle.[25]

Photosynthesis, the process by which green plants convert sunlight, carbon dioxide, and water into carbohydrates, is the basis of all life on earth, but it is not a very efficient process in terms of maximizing yields. A field of healthy, high-yielding corn, for example, is estimated to utilize only about 3 percent of the sunlight energy that strikes the leaf surface.[26] Any new development in plant genetics, farm management, or chemical modification that could help us capture some of the benefit of that remaining 97 percent would be a tremendous boon to humans. Even a small fraction of 1 percent improvement would translate into millions of tons of food each year.

Nitrogen is needed for protein development in plants, and there are wide variations in the ability of different crops to utilize nitrogen from the air and soil. Legumes, through a symbiotic relationship with nitrogen-fixing bacteria, capture nitrogen from the air. Grasses do not. If strains of wheat or corn could be developed that would fix atmospheric nitrogen as peas or beans do, they would produce grain without needing additional fertilizer, for a significant cost savings. Research has been searching for a suitable combination of crop strains and bacteria, but no commercial possibilities are yet in sight.

Bioregulators are natural or synthetic compounds that regulate the ripening of horticultural crops. If artificial bioregulators can be successfully developed, the ripening of fruit can be timed so that all the crop ripens at the same time to facilitate mechanical harvest. Similar compounds might keep fruit from deteriorating after harvest, increasing storage life and quality. The safety of these artificial hormones for human consumption will need to be proven before this technology finds wide acceptance, but it appears that they could raise productivity by making more food available from the same yield of crops.

Several approaches are being tested that may soon result in a significantly higher percentage of multiple births in beef cattle. These include breeding and selecting livestock that naturally produce twins, multiple ovulation through hormonal control, and embryo transfers. Embryo transfers involve the removal of fertilized ova from one cow, to be implanted in another cow for incubation. This allows superior animals to produce many more offspring than otherwise possible, with them borne and raised by common brood stock. Non-surgical methods of removing fertilized ova have been developed, but better methods of implanting the embryo are still being sought. Agricultural scientists have produced as many as 28 calves from one mother cow in one year through the use of hormonal treatment and embryo implants.[27]

This technology could make it easier and faster for the beef industry to breed animals with desirable traits such as a faster rate of feed conversion or the ability to produce table-ready beef from grass pastures. Both traits could reduce the need for feed grain and free grain supplies for direct consumption by humans. In addition, an ability to encourage multiple births would allow cattle numbers to build up faster in response to new demand, a process that now takes several years.

Research can also help combine several technologies into new systems that open up new agricultural possibilities. Agricultural scientists in Colorado, for example, have developed a new small-farm approach to growing apples in the western part of that state. Dwarf apple trees are planted in a hedgerow pattern that results in 20 times as many trees per acre as would normally be planted. A trickle irrigation system waters 40 trees with the same amount of water that it took to flood-irrigate 1 tree before. The result—4,000 bushels of apples per acre, compared to 1,200 before. The 20-acre farm now becomes a feasible commercial family operation, and land and water are used efficiently where they were probably too limited to allow commercial orchard production under conventional management systems.[28]

Wes Jackson of the Land Institute, in Salina, Kansas, raises the possibility of perennial crop plants that can be harvested year after year without cultivating the land or replanting the crop.[29] The advantages that could be gained in reducing both farm costs and topsoil erosion are self-evident, but the ease with which such crop varieties can be developed is not. Some scientists feel that a perennial growth habit may be impossible to link genetically with high seed yield, but others argue that the past history of science suggests that many "impossible" things have been converted into reality, so no possibility should be rejected without serious research effort. The latter point of view would seem to be supported by the experience with crops such as alfalfa, where the development of proper soil, water, vegetation, and fertilizer management has raised seed yields from 300-400 pounds per acre to a high of almost 2,000.

Idaho alfalfa seed grower Burtt Trueblood checks alfalfa residue being cultivated back into the soil to build up organic matter levels. Trueblood raises high yields of alfalfa seed, corn, and wheat from stony, fairly steep irrigated soils with no soil erosion through careful soil and water mangement. Photo by Neil Sampson.

High-yielding perennial varieties of wheat, corn, soybeans, or other major crops would be such a revolutionary, and valuable, contribution to agriculture that a great deal of investment in research on them would seem warranted. But they might not be the panacea they seem at first glance. Burtt Trueblood, of Wilder, Idaho, pushed alfalfa yields in perennial stands up five-six times over normal, but has been unable to hold those peak levels, despite a finely tuned system of soil and crop management. While there are no conclusive research results, Burtt feels that perhaps even alfalfa, with its soil-building characteristics, can only be grown under intensive high-yield management for a few years before some aspect of crop/soil interaction begins to be limiting. At that point, rotating the field to corn, beans, or wheat for a few years may be needed to break up the monoculture and rebalance the soil system.[30]

Buying Research

Since it is obvious that additional agricultural productivity is going to be vital in protecting the nation's lands from irreparable harm, the question that must be answered is "how much research is enough to give the productivity increases we will need?" To ensure new technologies, USDA scientists argue for a budget growing at a rate of 3-7 percent a year above inflation.[31] A 3 percent growth rate would spur a productivity rise of 1.1 percent per year, they predict; a 7 percent rise, if the emerging new technologies such as enhanced photosynthesis, bioregulators, and twinning in beef cattle become commercially available, could result in a 1.3 percent annual rate of productivity growth.[32] Neither rate, it is clear, would provide the 2 percent rate needed to keep land resources from being stressed.

To put those numbers into perspective, productivity grew at the rate of 2 percent per year in the 1950's and 1960's, but has now slowed to around a 1 percent growth rate. Public research budgets grew 3 percent per year between 1929 and 1972, but that rate has slowed to about 1 percent in the last decade. Total public expenditures for agricultural research now total $1 billion per year, with about 40 percent coming from state appropriations, the rest from federal funds.[33] Estimates of private research related to agriculture also fall in the $1 billion per year range, but about half of that research is directed toward food preparation, storage, and handling, so it contributes little toward expanding agricultural productivity.

The recent record of federal budgets, however, gives little hope of getting public research budgets up to the spending level that would, by itself, return the historic 2 percent annual growth rate in productivity. Given the climate for reduced public spending facing the nation today, it will not be easy to accelerate research rapidly enough to achieve the 1 percent goal set by USDA.

Researchers face a serious problem of credibility when they go to Congress and state legislatures to ask for ever-rising budgets. They can't guarantee that their research work will result in scientific breakthroughs that will have important commercial application. Research scientists don't always know for sure what they are looking for, let alone what they might find. The best argument to be mustered is that, without research, new scientific breakthroughs are almost certain not to happen. With research, the odds of finding new technologies go up. Legislators, facing tough demands to hold down public spending, have a hard time justifying immediate and real costs against future and somewhat uncertain benefits.

There are also built-in time delays that will prevent current research spending from having much real impact in the immediate future. Even if total spending were to increase in the next few years by the 3-7 percent annual rates that USDA claims are needed to assure new farm technologies, the chances are good that it would be 10 years or more before those new investments would begin to pay off. The contributions of science and technology to new agricultural productivity in the 1980's will reflect the tight research budgets of the 1970's, not those of the current decade. For the near-term, we need something besides research to provide the production that will be required.

Figure 9.2 sums up the importance of scientific discoveries in altering the pressure on the land. In 1977, it took about 381 million acres to produce food and fiber for domestic and export needs. In just 20 years, we will need 440 million acres to do the same job if per-acre yields continue to rise at the rate of 1.1 percent per year (line B). But what if we don't get those increases? If yields in 2000 are still about the same as we see today, meeting those same demands would require 575 million acres of cropland (line A). Either we will not be able to meet that need, or we will have to convert a great deal of land—much of it very marginal and expensive to farm—into crops. On the other hand, if dramatic scientific breakthroughs are immediately adopted for commercial use and the average rate of yield increase grows to 2 percent per year, crop yields will grow at the same rate as predicted demand, and the crop acres we currently farm would be all that would be needed in the future (line C).

Since the payoff from the new research effort may well not come until the mid-1990's, the outlook for the next two decades is serious. By the time current research begins paying off, given the current trends in domestic and foreign demand, the land:food:people crunch will be so severe that food production costs will have at least

Figure 9.2. Potential Demands
for Cropland, 1977-2000

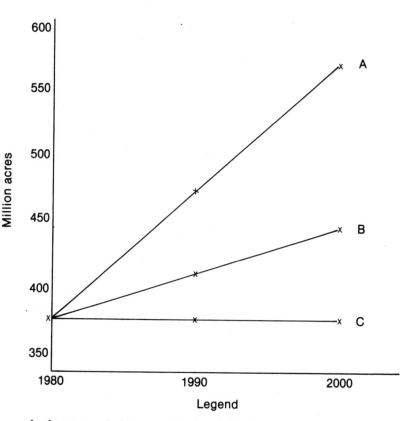

Legend

A. Acres needed to grow food and fiber if yields stay at 1977 levels.

B. Acres needed to grow food and fiber if yields rise at an annual rate of 1.1 percent per year.

C. Acres needed to grow food and fiber if yields rise at an annual rate of 2.0 percent per year.

SOURCE: Compiled by the author

doubled. That would mean serious difficulty in filling the demands for products people are able to buy, without consideration for feeding the people who can't afford to pay the higher prices.

There is another issue that is even more troubling. Part of our current dilemma may be the result of doing the wrong kinds of research and adopting the wrong kinds of technologies. Not all technological gains help maintain the productivity of soil and water; not all contribute to a sustainable agriculture. Some kinds of technology, such as those that allow high-intensity, clean-cultivated, one-crop monocultures are but a thinly masked method of mining topsoil. If that is the kind of new technology produced by research scientists, the nation will be wise in not purchasing too much.

Needed: New Directions

Agriculture is a man-made ecological system. Viewed through the eyes of an ecologist, agriculture can no more escape certain ecological laws than Newton's apple could ignore the law of gravity. Any ecosystem, whether man-made or natural, must in the long run achieve a steady state in regard to both energy and materials. Fertility, organic matter, energy, and water must flow into the system in relationship to outflow, or the potential productivity of the system will change. The soil is a living resource reservoir of constantly shifting matter, energy, and productivity capability. If the size or capability of that soil reservoir is steadily shrinking, it is only a matter of time until that becomes a problem to the animals or humans that depend on the production of the soil for their sustenance.

Agriculture and agricultural technology are, to that soil reservoir, like a pump on a well. The amount of food produced can be increased, as can the water from the well, by speeding up or improving the pump. But those "speed-up" efforts will only render a easy profit so long as the level in the reservoir is maintained. If we begin depleting the reservoir, improvements in the pump and ever-increasing energy inputs to run it faster become, instead of net gains, essential responses needed to maintain current levels of production.

American agriculture is in precisely that position today. Past productivity increases have come through the adoption of new technology and a huge increase in the amount of capital used, during a period when climatic conditions may have been unusually favorable. This has enabled a steadily dwindling number of farmers in America to stay in business by growing larger in size, adopting the latest in techniques, and absorbing narrow (or non-existent) profit margins. But is the agricultural "pump" speeded up about as fast as it will go? All the evidence indicates that the "water level" in the soil reservoir is continuing to drop.

We have essentially two ways to focus research. The first is to concentrate on plants and animals, breeding varieties that are more and more efficient at extracting minerals and water from the soil and converting them into useful products. That can make soil virtually unnecessary, since it is completely possible to grow plants in a medium of sterile sand, providing all the water and plant nutrients from artificial sources. Take the technology further, provide protection from adverse climatic conditions through construction and operation of a greenhouse, and crops can be grown in a completely artificial, subsidized environment. For some high-value food crops, that is a possibility. But for millions of acres of corn, wheat, and soybeans in America, it may cause more problems than it solves.

For those millions of acres that must be kept productive year after year, the search for more and more artificial growing conditions has resulted in the development of crop varieties that are more efficient at exploiting the soil. They flourish under high rates of nitrogen fertilizer, for example, which speeds up the depletion of other soil nutrients and, under heavy nitrogen fertilization, reduces the natural nitrogen-fixing capability of legumes. This may raise agricultural production in the short run, but may actually lower productivity over the longer term. As nitrogen fertilizer prices rise, farmers may try to go back to crop rotations, legumes, or other sources of nitrogen, but that will not be easy, as the soil may take several years to recover from the nitrogen "fix." When we look at the long-term sustainability of productive resources, including soil, water, and viable human communities, it is clear that the development of more and more exploitive agricultural technologies is harmful, not helpful.[34]

New agricultural research, if it is to contribute to the long-term strength of U.S. agriculture, must focus on a second type of technology that builds a sustainable agriculture without soil or water depletion. In our analogy of the pump, we must work at ways of keeping the well healthy and productive, as well as find all the ways we can to use efficiently the water we can extract at non-exploitive rates, not just concentrate on methods of turning the pump faster. There is no such thing as strength through resource exhaustion, regardless of the compelling arguments that are sometimes made for such a strategy as a necessary short-term expedient.

Organic farmers believe that if you take care of the soil, the soil will take care of you, an attitude that could greatly enhance the social impact of research and technology. The current interest in organic farming, occurring at the same time as a growing public concern for soil erosion and prime farmland conversion, is an indicator of a growing recognition that retention and enhancement of soil productivity is an important issue.

The current trends in agriculture's waste of the land must be turned around. That won't happen by some nostalgic return to some kind of mythical "good old days." It will only happen by an aggressive, innovative search for new agricultural technologies that give the *production* needed while building and maintaining *productivity.* That search must be led by the agricultural scientists in USDA, in State Agricultural Experiment Stations, and in other public sector agricultural research facilities, with strong support and cooperation from farmers. It won't happen in private industry research, because it is not the type of technology that will produce the salable commodities that will generate the profits needed to fuel private research. And it may not happen in the public research facilities unless they get the kind of public guidance, support, and funding that will be needed to push it in that direction.

American agriculture must not be allowed to follow other sectors of the economy such as the auto industry, the steel industry, the shoe industry and others in failing the test of change.[35] An all-out research effort on methods of enhancing *sustainable agricultural productivity* seems to be the best insurance against that disaster that is available today.

Unless we turn to the land that makes up the reservoir from which agriculture pumps food and find ways to reverse the damage and loss that is steadily depleting it, serious implications for American agriculture seems certain, and in the not-too-distant future. Our current wastefulness has important social consequences, for it passes the costs of today's production on to future generations. Those consequences have received little attention in the agricultural research efforts of the past, but they must now take priority if research is going to be a significant part of the action program to revitalize America's agricultural industry and place it on a permanent, sustainable, productive basis.

10.

New Crops from the Land

The future demands on America's land look formidable enough when the trends in topsoil erosion, farmland conversion, water competition, and research funding are compared to the growth of population and demand for traditional food and fiber products. But the crops of the past may be but part of the crops of the future as a totally new set of competitive forces are unleashed on the land in the declining decades of the petroleum era. America needs renewable sources of industrial materials and fuels to replace the increasingly expensive, scarce, or unreliable sources of the past, and we are turning to the land to get them.

The production of fuel ethanol as a means of stretching gasoline and diesel supplies will, to the extent that corn is used for the basic feedstock, compete most directly with common food and feed crops for the use of agricultural land and water. Another, less well recognized demand will come from the production of agricultural crops to be used in the rubber, plastics, paint, newsprint, and similar industries. Those industries have not used American farm crops to any great extent in the past, but that is likely to change. For the first time in U.S. history, farm-grown industrial products may seriously compete with food, feed, and fiber crops for the use of available farmland and water.

As this change occurs, the rural scene may look much the same to the casual observer, but far different economic forces will be at work. When corn is being used for commercial ethanol production, there will be a new kind of bidder in the market. This industrial user will not stop buying corn because the price moves up a few cents. He will need to have material to keep a multimillion dollar industrial plant in operation, and factors such as labor agreements and the need to fill contracts for alcohol will affect his willingness to pay for corn.

As grain prices climb, cattle feeders and other users who can make adjustments will drop out of the market, but the ethanol company will get the grain it needs. To grain farmers, these price rises will be welcome; to the cattle feeder, it will mean a higher price that he must pay in order to buy feed. Whether the cattle feeder will be able to withstand that kind of competition is a question.

The land will also feel the effects of this competition. Even in areas best suited for corn, there are many soils that cannot tolerate intensive corn cultivation without serious soil damage. A stronger price in the market will induce farmers to grow as much corn as possible, and this will require that they grow corn on land that is even less suited for such intensive use. The results will be more soil erosion, more water pollution, and an agriculture that is even less sustainable than that which we have today, since intensive cropping on those marginal lands will be a short-term expedient that must inevitably end when the topsoil is gone.

We may, as some argue, be entering a new era of agricultural progress and prosperity, but in that era there will be serious conflicts to resolve. The new economic and environmental forces at work may pit fuel against food, with the result that prices rise and poor people lose access to the necessities of survival. Critical public policy choices may make the difference, and they will be made with a new actor on the scene: energy.

The American people are not going to "starve to death in the dark." Neither is anyone else in the world, if they can help it. Fuel to power our devices and our lifestyle will be in demand, and when it begins to compete with food for our land and water, a new calculus must emerge, and new priorities must be set.

Biomass as a Fuel Source

The economics of petroleum are simple and deadly. Each barrel of oil pumped means that the next barrel must come from a deeper hole, at a higher cost. Only when there is a technical innovation that makes oil pumping cheaper, or a new oil field is discovered, will we get an economic reprieve. And that "reprieve" will be only temporary, as the costs will continue to climb from wherever the new "normal" cost level has been established.

In addition, we have the "energy multiplier effect." Extracting energy in the form of oil, coal, or other non-renewable resources consumes a great deal of energy in the process. As each barrel of oil gets more expensive, two kinds of new expense are added to the cost of getting the next barrel. Not only is the next barrel deeper and harder to pump; the fuel used for pumping is going up in price as well. In short, there is little long-term future in an economy based on non-renewable sources of energy. The best course is to turn to renewable sources of energy for many of our needs.

Burning fresh plant material (as opposed to burning the fossilized remains of prehistoric plants) is one way to shift from a non-renewable to a renewable energy base. All kinds of plant materials, ranging from trees to the corn stover and wheat stubble left after harvest, can be used for energy production. These plant products, along with organic materials such as animal manures and organic solid waste, are commonly called biomass. Despite the tremendous amount of biomass produced in the United States each year, we use it to produce only 1.5 percent of our energy. This leads to suggestions for converting more of these materials, often thought of as "waste," to a productive, economic use.

Generation of industrial power by wood burning is nothing new, of course. It was a feature of many early energy attempts, such as the wood-burning locomotive that carried its own fuel supply, or the sawmill that burned waste products to generate steam to run the mill. Now, with prices of other fuel sources going up, wood has been rediscovered, with the number of wood-burning stoves estimated to have increased from 1 million in 1974 to 5 million in 1976.[1] The managers of public forests, as well as private woodlot owners, are beseiged by homeowners searching for wood to help cut the cost of home heating.

Some forecasters predict that biomass use for energy could triple by the end of the century and double again by 2020.[2] Certainly, there have been developments to support that prediction. The Energy Security Act of 1980 authorizes federal expenditures of $1.45 billion through 1982 for financial assistance to synthetic fuel projects using biomass energy sources.[3] These funds have been held up by the Reagan administration, on the assumption that the momentum is already so strong that federal subsidies will not be needed to make biomass energy sources economically competitive.

One source of increased damage to current cropland could come from misguided attempts to use crop residues as an energy source. The nine leading crops produce wastes each year that, if burned, would provide about 5 percent of the nation's energy use.[4]

But that diversion would rob organic matter that is vital to the maintenance of soil fertility and tilth, leading to disastrous soil erosion levels.

This field of sugarcane looks like any other, but it will be harvested for energy, not food. Department of Energy (DOE) photo.

Some agricultural scientists feel that it may be feasible to use from 10 to 80 percent of the residues, depending on the crop, location, and soil type.[5] However, this calculation was based solely on the value of crop residues in preventing soil erosion and did not account for the value of organic matter in providing plant nutrients on contributing to soil tilth and structure. Also not considered was the importance of plant residues as a primary source of energy for soil microbial activity, a value that might best be assessed by conducting an energy analysis of the soil system.[6] Since soil scientists do not agree on the amount of crop residues that can be safely removed without adverse effects on soil productivity, the most prudent course, clearly, is to continue to recycle most crop residues back into the soil, where they are vital in keeping organic matter levels high enough to make the soil more open to air and water, more resistant to soil erosion, and more productive.

Farmers have much to gain by producing energy from farm products. At the Mason-Dixon Farms, of Gettysburg, Pennsylvania, for example, some 2.7 million tons of manure from 700 holstein cows are producing $30,000 worth of energy each year. In this system the manure is collected from the dairy barns and flushed into a pit where the solids are separated out and carried to a "digester," a specialized manure pit within a building. The pit is tightly covered by a black plastic bag that resembles a large balloon. In the oxygen-free air under the bag, bacteria convert the manure to a mixture of 60 percent methane and 40 percent carbon dioxide. The gas is captured in the bag and then pumped to an ordinary engine which drives a standard electrical generator.

With the addition of a second methane digester so that the system can utilize all of the farm's manure instead of only part of it, the operation should produce a net value of some $1.8 million over the next 20 years, its owners say. This farm will sell power to the power company instead of buying it, an interesting reversal.

The operation is simple and effective. With only 8 days of down time in the first year of operation, farm co-owner Dick Waybright is satisfied with the dependability of the system. He points out that it was not the methane digester, but the conventional generator, that caused the only trouble they have had so far.

Waybright points out several other advantages to his system. The by-product of the methane digestion process is a fluffy, odor-free material that can be dried and used as bedding in the dairy barns and loafing sheds. The waste heat from the engine running the generator could be used to power a small still to make ethanol from corn grown on the farm. In total, he feels it is completely feasible to turn his farm from an energy consumer to an energy producer.[7]

Such small-scale uses have grown rapidly in recent years. Farmers are lining up by the hundreds for non-commercial fuel alcohol permits at the U.S. Bureau of Alcohol, Tobacco and Firearms. In 1979, the Bureau of Alcohol, Tobacco and Firearms received over 4,000 applications for non-commercial distillery licenses, up from only 18 in 1978.[8]

Since the total energy use for all American farm production is only about 3-4 percent of total national consumption, farmers may find it advantageous to develop on-farm sources of some liquid fuel. Some studies have shown that the economics of most on-farm alcohol operations are still marginal, but with so many experimenters at work, the odds are good that economical methods of production will be developed. Besides, farmers don't always compute economics the same way that economists do. If they did, most of them would have quit farming a long time ago, because many aspects of their farm may not measure up to an economist's yardstick, in terms of profit and loss.

In rural Louden County, Virginia, farmer John Rocca has designed a solar-powered still that has been producing 25 gallons of ethanol per hour at a cost of 35 cents per gallon, far below the costs being estimated for commercial ethanol producers. His conversion rate of 2.3 gallons of ethanol per bushel of corn is similar to those achieved by commercial concerns, but his process is far different. Rocca is adding urea, a common farm fertilizer, to the corn paste. The result is starch formation without going through the energy-consuming process of boiling. In the process, the urea is reclaimed and the left-over distiller's grain is fed to the farm cattle. Rocca sees the production of ethanol as part of his small, general farm operation as an exercise in survival. "I think I see something here that's so simple and available—it answers so many problems for us," he says.[9]

Small-scale biomass technologies such as the Mason-Dixon and Rocca farms are using will be far less threatening in terms of committing the land base than if biomass energy is produced for large-scale commercial operations. But judging the land effects of a major increase in biomass energy utilization is not a simple matter. Even the small operations move nutrients off the soil in a one-way flow, and the fact that manure is used to produce methane means that a great deal of carbon and nitrogen is not being spread back onto the soils that produced the feed for the cows. With a system where all by-products do go back to the soil, however, less nutrients will need to be replaced to keep the land healthy than in a totally one-way system where the grain is sent many miles to an ethanol plant. Large industrial plants must treat their by-products as waste and often dispose of them where they are of no value in recycling fertility back to the soil that produced the grain.

What is feasible for a small farm with a full-time owner-manager may not be possible for large commercial farms, however. Even if large farmers needed to produce all their own fuel, they could not hope to without major readjustments. Wes Jackson calculates that, in order to replace totally the gasoline and diesel needed to run America's farms, 133 million acres of corn would need to be devoted to alcohol production.[10] That would be more than the total 1980 corn crop. Worldwide, it has been estimated that if *all* the grain grown in 1975 were converted to ethanol it would still only equal 6.9 percent of the crude oil used that year.[11]

The notion of growing crops to provide fuels for a large-scale commercial industry makes the numbers even larger. USDA has estimated that it would take about 300 million acres of corn to produce just 10 percent of the nation's current energy usage.[12] This is about triple the acreage currently in corn production, and 75 percent of the total cropland in use in the nation today. Such a production level from cropland is clearly impossible; the land just doesn't exist.

In addition to cropland, however, there are about 740 million acres of forestland in the nation, part of which could be devoted to the production of biomass for energy. In its Resources Planning Act report, the Forest Service estimated that more than half of this land is capable of commercial timber production.[13] Wood products can be used for energy by direct burning, through the production of methanol, or through "gasification." Most biomass-derived energy is now obtained by direct burning, but the development of improved technology for converting cellulosic materials into alcohol seems likely. At a meeting of the Bio-Energy World Congress and Exposition, held in Atlanta in April, 1980, most speakers agreed that, for economic reasons, wood would far surpass grain as the most economical source of biomass for energy production.[14]

This forecast suggests greater pressures on forestlands, pressures which will add to those created by the growing demand for wood products to be used in the building and paper industries, as well as the demand for forestland as a source of needed additions to the nation's cropland. One estimate is that 70 to 83 million acres of forest plantations would be required to produce just 10 percent of the nation's 1980 energy use.[15] That would be about equal to 10 percent of current forestland.

New energy crops will be grown on forestlands that are more accessible to communities, farms, and roads, where proximity as well as good soils will make them more attractive for highly managed energy plantations than the more remote or unproductive sites. Because of this factor, energy plantations will compete with traditional log and pulp production much more seriously than the acreage estimates would indicate. Energy crops will take the most productive soils, relegating the traditional forest crops to more marginal sites.

Since many new energy plantations will probably be established on rangeland and pastureland or marginal cropland, however, far less than 10 percent of today's forestland will probably be affected by such a biomass future. Instead, there will be a shifting of uses within several types of land, as marginal cropland, pastures and rangelands are planted to woody crops for energy production. In many areas, this can be both a conservation and an economic plus. It could provide an economic use for marginal croplands by encouraging people to plant them in trees for energy rather than letting them continue to erode away under row crops.

Gasohol:
Panacea or Problem?

There have been two major questions about the wisdom of producing crops for alcohol on America's farms. Will a national program encouraging ethanol production result in competition between food and fuel? Can a positive net energy balance be achieved if the energy consumed in crop production is included in the calculation?

The latter question, while not resolved, is generating less concern among policymakers than the issue of whether or not ethanol production burns up more liquid fuel than it manufactures. If a net liquid fuel balance *is* realized, gasohol may be said to have converted a type of energy (sun) that is not scarce to a type (gas or diesel) that is. Old distilleries, designed to produce consumption alcohol, often require almost a gallon of other types of liquid fuels in order to produce a gallon of alcohol, but the Department of Energy (DOE) estimates that new distilleries, designed especially to produce fuel-grade alcohol, could produce 4 gallons of liquid fuel for each gallon used in production and distillation.[16]

The food vs. fuel debate is less settled and less likely to be resolved through technology. One argument holds that, since the distiller's grain left over after the fermentation process contains all the original protein of the grain and can be used as a livestock feed, the net loss to the food supply is negligible. But this does not, in itself, resolve the question.

Lester Brown of Worldwatch Institute in Washington, D.C., points out that there may be a limit to the amount of distiller's grain that can be used as a feedstuff. Producing 2 billion gallons of ethanol from corn would yield 17 times as much distiller's grain as was consumed in 1976.[17] In light of the fact that distiller's grain is a less desirable feed supplement than soybean meal, it is doubtful that this much can be absorbed by the current U.S. animal industry.[18] If any significant amount is used, it will replace soybean meal as a protein supplement. The net effect on food supplies (or land use) that might result from shifting cattle from soybean meal to distiller's grain, and soybean land to corn for ethanol, is not subject to a simple calculation, even if the shift is technically feasible. In terms of total cropland needed, the difference would probably be negligible. Soil erosion might be reduced, since corn can be managed better for soil protection than soybeans, which have a limited amount of crop residues for soil protection over the winter season.

One possibility would be to export distiller's grain instead of whole grain. Since most of the grain going abroad is being sold for cattle feed rather than for food, there might be a good export market for distiller's grain, provided the demand for livestock feed abroad stays strong.

Even if the production and economic issues are solved, a major concern in any large-scale national gasohol program is the need for added cropland to grow the grain (mainly corn) likely to be needed. If marginal lands are brought into production as a result of new demands and higher prices for farm crops, added soil erosion is certain.

The Congressional Office of Technology Assessment expressed this concern in 1979 by pointing out that producing enough gasohol to meet most of America's automotive needs would mean putting approximately 30-70 million additional acres into intensive crop production. Such an expansion of new crop production would "accelerate erosion and sedimentation, increase pesticide and fertilizer use, replace unmanaged with managed ecosystems, and aggravate other environmental damages associated with American agriculture," their report said. They further noted that "a combination of ethanol subsidies and rising crude oil prices could drive up the price of farm commodities and ultimately the price of food."[19]

Even the DOE, trying to demonstrate both the viability of and the need for a national program to encourage ethanol production, could not keep from noting that major stresses might be created for both agriculture and the land. They pointed out that 4.7 billion gallons of ethanol per year could be produced using existing technologies. Such production, which would amount to around 2 percent or so of our petroleum use, would require, they said, "bringing into production all existing grain land and supplementing food processing wastes with sugar surpluses and fermentable municipal solid waste."[20] Just what DOE meant by the term *all existing grain land* is unclear, but my interpretation indicates that they were recognizing—correctly, I feel—that a major ethanol push would require marginal and unsuitable land to be pressed into service.

Wes Jackson brings the problem down to a more human dimension, arguing that the issue is primarily one of ethics:

> Keep in mind that the energy in the alcohol required to meet the demands of an average U.S. car for one year could alternately be used as food to feed 23½ people for an entire year. From our point of view, the issue is not whether the alcohol is there, but that massive alcohol production from our farms is an immoral use of our soils since it rapidly promotes their wasting away. *We must save these soils for an oil-less future.*[21]

Jackson's point should be considered in light of the situation outlined in Chapter 2. America can devote significant amounts of grain to ethanol without affecting our ability to produce enough food for domestic needs. We are inextricably tied to a global food system, however, and every bushel that goes into a fuel plant in the United States is a bushel that doesn't go abroad. There is no "extra" grain for ethanol.

Prior to 1979, the USDA expressed serious reservations about the ability of the land to absorb the added demand of a national gasohol program. Since then, however, the department's position has changed, and it now favors the production of gasohol. Analysts at USDA claim that the policy switch was due to new facts about gasohol, but there are also indications that pressure from farm groups, Congress and the White House was substantial. A positive stance on gasohol gave farmers some good news from Washington in an election year when the Russian grain embargo had been the dominant, and seriously negative, farm issue.

In describing the department's new program to Congress, Deputy Secretary Jim Williams noted:

> This alcohol fuels program represents a basic policy change. The USDA is now including production of farm commodities for alcohol feedstocks as a major objective of agricultural policy—alongside the production of food, feed, and fiber. Grain reserve targets, commodity price supports, acreage diversion, and other related agricultural policies are being managed to include the grain requirements for alcohol equally with other consumers of grain.[22]

In January, 1980, President Carter set a national goal of producing 500 million gallons of ethanol per year by the end of

1981. At USDA, Secretary Bergland estimated that such a goal could be reached by a 4 percent increase in the land devoted to corn (or a 4 percent increase in average national corn yields), and noted that "distillation capacity, not agricultural feedstocks, is currently the restraining factor on fuel alcohol production."[23]

That ethanol production is attractive to oil companies is in little doubt. In June, 1980, Martin Abel, of Schnittker Associates, a Washington-based consulting firm, told a conference sponsored by Resources for the Future that:

> Only recently Ashland Oil and Publicker announced plans for a 60-million-gallon plant at South Point, Ohio, and American Maize Products and Cities Service Corporation announced a 50-million-gallon plant at Hammond, Indiana. Furthermore, an Iowa cooperative is considering building a 50-million-gallon plant. We believe, therefore, that beginning in 1982, production capacity will rise rapidly, reaching 1.1-1.3 billion gallons by 1985-1986 and 1.5-2.0 billion gallons by 1990-1991. Thus, if U.S. and world energy prices evolve in the way we and others anticipate, there may be no shortage of incentives for investment in facilities to produce ethanol from grain.[24]

Gasohol is clearly not the only solution to diminishing supplies of petroleum fuel, nor is it without its costs. Whether it is more panacea than problem awaits an answer. One thing is certain—that answer is on its way. With the enthusiastic support of farm groups, who saw a new market that might boost farm prices, and Congress, who had been looking for something (anything?) to make farmers happier, the passage of the Energy Security Act of 1980 signaled a major political commitment to public support (and subsidies) for the ethanol industry.

If Abel is correct in his assessment for the future, and USDA was correct when they predicted that major investments in plant capacity would tend to lock the nation into the allocation of grain for fuel production up to plant capacity once conversion plants are constructed and operational, it seems that the die is cast.[25] The plants will be built; their buyers will begin to bid for grain. Contracts for future supplies will be purchased, so as to assure a steady supply at a predictable price, and the price of grain should go up. Farmers, in response to those price signals, will seek ways to grow more grain on their land. The challenge for conservationists will be to help farmers and ranchers find ways to integrate energy crop production into their land and water management systems without permanently damaging those basic resources in the process. Whether or not this can be done will depend largely on the way that the economic incentives are structured. If there are no economic rewards for conservation (a subject to which we shall return), then the efforts of conservationists will be frustrated and ineffective.

The challenge for national policymakers is more serious. Grain prices will be strongly affected by the existence of a grain alcohol industry, and the price that cattle feeders pay for feed will almost surely go up. If that happens, consumers will see a rise in meat prices. The whiplash effects common to agricultural supplies and prices, due to weather if nothing else, will be magnified by the new competition created by industrial purchases of grain. The effect of this on agriculture, and food prices, will make agricultural policy and farm programs even more difficult to manage in the future. A gasohol future holds more competition between food and fuel and more extensive topsoil damage.

Industrial Materials from the Land

American industry uses a wide variety of raw materials for its processes, but it is not widely known that many of them are agriculturally produced. In addition to the petrochemicals that are used by American industry to manufacture clothing, chemicals, paints, and plastics, the United States also imports a variety of agricultural materials. Included in the list are natural rubber, waxes, resins, newsprint, and adhesives. Many of these materials can be produced within the country, and this option is beginning to look more and more attractive. Economically, the stakes are large. The United States currently imports agriculturally produced industrial materials at the rate of an estimated $27.3 billion per year. In addition, another $8 billion is spent for petroleum products to be used as industrial feedstocks (the basic materials that feed into industrial processes).[26]

Recent political instability in many of the countries where this material is obtained has triggered a look at the potential for domestic agricultural production. Interest in the Congress, the Department of Defense, and the industrial community is centered on the possibility of achieving more self-reliance, at least in those products that are either "strategic" (critical to defense) or "essential" (required by industry to continue normal operations).

In assessing the situation, Howard Tankersley, Director of Land Use for the Soil Conservation Service, points out that most of the research for commercializing the use of these agricultural crops is either under way or could be done within 5-20 years, if producing them becomes a national goal. But a switch to domestic production of these products would require the use of about 55 million acres of land, about 22 percent of our current cropland base.

Table 10.1. **Products that Might Be Grown in the United States to Replace Current Imports**

Product Imported or Manufactured from Petro- chemicals	Percent of Products or Feedstocks Imported	Plant Species That Could Be Used to Replace Imports	Year Full Production Could Be Achieved	Land Needed for Full Production
				(million acres)
Natural Rubber	100	Guayule	1995	1.2
Synthetic Rubber	50	Guayule & Milkweed	2000	12.0
Plastics	50	Many Oilseed Crops	1990	50.0
Rubber and Plastic Additives	50	Same as Above	1990	(with above)
Coatings and Printing Inks	60	Flax, Caster, Soybean, Cottonseed, Safflower	1985	4.0
Adhesives	50	Stokes' Aster	1995	1.0
Lubricants	60	Jojoba, Crambe	1995	20.0
Detergents, etc.	60	Cuphea	2000	1.0
Newsprint, Paper	*	Kenaf	1988	0.9
Synthetic and Natural Fibers	50	Cellulose from Trees, Cotton, Flax	1990	1.0
Waxes	50-100	Jojoba, Crambe	1990	0.2

SOURCE: Table prepared from data developed by Dr. L. H. Princen and staff scientists at the USDA/SEA North Central Regional Research Laboratory, Peoria, Illinois, March, 1980.
*Information not available.

Not all the land for these new crops would need to come from current cropland. Jojoba, for example, is a desert shrub that can be grown in the southwestern deserts under conditions where little, if any, other agricultural production is possible. Jojoba (pronounced ho-ho-bah) seeds contain 45-60 percent of an unsaturated liquid wax similar in composition to sperm whale oil.[27] This high-quality

This jojoba bush (top) will produce a crop of seeds (bottom) that yield a high-quality oil good for special lubricating jobs, cosmetics, or other special uses. National Academy of Sciences (NAS) photos.

oil is used for specialized lubricating applications, such as those that must withstand very low temperatures, as well as being used as a base for cosmetics, pharmaceuticals, and other similar products. Commercial production of jojoba could not only make the nation more self-sufficient in a "strategic" material (sperm oil), but also reduce hunting pressure on whales, which might help prevent their extinction.

Jojoba must be grown in hot, low deserts where freezing is not a hazard. With small catch basins around each plant to concentrate the limited rainwater, it is possible to grow the crop with natural rainfall or limited supplemental irrigation. It takes from 5 to 7 years for the plants to mature and produce an economic yield, but a 60-year-old plant can produce up to 30 pounds of seed per year.[28]

Guayule (pronounced wy-oo-lee) is another desert shrub that is currently undergoing intensive research. It produces rubber of a quality nearly identical to that of the Hevea rubber tree, with the foliage containing up to 25 percent rubber by weight.[29] The guayule plant can be harvested every two to five years, after which it will regenerate from its perennial rootstock. Because much of the rubber is contained in the root, however, USDA scientists feel it may sometimes be preferable to pull the entire plant and reestablish the stand on a four-year cycle. Yields now run from 200 to 1,000 pounds of rubber per acre per year, and researchers are testing varieties that will yield even better.

Guayule is no newcomer to the American scene. In 1942, the Emergency Rubber Project involved some 10,000 people in an effort to produce natural rubber from the plant. By 1945, some 32,000 acres were growing in Texas, Arizona, and California, and 15 tons of guayule rubber a day could be processed at a plant in California.[30] But at the end of the war, with overseas sources of natural rubber once again available and the new synthetic rubber promising to make natural rubber obsolete, the guayule plantations were burned and all research stopped.

But synthetic rubber did not completely replace the natural product, and today, with the new radial tires that require high elasticity and low heat buildup, natural rubber use is rising rapidly. Demand seems likely to outrun supply by the late 1980's, and the overseas sources could complicate that with an OPEC-type cartel, so, once again, guayule is a topic for serious consideration. USDA, along with the Departments of Commerce and Defense, is studying the feasibility of a government-sponsored project to demonstrate the

This 1931 crop of guayule was grown in California as part of early efforts to mechanize the handling of the natural rubber crop. The rubber produced by the guayule plant has proven itself equal in quality with natural rubber imported from abroad. As we struggle for more resource independence, more and more land may be needed for plants such as this. National Academy of Sciences (NAS) photos.

potential for growing and manufacturing guayule rubber. If the project moves ahead and is successful in showing farmers how to grow the crop and in demonstrating that the process is economical, a new agricultural industry could be in sight.

Two other crops, buffalo gourd and devil's claw, also show promise for semi-arid agriculture in the southwest. Buffalo gourd grows well on disturbed soils and can survive on as little as ten inches of rain annually. An acre yields up to 3,000 pounds of seed, containing over 1,000 pounds of vegetable oil and 1,000 pounds of protein meal. In addition, the roots can be harvested for starch, yielding up to six-seven tons of starch per acre.[31] Devil's claw seeds contain up to 40 percent oil and 27 percent protein, with the oil similar to safflower. The plant is adapted to both dryland and irrigated farming, and work to collect and improve seed stocks is now under way.[32]

These new plants offer a potential crop for land that could not grow normal crops, and may also extend the agricultural future for a great deal of cropland in the southwest that might otherwise revert to desert because of the loss of irrigation water.

The promising new crops also include some varieties that could become competitors for agricultural land in the more humid climates. Among these are Crambe, an oilseed crop that contains 30-40 percent of an oil that can be used for the manufacture of lubricants, plasticizers, nylon, and other products.[33]

Crambe can be profitably grown from North Dakota to Texas, and from California to Connecticut, with yields of up to 4,000 pounds per acre. The basic information needed to grow Crambe successfully is available from USDA, and the crop is competitive with all traditional crops except corn at a 1980 sales price of 8 cents per pound.[34] Chances are strong that Crambe has a part in America's farm future.

Kenaf, a source of cellulose for newsprint or other paper products, is another crop that can compete for agricultural land, particularly in the warm, humid zone. It has produced yields of five-ten tons of dry harvested matter per acre, about twice the production that can be obtained from normal tree farming operations.[35]

Growing domestic crops to replace imported materials is, indeed, a feasible option. Success will demand coordination and timing of all facets of the research and development program, however. When the knowledge is on hand for growing the crop, the industrial capacity to use it must also be ready. If one gets ahead of the other, there is a problem. With most of the crop research in USDA and most of the utilization research in industry, timing is difficult.[36] Compared with normal crop research, where USDA laboratories can fairly easily simulate a housewife's kitchen or a baker's oven, these crops require complex industrial processing that can be replicated only under industrial conditions.

In the past, American industry has always found it easier to buy imports than to deal with the necessary problems of domestic production. In this way, they have left all the headaches of production, processing, by-product utilization, and waste disposal to the countries of origin. If we start growing our own supplies, industries will be forced to deal with farmers, contract for supplies, and make contingency plans for years when the crop is less than normal.

Regardless of the problems, the nation's interest in becoming more self-reliant in these products, in decreasing consumption of imported petroleum, and in reducing our import costs by billions of dollars each year, is likely to lead to added research, testing, and production. These new crops could easily become an important part of our agricultural picture in the near future.

The production of new industrial crops could take up to 55 million more farming acres, although well over half of those crops might be grown on lands not now in cropland or considered to be potential cropland for ordinary crops. Without a great deal of data to rely on, it seems reasonable to estimate that the new farmland demand for these crops could be in the range of 10-20 million acres by the year 2000, with the remainder of the production coming from semi-arid or desert lands. The question is, for those who contemplate the future of the American land, where will those new crops be grown? What other crops will they displace in the process? Will the acreage requirements, in a new age of limits, create one more pressure on the people:food:land equation?

11.

Total Demands on the Land

In order to convert numbers of people, consumption of food, and production of energy into a coherent view of the demands on the American land, we must bring all those factors together into an estimate of total demand. The estimate is a complicated one. An oversimplified version could be prepared by making a list of all the farm products we think will be needed in some future year and then calculating the acres it will take to produce those crops.

The answer would be hard enough to calculate if all the products came from the same kind of land, and if the annual yields were fairly predictable, but neither is the case. American farmers produce from a fantastic variety of soils and types of land. They make individual planting decisions based on their judgments about weather, prices, and other factors that may—or may not—turn out to be right. Nobody tells them what to plant, or where to plant it, so trying to guess how much they will plant—and where—in some future year requires some bold assumptions and a sizable dose of humility about the accuracy of predictions based on hypothetical conclusions.

As one approach to making the kind of estimates required, USDA, in cooperation with Iowa State University, has developed a complex computer program or "model." The computer is filled with information about the extent and quality of the soils in the United States. For each of 109 geographic regions that contain similiar soil and water conditions, the computer holds data on farm

214

budgets showing the cost of producing each major crop, based on 1977 conditions.

In addition to the amount of cropland currently used in each region, the computer program contains estimates of the amount of new cropland that might be developed, including estimates of the cost of developing it. The rate of cropland loss to urban use is estimated, and the future acreage in each region adjusted accordingly.

Finally, an estimate of future yields is entered into the computer so that it will reflect the predictions of USDA scientists about the conditions most likely to exist in some future "target" year.

As part of the land supply appraisal required under the Resources Conservation Act (RCA), USDA selected 2000 and 2030 as the two target years to be used in the model. For each of those years, they used the projections about population, consumption, and exports that we have already reviewed to develop an estimate of the total amount of the nation's major farm crops that would be needed. The computer was then asked to solve the problem of how many acres would be needed to produce this crop at the lowest cost.

To illustrate: If we will need 10 million bushels of corn in 2000, and corn yields at that time are expected to be 1.3 times higher than they are today, the computer searches the regions for the lowest-cost soils to grow corn. It subtracts all those acres from the land balance, then searches for the next-cheapest land and removes it. By the time it gets to the 10-millionth bushel, it will have selected the lowest number of acres that would be required to grow that whole corn crop at the lowest cost.

If corn were the only crop, that would be hard enough, but solving for ten crops simultaneously makes this a fantastically complicated analysis. The result of all this computerized wizardry is that USDA has prepared some of the most sophisticated projections about the future land and water needs for agriculture that have ever been available. But in that process, many key assumptions were required. If any one of them turns out to be either too high or too low, the projections will suffer accordingly.

It is well to keep in mind that the results of all the USDA calculations are projections, not predictions. They don't try to predict the future, only show where the current trends and most likely futures may lead us. Many of the past trends in agriculture might, in fact, change; some could be changed on purpose if the nation were to decide that the current trends are counterproductive. To make that kind of judgment, however, we need to understand where these trends might lead us if they are allowed to continue into the future. That is the value of the computer model, which is, after all, a very, very oversimplified version of the real world and how things will really change on the American land.

The cropland requirements to produce the basic food and fiber needs that we might face in the future are illustrated in Table 11.1. As can be quickly seen, the cropland required is expected to rise by about 70 million acres over the next 20 years in order to meet the anticipated domestic and foreign demands, then hold fairly steady until 2030. Leaving aside, for the moment, the question of whether, and at what price, that much cropland can be made available, let us concentrate on some of the underlying assumptions that were used to make the projection.

Table 11.1. **Cropland Requirements Based on Domestic and Export Needs, Adjusted for Technological Advancement**

Year	Domestic Needs	Export Needs	Total Needs	Yield Technology	Total Cropland Required
	(index)	(index)	(index)	(index)	(million acres)
1977	100	42	142	100	381
1990	129	55	184	115	428
2000	152	64	216	126	451
2010	156	74	230	134	453
2020	160	84	244	142	452
2030	165	94	259	150	454

SOURCE: Developed by the Soil Conservation Service from data presented in the *RCA Appraisal 1980,* Part II.
(Columns 2-5 are shown as indexes. What that means is that the domestic food needs and the yields experienced in 1977 are given a value of 100 and all the other years are related to that 1977 index. Domestic needs in 2000 are estimated to be 52 percent greater than 1977, so the index is 152.)

Domestic needs are based on the following: Population will climb to 260 million in 2000 and 300 million in 2030. Per capita disposable income (in 1972 dollars) will increase from $4,148 in 1977 to $7,640 in 2000 and $13,779 in 2030, and the percentage of disposable income spent on food would stay about the same, increasing from 16.8 percent today to 17.5 percent in the future.[1] On the basis of these assumptions, which are commonly used in government studies today, Table 11.1 shows that domestic needs could rise by over 50 percent by 2000 and 65 percent by 2030. Should any of these assumptions prove to be too high or too low, the projections will suffer accordingly.

The export index in Table 11.1 was calculated on the basis of USDA's "moderate" projections for future exports, which foresee an increase of 2.3 percent annually between 1977 and 2000, slowing to a 0.6 percent annual increase between 2000 and 2030. On this basis, roughly 30 percent of the total crop in 2000 would be exported. But in 1980, we were already sending some 40 percent of the grain crop abroad and well over 33 percent of *all* crop acres produced for export. The 1977-based USDA export projections indicate a doubling in the amount of exports between 1977 and 2020, but they are unrealistically conservative, having already been overtaken by real events.

These assumptions are important in evaluating current national policies for expansion of U.S. farm exports. This is a popular topic with farm groups, industry, and state governments who are pushing the federal government to accelerate an aggressive campaign to promote overseas sales. Secretary of Agriculture John Block supports that idea and has pledged to do everything possible to promote export sales. But comparing the already-explosive growth of exports with the acreage needs shown in Table 11.1 leads quickly to the question of how much more we will be able to expand production for export markets. Perhaps the nation's energy should go into keeping the land productive so we can have a better chance of meeting the demands we have already created in the world, rather than creating new demands that we may find difficult, if not impossible to meet.

The situation is reminiscent of the mid-1970's, when many electric utilities in America were heavily advertising the virtues of all-electric homes. Now that those same companies have switched their advertising emphasis to conservation of existing energy and support for research on new or alternate sources, the effect of those old demand-inducing ads is to make what must be done much more difficult than it should have been. Unfortunately, many Americans are still hooked on the "more is better" syndrome, and would like to believe that there is no limit to our ability to produce, whether the commodity is electric energy or food. If the land and water resources aren't there, we think we can turn to science and keep increasing yields indefinitely in order to meet the ever-growing demands created by our sales pitch. That may be a false hope.

Future yield increases, if we can realize them, could indeed help us meet the vastly larger demands that may be placed on the land. USDA projects a future yield increase of 1.1 percent per year between now and 2000, slowing to 0.8 percent between 2000 and 2030. This compares to a historic rate of growth in the range of 1.6 percent and reflects the judgment that the rapid rate of increases enjoyed in the past may not be so easily attained in the future.

The importance of this assumption is not difficult to illustrate, using the numbers in Table 11.1. The Soil Conservation Service (SCS) calculated that it took 381 million acres to produce a total "needs index" of 142 in 1977. If yields in 2000 were the same as in 1977 (index of 100), a simple mathematical calculation shows it would take 580 million acres to produce a total "needs index" of 216 in 2000. That means that USDA, even in its conservative view of the potential of technology on future yield increases, is projecting new technical breakthroughs in the next 20 years that will have the effect of reducing the need for cropland by over *120 million acres.* Chapter 9 discussed the various yield-raising technologies that may become available over the next few decades if research is given a high priority. This simple calculation shows the urgency and significance of moving toward that goal — even without a great deal of *new* demand created by an accelerated U.S. sales campaign abroad.

There is one other important factor in these assumptions: The 70 million acres of new cropland needed to meet food and fiber needs over the next 20-50 years *must* be acres that *contribute* to an overall yield increase of 50 percent during that period. This means that the new soils must be top-yielding, not marginal. However, most of the top-yielding soils are already in crops, and much of what remains available is marginal. In addition, a considerable amount of the new productivity achieved in the past few decades has been due to the expansion of irrigation. Regional water shortages, rising energy costs, and competition for agricultural water do not auger well for greatly expanding irrigated agriculture.

Even when the conservative projections made by USDA signal a potential land shortage within 10-20 years, they still do not adequately reflect all of the demands for agricultural crops. The domestic and export needs shown in Table 11.1 are for traditional food and fiber crops only. Before we get to the question of whether meeting those demands is a realistic expectation, we need to add the demands on the land likely to be created by the new crops discussed in Chapter 10. Those new crops are on the horizon, and the extent to which they compete for cropland will add a new, and very important, dimension to the demands of the future.

In spite of the fact that we don't know a great deal about the rate at which American farmers will adopt new energy crops, or exactly what percentage of them will be grown on existing cropland, I estimate that the total demand on U.S. cropland for new kinds of energy and industrial products by 2000 will range from 18 million acres upward to 50 million. Grain for gasohol could add from 8 to 30 million acres of new demand in the next two decades, just on the basis of the plants now under construction. These would be acres in addition to those needed for food, feed, and fiber production as we now know it, allowing for some substitution and double-usage such as a gasohol-cattle feed dual use or double cropping that combines an energy crop with a food crop.

Figure 11.1. Potential Demands for Cropland, 1977-2030

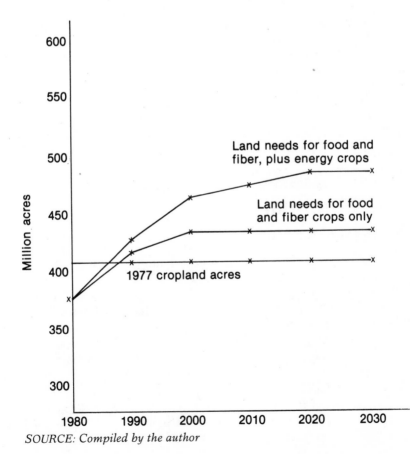

SOURCE: Compiled by the author

The production of new crops for rubber, pulp, and oil could add up to 55 million more acres, but it appears that well over half of those acres could come from lands that are not used for traditional crops, so they would not affect our cropland resource pool. Without a great deal of data to rely on, it might be reasonable to estimate the new demand for cropland in the range of 10-20 million acres by the year 2000, with the remainder coming from semi-arid or desert lands. Adding the two types of demand together gives a range of 18-50 million acres. That the range is very wide indicates the uncertainty of some of these crops, but does not make their demands insignificant. Even though I feel these estimates are conservative, they could mean, at the upper limit, adding a net area to our current cropland that would be equal to the total size of the state of Idaho.

It is clear that attempts to grow energy or industrial crops will compound an already difficult situation. Figure 11.1 displays the USDA data taken from Table 11.1, plus my projections for land needed to produce fuel and industrial crops. The projections are based on the assumption that yields will continue to rise at about 1 percent per year, so many millions more acres of cropland—high-yielding cropland—must be brought into production. The other factor shown by Figure 11.1 is the importance of keeping the acres now in crops available for future cropping. Every time another crop acre is covered over with concrete, or pulverized by a bulldozer, the land supply drops. The facts are clear: This nation can't afford to *lose* good cropland; what we have to do is *find more.*

The Potential
for New Cropland

Throughout history, the United States has always had a reservoir of fertile cropland that could be used to meet new demands. The existence of this land surplus has dulled arguments for improved methods of soil conservation or farmland protection. It has always been too easy to argue that, with more good land just waiting for the plow, there is little need to worry about some topsoil loss or conversion of some land to urban uses. But a commonsense look at the landscape tells us there is a limit to the amount of land that can be feasibly used for crop production.

As part of the 1975 Potential Cropland Study, and again in the 1977 National Resource Inventories, SCS tried to estimate this cropland limit by identifying land that had potential for conversion to cropland. On lands that were not being cropped, field samplers estimated how feasible it appeared, under current economic conditions, to convert the land. Comparisons with similar land nearby that may or may not have been converted were part of the judgment, as were the local experience of SCS technicians, county extension agents, and farmers. Each sample was rated on the basis of high, medium, low, or zero potential for conversion to cropland in the near future. The results, depicted in Figure 11.2, indicate that there are around 127 million acres that have high and medium potential for such conversion. (Appendix E shows where these lands are located, as well as their current land-use status.)

Figure 11.2. Potential of Rural Land for Conversion to Cropland, 1977

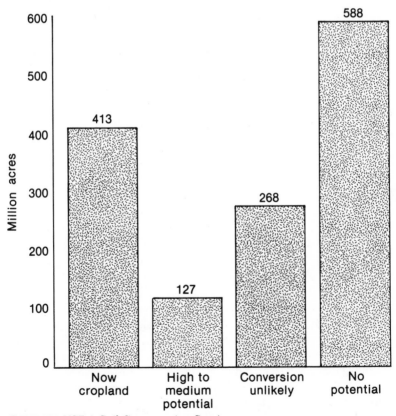

SOURCE: USDA Soil Conservation Service

The concept of a cropland resource pool, formed by adding the potential cropland acres (127 million) to the land currently used for cropland (413 million acres), gives a measure of the nation's total cropland capability (see Appendix C). The size of the total pool may change with new economic or technological conditions, but it is the best measurement now available for making predictions of future land productivity. If exploding needs require that we press all available cropland into service fairly soon, 540 million acres is the best estimate of the limits to our practical capability.

But that resource may not be as generous as it seems at first glance. For one thing, that total is the starting point from which we can begin subtracting the land that is being irreversibly converted out of cropland use. These losses, some of them caused because soil erosion has stolen all the topsoil, and some caused as bulldozers rip into fertile topsoil to prepare land for houses, roads, strip mines, and shopping centers, continue to cut into the total cropland resource pool. As long as that pool was dramatically bigger than could ever possibly be needed, the true damage caused by these losses stayed largely hidden. In the face of the demands for land in the future, however, many people in the United States are beginning to question the wisdom of allowing cropland loss to go unchecked.

Further, those 127 million potential crop acres are not just "laying there," waiting for the plow. The land is being used now, mainly for the production of meat, milk, and wood products. If the 127 million acres were plowed up, the result would be the loss of 51 million acres of pastureland, 39 million acres of rangeland, and 31 million acres of forestland. So, while the additional production of crops is possible, it is at a loss in our ability to produce other valuable commodities.

Just what those losses might be is difficult to assess, but some rough estimates are possible. Studies have shown that commercial cattle ranches in Colorado and New Mexico market from 9 to 14 pounds of cattle per acre, depending on the quality and condition of the forage on their rangelands.[2] It is reasonable to assume that pastureland, being more liberally endowed with moisture, will usually produce more beef than rangeland, and it is also true that most of the pastureland and rangeland identified as potential cropland is the more productive grassland, since it has the best soil conditions. At 14 pounds of cattle per acre, converting 90 million acres of pastureland and rangeland to crops represents a loss of beef production of over 1.25 *billion* pounds a year. That is, it could be said, a lot of hamburger!

Conversion of new lands into crops will entail other costs as well—costs for clearing trees, preparing the land, providing needed drainage or irrigation, and making necessary soil amendments, in addition to losing the products from the land's earlier use. Farmers and ranchers will weigh all these factors before they change what they are now doing on their land. Just how they will make these decisions is hard to guess, but the resource data at hand suggest that they are not likely, under any set of reasonably predictable economic circumstances, to add much more than 127 million new acres to the cropland base.

Even with all the qualifications, that 127 million acres still looks like a tremendous resource opportunity. If it were all brought into production reasonably soon, it would represent a 25 percent jump in the amount of land under the plow in the United States. But the impact of that may not be as great as it might seem. Over 50 million acres of cropland were added back into intensive cropping from set-aside lands between 1967 and 1977, most of it between 1973 and 1975. And, if we don't continue to realize a steady rise in the yields per acre—something on the order of a 1.1 percent yield increase every year—the 127 million acres of potential cropland will be used up before the end of this century, even at today's levels of demand for crops.

In sum, this nation has a clearly limited ability to add new land to its cropland base, and there will be significant—and rising —costs of doing it. In a world of virtually unlimited demand, we cannot totally depend on new acres as the source of future food expansion.[3]

Three Future Projections

Despite the powerful information from the cropland and resource inventories, arguments among land analysts over the rate of farmland loss still rage. USDA economists, who have long insisted that cropland loss is no problem, continue to defend that point of view. Meanwhile, many other analysts, including myself, look at the data and come out with clear indications that land scarcity is an important concern lying not too far in the future.[4] How can such divergent views flow from the same information?

For one thing, USDA and I have used different portions of the data to indicate how much land is being lost to agriculture. The USDA analysts, in preparing estimates of cropland availability for the RCA analysis, focused on the acreage of cropland being converted directly to urban and water uses as shown by the 1975 Potential Cropland Study and the 1977 National Resource Inventories. Those estimates suggest that, out of roughly 3 million acres being converted to urban and water uses each year, somewhat less than one-third was coming directly from cropland. As a result, the USDA analysts picked the rate at which cropland is being converted directly to urban uses—675,000 acres per year—and projected it on a straight-line basis into the future. A similar estimate was used by the analysts for the National Agricultural Lands Study (NALS).[5]

In response to my criticism that their estimates were far too low, the USDA analysts restudied their figures and, not too surprisingly, reaffirmed their original conclusion: "Given the available data and other factors, there is not much way to support a projection of cropland and potential cropland conversion to other uses of over about 40-50 million acres between 1980 and 2030."[6] When I continued to press my case, however, several salient points emerged, the most important of which is that, while the direct conversions of cropland to urban and water uses are an important factor in reducing cropland availability in the future, they are far from the only factors at work. In addition to the direct loss of cropland, there is a continuing loss from the 127 million acres of potential cropland. If those acres get co-opted by urban uses between now and some future time, they are no longer "potential" for cropland. There is also the cropland that moves out of crop use into some kind of "holding" pattern while its owners await the proper time for development. Much of this land is identified in the "other" category in the USDA data, but some of it continues to be used as pasture or some other agriculturally related use during the transition period.

USDA's crop-to-urban estimate totally ignores the indirect effects of cropland conversion on adjoining lands. The most significant effect on agricultural productivity involved in the farmland conversion process is the "idling" of leapfrogged or adjacent lands, coupled with the reduced investments in productivity by nearby farmers who lose faith in the future of agriculture in the area. None of these factors—vital as they are to continued agricultural productivity—are included in the data set used by USDA and NALS.

Finally, there are many acres of cropland that are not urbanized, but are, in fact, permanently lost for future cropland use. They will show up in the land-use data as pastureland, rangeland, forestland or "other" land, but they are no longer feasible for cropland. Some of this is due to soil erosion, salinization, or desertification. Some is due to the mining of groundwater. Some may be isolated by an adjacent land use such as urbanization or surface mining. A study of the 1975 Potential Cropland Study (see Appendix C) revealed that, of the cropland that went into other land uses between 1967 and 1975, only about one-third was rated as having a high or medium potential for return to crop use under 1974 economic conditions.

Surprisingly, that study revealed that even where cropland was converted to pastureland, less than half of it was given much chance of ever being cropped again. The figure dropped to 17 percent when trees had been planted on the land, and 14 percent when the land was placed in "other" uses (this included urban and water as well as farmsteads, idled land, and a variety of miscellaneous uses). In other words, there is considerable evidence that the conversion rates used by the USDA and NALS analysts are far too conservative since they miss several important aspects of the cropland conversion problem.

Since different people have looked at the same data and come to such different conclusions, it may be best to show the most likely range of projections for our land supply. Figure 11.3 shows what such a land supply forecast might look like in graphic form, with the USDA-NALS projection showing the slowest rate of conversion (line A), my interpretation of the 1967-1977 trends projected into the future being by far the fastest rate of loss (line C) and a "middle" estimate derived from a rather arbitrary assumption that rising pressures on land resources for agriculture would serve to slow down the conversion trends in the future (line B). Appendix C explains in detail how the projections for lines B and C were derived.

The basic message in Figure 11.3 is that cropland supplies are dropping, and will continue to drop, even though we find it difficult, if not impossible, to agree on how fast that process is proceeding and how it might look in the future. My best guess as to the accuracy of the graph is that line C is clearly too pessimistic. Even though I think it captures, as well as the current data base will allow, the total losses of cropland and potential cropland in the 1967-1977 period and extends them into the future, I do not think that such a rate of loss will continue. The reason seems obvious: It can't. A rate of loss such as that would make agricultural land loss a crisis issue in a very short time, and I can't help but believe that there are ample economic pressures in the system to slow down any conversion trend long before it hits that kind of crisis stage.

Line A is, I believe, too optimistic. It does not take into account many factors that are affecting land supply today, as discussed earlier. That leaves line B, or any one of thousands of other lines that could just as easily be drawn in the wide gap between A and C. For the moment, however, my best guess is that the supply of good cropland available to U.S. farmers under foreseeable economic conditions will drop from the current 540 million acres to something under 500 million acres by 2000 and around 450 by 2030.

Figure 11.3. Three Projections of Total Cropland Availability, 1977-2030

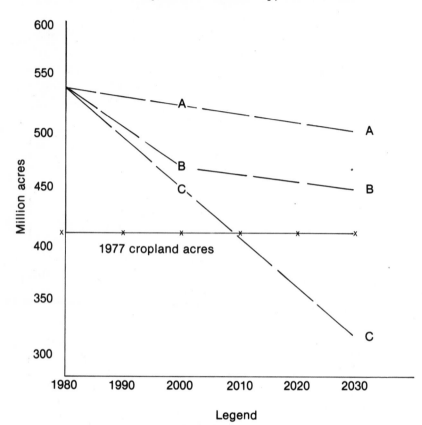

Legend

Line A. USDA-NALS projections of cropland converted to urban and water uses (675,000 acres per year).

Line B. Cropland losses that might occur if the 1967-1977 rate calculated by the author slows down dramatically (see Appendix C).

Line C. Cropland losses that might occur if the 1967-1977 rate calculated by the author were to continue unabated (see Appendix C).

SOURCE: Compiled by the author

**Figure 11.4. Comparison of Three
Potential Levels of Demand for Cropland
in the Future with
Three Possible Levels of Land Supply**

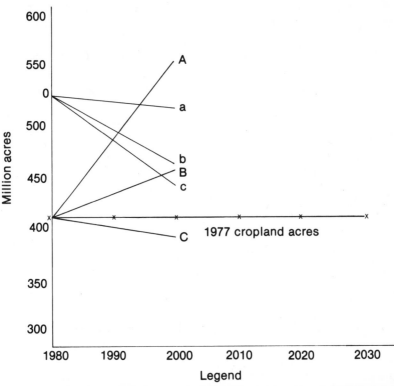

Legend

A. Land that might be needed if crop yields stay at 1977 levels.
B. Land that might be needed if yields increase at 1.1 percent per year.
C. Land that might be needed if yields increase at 2.0 percent per year.

a. Land that might be available under USDA projections of cropland converted to urban and water uses.
b. Land that might be available if current rates of cropland conversion slow down in the future.
c. Land that might be available if current rates of cropland loss continue unabated into the future.

SOURCE: Compiled by the author

To keep the land supply possibilities in perspective, it may be helpful to superimpose the lines from Figure 11.3 onto the different potentials for land demand that were shown in Figure 11.1. The results are shown in Figure 11.4. All sorts of possibilities emerge, from an emerging crunch that will signal the full utilization of all U.S. cropland, to a future where land demands are easily met by the supply at hand. Depending on the assumptions one wishes to make about future yield increases and future land conversions, reasonable arguments can be marshalled to predict that any one scenario is as likely as the others. What is startling about the graph, however, is that the two worst possibilities—the supply and demand lines that would run together the fastest—are the two lines that extend 1977 conditions and trends. Equally startling is the fact that, without continued yield increases, the land conversion argument is moot. The land base would need, in that case, to be all used far before the year 2000.

Expanding the Equation

The previous discussion of the probable future trends in the land available for agriculture is not optimistic. Maybe it is too cautious, overlooking some possibilities, or perhaps it overstates the degree to which higher energy costs may drive up the cost of agricultural inputs. But there is the strong chance that it is right, and the penalties for erring on the side of caution are far less than those of proceeding blindly onward until a serious crisis emerges.

Agricultural production results from a combination of inputs —land, water, capital, technology, and labor. If any of the input factors diminish in effectiveness, it will take more of the others to compensate. That is the simple economic fact—the productivity of the agricultural system will be held down by the limiting factor. The data today shows that the quality of the nation's topsoil is diminishing, and that means something is going to have to take up the slack. What will make up for these losses, and will it happen as a normal part of the economic marketplace or will it require public intervention?

New cropland is available, but in limited supply, and its quality is not likely to be as good as what is now being farmed; therefore, addition of new acres may serve to lower average per-acre yields rather than raise them. Water is limited and irrigation expansion will be slight as a result. That leaves us with the need to increase the inputs of capital and technology as our major means of expanding output. Neither will come to pass unless people in the United States are willing to pay the price.

Agriculture probably can't compete for capital on an open money market; the subsidized loan programs of USDA have already pumped billions of public dollars into farm loans to help farmers through difficult times in the 1970's. Now, the Reagan administration proposes to cut back farm loans drastically. New capital for land, machinery, and buildings may be limited, and operating money for fertilizers, pesticides, and other high-cost yield-raisers may suffer as well.

New technology means an intensive thrust on agricultural research and that means higher budgets for research and development, at both the state and federal levels — in a period when holding down public spending is a growing political priority.

That leads to little confidence that massive new infusions of capital or technology in agriculture are in the offing. So we are back to dependence on the land base. Is it in the best interests of the American people to ignore the inevitable costs of letting the nation's topsoil wash and blow away, and its best farmland be paved over? I would argue that it is not, and the results of the Louis Harris Poll taken in 1979 indicate that many Americans agree. In that poll, 60 percent of the respondents indicated that this nation should be moving toward policies that save resources for future generations, while 44 percent supported public policies aimed at conserving the natural productivity of the soil.[7]

The RCA analysis carried out by the Department of Agriculture led to the same conclusion. That study demonstrated, perhaps for the first time, that the nation will pay a prohibitive cost in the future if current rates of land loss are allowed to continue. As a result, USDA has developed conservation objectives that would halt the downward slide in soil productivity. There are solutions available, if we decide to use them.

The formula for slowing land losses dramatically is deceptively simple. Soil erosion *can* be cut 50-75 percent by the adoption of economically feasible conservation systems on the lands that need them, plus the removal of unsuited and marginal lands from cropping. The conversion of prime farmlands to non-agricultural uses can be slowed through a variety of means. But, like solutions to every other environmental or social issue, even feasible, practical measures can be very difficult to implement.

Before we devise new land policies, however, it is important to look at what can actually be done on the land to protect it for future farm use and to prevent the levels of soil loss currently occurring. We need to be convinced that feasible solutions exist, and that they won't be as painful as the ailments they are meant to cure.

12.

Protecting Agricultural Lands

"Conservation," Aldo Leopold said, "is a state of harmony between men and land."[1] Before people arrived on the scene, nature went through a complex series of "building" and "holding" actions that, through the weathering, grinding, and transportation of rock particles, the growth and decay of plants and animals, and the actions of microorganisms, resulted in the soil as we know it. The system of soils, plants, and animals that emerged form a dynamic *ecosystem* where the forces that create and build up the soil are roughly balanced out by those that destroy it. Each individual soil is a unique product of the rocks, plants, animals, climate, and history of geologic events that formed it.

Human intervention, usually in the form of a plow, changes the plant and animal relationships that have developed over centuries and imposes a new set of man-animal-plant-soil relationships known as agriculture. Instead of a complex plant community composed of many species of grasses, brush, or trees, there is now a simple community, often composed of only one or two annual plants. Instead of long periods of stability where the soil lies undisturbed, and permanent internal soil structures such as pores and aggregates can form, the soil is now cultivated on a regular basis. An entirely new set of air, water, and plant conditions is created. This new ecosystem is, in many ways, more difficult to manage properly than the natural one it replaced. Gone is much of the complexity, and many of the checks and balances, that helped natural ecosystems stay stable. Agriculture can get out of balance very easily, if not constantly tended by the skills developed in farmers over thousands of years.

The question is: "Can farm ecosystems be managed in such a way that the forces tearing down the soil are still being offset by soil building, much as they were under natural conditions?" The record, as we have reviewed it, indicates that in too much of agriculture today this is not the case. Soil destruction and waste are proceeding far too rapidly. Those forces, unless stopped, will gradually ruin the soil itself, and man's inattention to proper stewardship as an essential part of agriculture will be to blame. Today, knowing the history of man's waste of the land over the last 10,000 years, and having the scientific sophistication of the Space Age, we ought to do better.

In the interests of human survival, we must do better. It is not only foolish, but dangerous to think that we can continue the current patterns of waste, for humans are neither powerful nor inventive enough to ignore nature with impunity. When sufficiently violated, nature strikes back hard, with no misgivings whatsoever about genocide. All the technological and economic trappings of civilization are but temporary unless we can reach a state of balanced harmony with the land.

Achieving that harmony is not an impossible task, but it is very complex. Because each soil is a unique product of its original ecosystem, each requires the design of an appropriate agricultural ecosystem compatible with the soil, climate, and human factors at hand. The system must not only be adapted to each individual soil situation, it must be able to adapt or be changed to meet new forces created by people, animals, or important shifts in climatic conditions. When a system can meet these criteria and create a new ecosystem in which soil-building and soil-destroying forces are once more in balance, it is a soil conservation system.

It may be important to think for a moment about what such a system is—and isn't. It is not, for example, a system designed to preserve the "status quo" as created in nature. It does not require that humans be eliminated from the scheme of nature, only that they manage their activities in a way harmonious with nature's demands. It does not prevent the use of the land for producing the agricultural products required by people; it establishes a way in which people can *use* the land, but *save* the soil. It is not something that can be "patched on" to any farming scheme that people wish to dream up, on any land, anywhere. It *is* a *total* farming scheme itself, embodying all elements of agricultural production and management, together with ways of meeting the needs of the soil for protection and rebuilding.

Soil conservation is not an unreasonable or impossible standard to ask of the farmers who manage agricultural lands, and thus hold the fate of the whole society in their hands. It is not an economic burden; it is, in fact, the *only* truly economical way in which lands can be used, if we extend the economic calculation beyond a single year or two and calculate the full range of costs and benefits involved.

There is little doubt that soil conservation, as an integral part of a sustainable agriculture, can be accomplished in the United States. There are areas in western Europe that have been farmed for centuries without appreciable soil destruction. There are soils being intensively farmed in the United States today that are as good or better than they were when they were first put into agriculture. There are farms in the United States that produce bountiful crops, bring good financial and social rewards to their owners, and *lose no topsoil whatsoever*. These "success stories" occur in every state, in virtually every county. Their achievements are no secret; the methods they utilize involve no magic or mystery. But the people of the United States, as a society, have not insisted that this is the type of responsible performance that marks agricultural success.

That can be changed, if the American people decide it must be, but it will not change of its own momentum. The forces causing people to abuse the land today are strong, and will not be changed by anything short of intensive reforms or the collapse of the whole system. Given those options, it is clearly preferable to try the reforms.

Before the public will insist that farmers follow soil conservation systems on their land, however, both farmers and the general public must be convinced that such systems are feasible as well as urgently needed. A brief survey of how soil conservation works in practice may help that understanding.

Constant-Care
Maintenance of the Land

A soil conservation system that provides a rough balance between the forces that tear down the soil and those that build it up does not need to stop erosion completely. It must, however, create a balance that allows the soil to maintain itself as part of the agricultural ecosystem on the land. The basic elements of that balance involve three things: maintaining the soil's ability to absorb water from rainfall and snowmelt; creating a system that can safely handle the water that is excess to the soil's capacity or needs; and helping soil-building processes work so that any topsoil that is lost will be replaced.

Vegetation is the key to soil conservation management. The major function of plant cover in preventing soil erosion is to protect the soil from the force of falling raindrops, the primary cause of erosion on cultivated land.[2] When raindrops strike unprotected soil, they dislodge soil particles, create turbulence and energy that helps lift and move soil downhill, and compact the surface of the soil so that less water can enter, forcing more to run off and increasing the soil-eroding force of the rain storm. Where the water's force is intercepted by plants, most of the energy is dissipated before the water reaches the soil. Where soils are fully covered with a dense mat of vegetation, soil erosion is virtually non-existent, and water runoff from the land is greatly reduced as well. Most of the water that falls can soak into the soil, there to be used for plant growth or to soak down and replenish underground water supplies.

Under natural conditions, year-round plant cover provided this soil protection as part of the total soil-water-plant ecosystem. Under agricultural conditions, it is sometimes possible to copy nature. Where the land is kept in pasture or trees, and sufficient plant growth is left on the soil for protection, a natural soil system can be reproduced and soil erosion controlled. Under crop management, however, soils are often left bare for portions of the year which may coincide with periods of rainfall or snowmelt. Consequently, some other form of protective soil covering must be provided.

Often, the easiest source of vegetative material to cover the soil is provided by the plant residues available after the crop has been harvested. The residues from some crops, such as wheat or grain corn, may be adequate protection for certain soil and climatic conditions, provided they are left on the soil surface over winter. Once the erosion season is over, the residues can be worked into the soil, providing a source of new organic material. However, if crop residues are removed from the field, burned off, or plowed under immediately after harvest, their value in reducing erosion is lost.

With some crops, such as silage corn, farmers remove most, if not all, of the material from the field during harvest. Many crops, such as soybeans and peanuts, do not produce much plant growth, and their vines and leaves are succulent and deteriorate rapidly following harvest. With these crops, there are few residues to protect the soil after harvest, regardless of how they are managed.

Where crop residues will not provide adequate protection through the critical times of the year, other options are sometimes available. A quick-growing crop such as rye may be planted immediately after harvest, allowed to grow until the following spring, then plowed into the soil as part of seedbed preparation. This "cover crop" not only protects the soil over the winter months, but also provides a supply of organic matter when it is plowed down.

Another option is to schedule different crops so that some form of protection is available through most of the year. For instance, in some areas a crop of soybeans might be followed by fall-seeded winter wheat that provides soil protection over the winter and spring months. After the wheat is harvested, the stubble may be left standing for soil protection over winter, then the land planted back to soybeans the following spring.

There may be certain types of fall tillage that can reduce runoff and erosion. A chisel, for example, is a tool with long, pointed shanks that can be pulled through the soil to open up channels from the surface into the subsoil, allowing faster water intake during the winter and spring months. This, combined with good crop residue management, may provide good soil protection.

In one pass, this Minnesota wheat straw is chopped up by a flail chopper, then mixed into the topsoil by a chisel plow. The chisel plow does not turn the soil layer upside down like a conventional plow, so some of the straw remains on the surface of the soil to provide erosion protection. SCS photo by Arden Lynne.

There are many different systems of crop mangement that can be used on the land, but all should be designed to help with two of the goals mentioned earlier: providing plant material back to the soil and vegetative protection for erosion control during the peak rainfall or snowmelt periods of the year.[3]

Many conservation practices involve an investment on the part of the farmer, and often these investments return little immediate cash income. The possible exception is conservation tillage, which combines sophisticated machinery and timing with equally sophisticated chemical weed and pest control to result in less trips over the land, less fuel burned, and less soil erosion. By offering

This no-till corn in Missouri was planted directly in fescue sod after the grass had been killed with an herbicide. At the price of higher herbicide use, no-till planting is one way to cut soil erosion and energy use. SCS photo by Al Nottorf.

reduced operating costs as well as soil conservation, this practice becomes one of the few that ties together the farmer's need to see immediate profit with the practice of soil conservation. As a result, conservation tillage is the fastest-growing soil conservation practice today.

When minimum tillage (or a variation, no-till, where the crop is planted directly in grass or the stubble from the prior crop) was first tested, there were many difficulties. Crop varieties bred to perform well under clean-tilled conditions often did not do well in the cooler, moister soil conditions created by leaving crop residues on the surface. Tillage equipment that was not designed to handle large quantities of crop residue either didn't work or broke under the new strains. Those deficiencies have been rapidly overcome, however, as the popularity of conservation tillage systems have spurred both researchers and equipment manufacturers to develop new strains and equipment.

A continuing frustration is the lack of specialized equipment for farmers to use in testing conservation tillage systems. Beginning one of these new systems is a significant change for any farmer, and most want to test for a year or two to assure themselves that they can be successful. Purchasing all new equipment for that testing period is an expense that many feel they cannot justify. Local conservation districts have been obtaining special tillers, drills, and other equipment from equipment dealers, then leasing them to farmers for this testing. In 1981, the John Deere Company offered special incentives to dealers who would enter into such an arrangement with their conservation district. Through the intensive efforts being made, spurred by the strong demand from farmers for a soil-conserving system that also saves time, money, and energy, the future of the limited-tillage systems looks very bright.

In spite of all the soil-building efforts and soil protection provided, it often happens that the amount of falling moisture cannot be absorbed by the soil, and some runoff is inevitable. This leads to the second phase of a soil conservation system—the water management system. The goal of water management is to get excess water off the land without damaging the soil in the process.

Terraces are one of the methods that have been used to handle excess water safely. These are banks built across the slope of the land to intercept surface water and carry it slowly and safely to the edge of the field. Here the water is released into a grassed waterway or a pipe outlet where it can run down the slope without causing soil damage. Terraces have been built in some parts of the world for thousands of years, and their use has been a common feature of conservation systems in much of the United States for the past 40 years. They are expensive to construct, however, and often interfere with the operation of large farm equipment, so farmers are electing other types of erosion controls where possible.

Terraces protect a 12 to 18 percent slope in Iowa from soil erosion by breaking up long slopes into short segments where runoff water is stopped before it can build up large streams that would cut rills or gullies. The terraces also help align the corn rows on the contour, so that each row acts as a miniature terrace to hold rainfall on the land rather than let it run off. SCS photo by Erwin Cole.

Grassed waterways, with or without the assistance of concrete structures, can be used to carry water down a slope safely. As long as the slope is not too steep, or the amount of water too great, a good sod mat will stop soil erosion and carry the water without gully formation. If the water has too much energy for the grass to withstand, a mechanical "drop structure" may need to be built in the stream to get the water to run off safely. This involves building an artificial waterfall where the water can drop from one to several feet within an erosion-resistant structure made of concrete, rocks, logs, or other solid materials. After the energy of the falling water has been absorbed by the structure, the water can flow safely back into the grassed waterway. Several such structures may need to be constructed in a long waterway down a steep grade. In some places, it is necessary to construct a concrete or rock ditch, or a buried pipeline, that can let the water run rapidly down the hill without soil erosion.

Where water soaks into the soil but builds up a waterlogged condition instead of moving down into an underground water supply, the excess water may need to be removed by some form of soil drainage. This may be provided with a series of open ditches that collect the soil water and drain it safely off the field, a system of buried tile or plastic drain lines that intercept and collect water flowing through underground soil layers, or some combination that involves both pipelines and channels. Clay, concrete, or plastic tile is laid in the bottom of a trench which is carefully graded to provide the right slope for water to flow through the finished line. Water enters the tile line through open cracks between the tile sections or through holes in the case of plastic drain pipe.

Contour strip-cropping divides these Minnesota slopes so that strips of cultivated crop are protected by alternate strips of close-grown crop. This slows soil erosion but also creates oddly shaped fields that are difficult to farm efficiently with modern machinery. SCS photo by Erwin Cole.

Drainage of wet soils provides better growth conditions for plants, since the removal of excess water allows a proper air-water balance to exist in the soil. It also can reduce water runoff and erosion by keeping the soil in condition to absorb additional rainfall as it occurs. Drained soils can be cultivated with less compaction damage to the soil and with less energy.

Where supplemental water is brought to the soil from an outside source, irrigation systems must be designed to apply the water in a non-erosive manner if soil damage is to be avoided. This is no small matter, since it is common for irrigation to involve the application of four feet or more of water to the land each year, far in excess of the normal precipitation even in many humid regions. Irrigation water may be applied to the land through several kinds of systems, each of which must be designed to fit the soil, topography, and crops involved. The most recent kinds of systems involve sprinklers that spray the water onto the soil in a manner similar to normal rainfall. As long as the application rates are within the range that the soil can absorb, these systems cause little, if any, soil damage. If the water is applied faster than the soil can absorb it, however, sprinkler systems can cause serious soil erosion.

Fitting the System to the Problem

When I was working directly with farmers, it was common to be called out to a farm and asked to help fix the wrong thing. I would be shown a field with an active and growing gully that had finally gotten to the point where farm machinery couldn't cross it easily any more.

"See that gully?" the farmer would say. "I want to smooth it out and fix it up so I can cross it."

"That won't solve the problem," I would point out as tactfully as possible. "The hillsides above the gully are not being protected against soil erosion, and the water is running off instead of soaking into the ground for your crops. What you need to fix is the soil in the field. The gully is not the problem. . .it is just the symptom."

"But the field is fine," would come back the rejoinder. "I got 60 bushels of wheat off it last year, the best yield ever. There are a few little rills, but nothing I can't cross easily. The only thing that bothers me is this darn gully."

And so the negotiations would continue, I trying to convince him to treat causes instead of just symptoms and he to convince me that I was trying to get him to do things to his land that were not needed and not directly connected to the situation that was his real concern. Sometimes, if I was either skillful or lucky enough, I would convince him that total treatment of the whole field was all that was really going to work in the long run. At times, I just couldn't find the right argument, and he wouldn't budge.

My best tactic then was to try to use the gully as an experiment to teach him what he needed to know about the field. I'd try to make a deal with him.

"Let's fix this gully this year," I'd say. "I can help you grade and shape it right and seed it to grass so you can cross it easily, just like you want to do. But let's watch it for a year or two. If I am right about the source of your trouble, that gully will open up again. If that happens, we'll have to re-do this year's work, but then let's also fix the field so that it won't happen again."

Generally, that would work. Farmers don't want to be argued down by some college-educated kid just because he calls himself a soil conservationist, but they are also interested in learning something new if it will help them. By making the soil erosion problem in his field the subject of a wager between us, I accomplished two things: He watched the field more closely to try to prove that I was wrong, and I had a good excuse to come back later and check up on the progress of his conservation work. If the next year featured normal rainfall and runoff, chances were excellent that we would be back out there, developing a conservation plan to fit the whole field and remedy the basic cause of his problem.

The story illustrates a common misconception held by people who try to develop soil conservation systems. Soil erosion, excessive water runoff, salt buildup, gullies, or worn-out soils are not the basic ailments that must be addressed. Those are the symptoms—the external signs that something is wrong inside the soil system. They are like the cough associated with a respiratory disease. If all one wants to do is suppress the cough, there are any number of ways to go about it that will do nothing to address the basic infection. On the land, the basic situation is that the system is out of balance. The soil is not being properly maintained in a strong and resilient condition so that it can produce high yields of plants to help protect itself from the elements, or else its protection is being mismanaged or removed.

The cure may or may not involve treating the symptoms. Often, it is just as well to ignore them and go straight to the cause. Designing a whole new system that incorporates the kinds of crops grown, their rotation, the layout of cultivation patterns on the land, the methods of cultivation, the management of crop residues after harvest, and the handling of surplus water is what is needed. Included in that system, if it is to be useful to the farmer, must be considerations for weed, insect, and disease control; livestock management; manure handling; and all the minute details that must be managed in a successful agricultural enterprise. It is not an easy task and will test both the practical knowledge and experience of the farmer and the scientific knowledge and skill of a trained conservation planner. But it can be done, and it can be done in such a way that the farmer himself knows the reason behind every aspect of the system, understands why each part is essential to the success of the whole, and is able to make the minor adjustments that inevitably come up as he manages his farm through the annual cycle of growth and production.

Fitting a conservation system to the soil begins with a knowledge of the soil itself. That includes such characteristics as the soil texture and structure, the rate at which water can enter, the organic matter content, and the slope of the land. On soils that are subject to little, if any, erosion damage from water, it may not be necessary to be concerned about vegetative protection for erosion control or water management systems to convey excess water. In that case, a system that maintains soil quality by keeping organic matter levels adequate and preventing salt buildup may be adequate. Most such soils would fall in Class I* and would be highly prized as cropland.

Moving the same crops, management techniques, and conservation practices to steeper or stonier Class III soil, however, would be totally inadequate. Even though the Class III soil might be on the same farm—or even in the same field—its needs for proper treatment would be far different from those of the Class I soil. This creates serious difficulty in the intensive agriculture of today, where farmers want to make fields as large as possible and use a standard management system over the entire field. "Efficiency," measured by the speed at which farm operations can be completed with the least cost per acre, dictates that little time be spent to differentiate the subtle soil differences and manage each soil accordingly. Too often, that means designing a system suitable for the best soils and letting the marginal soils suffer in the process.

In many cases, erosion on marginal croplands can be curbed with the right combination of conservation practices. In one case, an Iowa soil that was eroding at the rate of 25 tons per acre per year

*See Appendix A for explanation of the soil classes.

under conventional tillage practices was reduced to less than 4 tons per acre of soil loss when conservation tillage, contour farming, and terraces were applied on the land.[4]

But not all systems can be adapted to all soils. There are serious losses today where farmers have initiated intensive corn-soybean crop systems on marginal Class IV and Class VI lands throughout the Corn Belt and the Southeast. This was started during the 1970's as high prices for those commodities encouraged additional production. Soil loss rates are now running as high as 50-60 tons per acre per year on those lands, an intolerable erosion rate that will soon render the soils unproductive. Where farmers have simply put the wrong crop system on the soil, there is no way soil erosion can be brought within tolerable limits without changing the entire cropping system. The nature of the soil requires that the land be put back into grass cover, either permanently or at least for a fair percentage of the time. The point is, no one can decide first that he will run an intensive row-crop system regardless of the soil involved, then find some conservation "magic" that will prevent soil damage.

The first step in designing a conservation system for the land is to pick a cropping system that is within the soil's capability. Once that is done, there is usually a wide variety of soil-conserving and soil-building techniques that will round out the system.

Developing this kind of a conservation system requires several ingredients for success. First, it takes a committed farmer, one who has decided that he will learn what it takes to carry out a soil conservation program on his land and then do it. It takes a farmer who will see his land often enough to recognize when it is showing stress and have enough time to make the management changes that will be needed to meet new symptoms as they show up. It takes some skilled farm advisors, professional agriculturalists, or soil conservationists, who understand how to design sustainable soil management systems fitted to specific soil and farm needs, and who have time to work with the farmer to thoroughly diagnose his soils and their ailments, help him decide on feasible ways to handle them, and teach him the skills he will need for success.

If any of these ingredients is lacking, the chances of success are slim. Many farmers today would like to carry out a good conservation program, but they have expanded their farming operation to the point where they may not get over the land as often as they should. Most of the actual farming may be done by hired hands, who are more interested in getting the work done than in checking for subtle symptoms that the soil may be wearing out.

During the growing season, many farmers are so busy that they have little if any time to make the "little" adjustments that are needed to minimize soil waste.

˙ The availability of skilled farm advisors may be limited, either because limited budgets at the state and federal level have cut back on University, Extension Service, or Soil Conservation Service personnel, or because those people have not been trained to understand sustainable agriculture. Too many times, the farm adviser shows the farmer how to extract more yield from the land, but does not recognize the need for maintaining topsoil quality in the process. As a result, agricultural professionals may disagree publicly on the correct methods for handling soils, and farmers are left to choose between the opposing views of two scientists who both proclaim loudly that they know what is best but who promote conflicting approaches.

The famous potatoes of Aroostook County, Maine, are no longer of the quality they used to be, for example, and yields have slipped as well. Farmers plant potatoes year after year, and even with massive doses of fertilizer, yields are lower than they were 50 years ago, according to one author. [5] Soil erosion robs some nine tons of topsoil per acre per year—three times faster than topsoil can be replaced in that northern climate even with good organic farming methods, and very few farmers practice any crop rotations or plant any green manure crops that would help rebuild soil strength. The economic pressures that are pressing Aroostook farmers to "mine" their soils are the same as farmers face all over America, as we pointed out in Chapter 3.

Yet dozens of technical reports and farm press stories printed through the 1960's and 1970's assured Maine farmers that it was perfectly all right to grow nothing but potatoes.

"Just use the latest in fertilizer and herbicide techniques," they were assured, "and you will be in the vanguard of modern agricultural practice."

"Crop rotations are old-fashioned and outmoded."

This belief that modern science could find "magic potions" so that it was no longer necessary to maintain a balanced soil-water-plant ecosystem has not been limited to Maine—it has been a pervasive feature of agricultural science since World War II. From USDA right down to the local county agent, farmers have been sold "progress," but with a price. Today, for Aroostook County farmers, and for many others around the nation as well, that price is coming due.

In those areas, where soil conditions are beginning to add up to real trouble, telling farmers to turn to the "experts" for advice is a tricky business, because in too many cases it was "experts" who got them where they are now. The survival of American agriculture is too important an issue to leave to such "experts," and farmers themselves need to ask the kinds of questions that sort out the good advice from the bad. "Will it build up the soil or tear it down?" should be asked about every new production suggestion.

If the answer is that the soil will be damaged, or if the scientist hasn't made certain about the effect of his suggestions on soil quality, the farmer ought to reject the advice out of hand, or ask the scientist to come back when he knows the real effect of his new technology. The same could be said for the "experts" who come from equipment or fertilizer dealerships. They know a great deal about improving a farm's production. Nearly all of their recommendations require the purchase of the product they are hired to sell. Farmers need to separate out the sales pitch from the valid and helpful technical advice. Or, as one farmer put it, "take all the advice you can get, but then make up your own mind."

That many farmers have, indeed, "made up their own mind" is evident in two kinds of farming that are definitely not in step with the "modern" version of American agriculture, but which stubbornly refuse to be relegated to the scrap heap of history. Although often decried as old-fashioned and unprofitable, both organic and grassland farming still attract significant numbers of farmers. Some, to the great surprise of many agricultural "experts," not only survive but prosper.

Organic Farming

It may be startling to many people in this age of "get big or get out" commercial agriculture that there are thousands of farmers running profitable family farms either totally without the use of chemical fertilizers and pesticides or with a minimum use of them. Organic farmers* believe that sustainable, successful farming must

*Although organic farming is difficult to define, a recent USDA study settled on the following definition:

> Organic farming is a production system which avoids or largely excludes the use of synthetically compounded fertilizers, pesticides, growth regulators, and livestock feed additives. To the maximum extent feasible, organic farming systems rely upon crop rotations, crop residues, animal manures, legumes, green manures, off-farm organic wastes, mechanical cultivation, mineral-bearing rocks, and aspects of biological pest control to maintain soil productivity and tilth, to supply plant nutrients, and to control insects, weeds, and other pests.[7]

be based on a healthy soil, and that good crops result from good soils. Instead of concentrating on crop yield as their first priority, they strive to see that the soil is kept in top condition.

Organic farmers have not made a great deal of news, and there is no big movement toward organic farming that is shown by any data on agriculture, but there is a growing interest in the idea, both in USDA and around the nation. Bob Rodale, editor and publisher of *The New Farm* and *Organic Gardening* magazines, tells what is happening with a touch of wonder in his voice. "We went along for many years, trading information about organic methods among a small group of people," he says. "Suddenly, the number of people interested in organic methods began to grow dramatically. We didn't do anything different, but they started coming to us."

For several years now, agricultural scientists and researchers have been going to organic farms, trying to analyze what they were, whether or not they really worked, and what there was about organic farming that could be usefully employed in agriculture as a whole. In 1974, ecologist Barry Commoner, at the Center for the Biology of Natural Systems at Washington University in St. Louis, began comparing the performance of 14 organic farms with that of 14 conventional farms.[6]

The organic farmers were selected wherever they were found. A conventional farm which operated a similar crop-livestock system on comparable soils was then picked to "match" each organic farm chosen for the study. Each conventional farmer was picked on the recommendation of local USDA personnel, who identified them as "top management" farmers. The average size of the organic farms was 425 acres; that of the conventional farms was 480 acres.

The preliminary results, which indicated that the organic farms fared nearly as well as their conventional counterparts, were met with surprise and skepticism from agricultural experts. One or two year's data does not prove anything in a business where weather and a host of other variables are constantly affecting the results, they said. But as the research has continued, the findings continue to be consistent: Organic farms produce less total product—about 11 percent less from 1974 to 1976—but the net income per acre is about the same. The reason: less outlay for inputs on the organic farm. For example, organic farmers used 60 percent less fossil energy to produce a yield of cash crops that was only 3 percent lower for corn.[8]

Some other results were equally interesting. Organic farmers are not bound to methods used 40 years ago, as most people imagined. They use, for the most part, modern tractors, harvesters, crop varieties, and other methods. The farms in the study were run as businesses by owner-operators, with labor provided mainly by the family and some hired help. In other words, these farms were surprisingly like the neighbors', except in the ways they dealt with soil-crop relationships.

The early results from the studies of organic farms were of interest to soil conservationists and agricultural scientists for a different reason, perhaps, than that of the organic farmers themselves. The organic advocates were largely concerned about the long-term effect of chemicals on soil, plant, animal, and human health. Most soil conservationists have no built-in reservations about the use of chemical fertilizers and pesticides, accepting them as part of the modern technology of agriculture and assuming that, if they are properly and safely handled and applied, there is no hazard to either soil quality or human health in using them. That is, it is fair to say, the "conventional technical wisdom" among most agronomists and agriculturalists in the United States today.

The methods espoused by the organic farmers are the same as those advocated in the "old days" by the early soil conservationists: crop rotations and green manures to build soil quality, improve soil tilth, and maintain high soil organic matter. Could it be that those basic ideas, so "out of style" in modern agriculture, still have some merit? What can be learned about "mid-way" techniques that pick up on some of the soil-building and cost-reducing ideas of the organic farmer, yet retain some of the technology that the conventional farmer still feels is economic or necessary to use?

Those questions, and many like them, led to an intensive study of organic farming by a team of USDA scientists led by Dr. Richard Pappendick.[9] The USDA study results basically agreed with those identified in the Washington University research. Organic farmers have not regressed to the past; they use modern methods except for minimizing chemical usage. They use animals as an essential part of their operation in most cases, recycling the manure back to the soil to help maintain nutrient balance. They do a surprisingly good job of controlling weeds and insects through crop rotations; timing of planting, cultivation, and other operations; and biological controls. They use a little more labor, but less energy.

In the area of soil conservation, to no one's great surprise, the organic farmers received high marks. "Organic farmers are strongly committed to soil and water conservation," says Dr. Garth Youngberg, a member of the study team and now Coordinator of Organic Farming for USDA. "Heavy use is made of terraces, strip-cropping, grassed waterways, and contour farming. Most organic farmers in the USDA case studies appeared to be highly successful in control-ling erosion."[10] By emphasizing the use of cover crops, green manure crops, legumes in rotation, and animal manures to keep soil organic matter high, these farmers increase soil infiltration rates and water-holding capacity in their soils. This, in turn, reduces runoff, soil erosion, and subsequent losses of water, topsoil, and nutrients.

But there are limits in organic farming, too. Organic plant nutrients don't always offset the slow decline of phosphorus and potash in the soil. The amount of nitrogen that can be produced by legumes in a crop rotation will not always give top yields of high-nitrogen-using crops like corn. The total production of organic farms is reduced because of the acres committed to soil-building crops. (Whether the total production from an organic farm repre-sents a long-term sustainable level of soil output, as compared to a conventional farm which may be producing at levels the soil is unable to sustain, was not addressed in the USDA study.) There is less gross income, with organic farms growing less acres of high-value crops than their conventional counterparts. There is also a definite three- to five-year transition period in switching from con-ventional to organic methods during which, as the soil and crop system gets fully established, yields may be down and losses to insects and weeds may be very serious.[11]

In a third major study comparing organic and conventional farming systems, the Council on Agricultural Science and Technol-ogy (CAST) points out that the decision whether to use legumes or chemical fertilizers is, to the conventional farmer, largely a matter of economics. From 1950 until the 1970's, as fertilizer costs dropped, legumes were abandoned for more profitable crops. The inference, not expressed by CAST, is that, if the cost of fertilizers were to rise relative to other farm costs, additional use of legumes as a source of soil nitrogen would be the natural economic response of conven-tional farmers. The same idea would hold for energy use in the farming system.

But CAST differed from USDA on the probable results of a move toward organic farming. Widespread adoption of organic techniques in the United States would cause, CAST said, an increase in soil erosion.[12] That conclusion is somewhat surprising, given the record of organic farming in soil conservation. It was apparently based on CAST's judgment that more acres of marginal land would need to be cultivated to meet the total crop production needs of the nation, causing enough added soil erosion to more than offset the reduced soil losses on the current cropland base. That could be a matter of dispute, because there is no law (either natural or man-made) that forces the United States to try to produce so much that we destroy our resource base in the process, by either organic or conventional means.

The organic farming movement is very closely allied to many of the central agricultural policy issues of the coming years, and it looks as though it is here to stay. Although there may not be a wholesale switch to "pure" organic methods, it seems likely that more and more farmers will move toward some of the organic techniques as they become better known and more economically advantageous. Many conservation farmers would not be (or like to be) labeled as organic farmers, but some of the soil-conserving techniques they use would put them in the middle of a broad spectrum of methods ranging from the pure organic farmer on one end to the all-out, full-production, mine-the-soil industrial farmer on the other. Conservation farmers would fall somewhere in the middle, and may share more common values with the organic farmer than either recognizes today.

Grassland Farming

Grass is the forgiveness of nature — her constant benediction. Fields trampled with battle, saturated with blood, torn with the ruts of cannon, grow green again with grass, and carnage is forgotten.... It yields no fruit in earth or air, and yet should its harvest fail for a single year, famine would depopulate the world.[13]

Thus did John James Ingalls, a senator from Kansas, extol the virtues of bluegrass in 1872. Ingalls was a politician, not a scientist, and knew what could touch the sensitivities of a Kansas homesteader. His eloquence, often quoted even today, touches a chord in all who know the land and feel its needs. For grass is, indeed, nature's healing cover. When the soil has been torn, ripped, and scarred, grass is the remedy. Properly planted — either by wind, birds, animal droppings, or the machines of men — the tiny shoots of grass soon become strong crowns with millions of fibrous roots to open up the soil and extract and transfer nutrients from deeply buried layers to the surface, as well as stems and leaves to manufacture sugar and cellulose from the air, water, and sun that nature provides.

Properly managed, grass produces food for domestic livestock while simultaneously building up the soil, utilizing and transferring water safely from rainfall to underground pool or bubbling stream, and providing food, cover, and shelter to a broad array of tiny, medium, and large animals that can live in the grassland ecosystem. It is small wonder that this marvelous and widely diverse plant species has been highly valued by people down through history.

In the years following World War II, there appeared to be a great move in the United States toward grassland farming—a type of agriculture that depends on the ability of grass and its leguminous cousins to convert sunlight, air, and water into food that can be harvested by livestock. The benefits are obvious—grassland farming can lead to the goal that all societies must eventually attain, a permanent agriculture. As P. V. Cardon said in the introduction to the *1948 Yearbook of Agriculture*,". . .around grass, farmers can organize general crop production so as to promote efficient practices that lead to permanency in agriculture."[14]

There were the apostles, some of them more than a little eloquent. Louis Bromfield, in his books *Pleasant Valley* and *Malibar Farm* described in engrossing prose how he used a grassland farming system to take a worn-out farm that was known as "the thinnest farm between Newville and Little Washington," and make it into a productive, profitable, pleasurable farm.[15] Channing Cope told how easy it was to take gullied, impoverished, abandoned Georgia soil and make it come to life again through a system of farming that started with building the basic foundation of all agriculture—the soil—then letting grass do the work of manufacturing food and having livestock do the work of harvesting it and converting it into usable forms such as meat, milk, or wool. His system of farming not only stopped all soil erosion, it was also kind to the people who inhabited it. It was so easy, in fact, that he called it "*front porch farming*," on the basis that he could manage most of the work required from an easy chair on the front porch of his farm house.[16]

Easy chair or not, many farmers have found that grassland farming is a sensible (if not always easy) way to extract the bounty of the soil without robbing the soil itself, and thousands of sound economical farm units are based, in whole or in part, on the theories of grassland farming. Those theories are all grounded in an ecological approach to agriculture: Harvest the excess production of each individual soil-plant ecosystem, but leave the essential resources to keep the ecosystem flourishing. In other words, as I used to tell ranchers on Idaho ranges, "if you harvest half the grass and leave the other half to build up the soil and strengthen the grass stand, the half you take will get larger and larger each year. If you try to take it all, the harvest will continue to shrink each year as the grass stand dies out, the soil washes away, and the productivity of the whole system falls off." The principles are basically the same, whether farming the humid soils of Ohio or Georgia or ranching on the arid plains of West Texas. A balanced soil-water-plant-animal ecosystem will produce the maximum sustainable amount of usable food from any type of soil situation. The names of the grasses to be used will vary with soil and climatic situations, but the need to balance the harvest of the grass with its annual production, so that the basic grass stand remains healthy and productive, stays the same.

In the humid areas, like Bromfield's Ohio and Cope's Georgia, the secret lies with planting the right kinds of grasses for the soil type and management system planned. Channing Cope described it in very simple terms:

> "Front porch farmers are using the following crops: kudzu, sericea lespedeza or Coastal Bermuda grass, ladino clover, and Kentucky 31 fescue grass.
> The kudzu occupies the land most easily eroded.
> The sericea lespedeza is planted on the next most erodible lands.
> The ladino and the fescue are seeded on the best lands.
> The Coastal Bermuda replaces sericea lespedeza in sections where best adapted.
> These crops, in proper combination and rotation, provide grazing throughout every day of the year. The animals move from pasture to pasture as the seasons progress and they are so arranged as to provide grazing. . . throughout the year.[17]

Cope, of course, had something not every farmer can depend on—a climate that allowed animals to graze outside all year. But even with the need to provide winter shelter or supplemental winter hay feeding, as is common in most of the United States, the system he described will work on many soils, under many different kinds of conditions. It is, when properly applied, an "erosion-proof, rain-held, permanent, continuous, weatherproof farming" method, as he called it, that, if applied to entire watersheds would reduce flooding and other soil waste while providing a sound agricultural economic base for the entire region.[18]

Grassland farmers differ, at times, from organic farmers in that they may have little or no reluctance to use chemical fertilizers or herbicides if they feel them needed. They are very similar, however, in recognizing that it is the strength of the soil, translated into usable food products by plants, that is the source of all life. Both organic farmers and grassland farmers know that if the soil is strong and productive, the crop produced will be as bountiful as the weather and other conditions will allow, and the production of that crop will not deplete the soil or take away the chance for a bountiful harvest next year.

Good farmers, whether they are labeled "conservation farmers," "grassland farmers," or "organic farmers," all put a great deal of store in the ability to produce a crop next year. But that is not the whole story. No kind of farming that ignores the ability to pay this year's bills is going to survive until next year. Louis Bromfield, in the introduction to Channing Cope's book *Front Porch Farmer*, dispels the notion that conservation means passing up a profitable farming operation:

> Too often reports of our farming efforts lead to the conclusion that we are altruists who have little concern for current income. This is not a correct deduction. We farm for today as well as for tomorrow and, like the vast majority of farmers, we are very concerned with labor income and profits.
>
> In seven years I have doubled the livestock-carrying capacity of my farm five times — that is, the carrying capacity without buying feed. In four years on one farm, I increased corn yields per acre from 30 to 75 baskets and oats from almost nothing to 75 bushels per acre.
>
> These facts, largely the results of grassland farming, speak for themselves. They show, among other things, that we farm for today as well as for all the tomorrows that we will live to enjoy.[19]

Today's grassland farmers differ very little from Bromfield and Cope. They can profitably use soils that are simply not capable of handling the intensive row-crop treatment of "big agriculture." There are thousands — perhaps millions — of acres of productive farmland in Appalachia, the Northeast, and the Southeast, not even counted in the current inventory of "potential cropland," that could be made into highly productive farms with the techniques so eloquently explained by Channing Cope.

Those acres are not counted as potential cropland because they do not fit the needs of modern, big-equipment agriculture. The fields are too small, the topography too broken, the hills too steep. But that does not say that the land is not good for agriculture. As the nation begins to push to the margin of the land suitable for "big-machine" agriculture, and as costs begin to readjust so that farm labor is once again a worthwhile input, many acres that are today lying idle will be "rediscovered" by enterprising farmers. When that occurs, the scientific techniques of grassland farming—made as sophisticated as modern knowledge of soils, grasses, and animals can make it—will once more be a vital part of the American agricultural scene.

Conservation and Self-Reliance

The American farmer is a rugged individualist who takes risks in the face of unknown weather and market conditions, depending as little as possible on other people (or government) for support and vigilant in his opposition to outside interference in his activities. He clings tightly to the idea that, if times really get rough, he and his family will survive. They will make it—even when people in cities starve—for they have one commodity no one else has—the land—and the skills to grow food on it. Many of today's farmers are first-generation descendents of farmers who went through the 1930's. Many experienced hard times as children, or were raised on stories about years when wheat was 10 cents a bushel and farmers burned corn to keep the house warm because they couldn't sell it for enough to buy coal. Those farmers who survived that terrible time became solidly established in the World War II era, and their legacy lives on today.

The situation has changed, however, and in the process, the survival opportunities open to farmers have dwindled. Many have been dosing their land so heavily with nitrogen fertilizers for so many years that, if no fertilizer was available next year (or if price were to jump dramatically), they would be hard pressed to grow a profitable crop. Many are fully equipped with huge, complex machinery that works beautifully and efficiently so long as petroleum and spare parts are available. If either gets in short supply—or experiences a dramatic price increase—that machinery could become very hard to keep working.

In 1980, for example, the International Harvester Company suffered through a long shut-down due to a strike. My friends who had IH harvesters became concerned as harvest drew nearer, for there was no assurance that ample supplies of spare parts would be available. Those machines are just like modern automobiles: complicated. They can't be "patched up" with baling wire like the older models. It takes the right parts and, often, specialized tools and skills not available to all farmers to keep them running. The strike ended, and the IH parts became available, but not before many farmers spent a few anxious moments, and some spent many days searching for and stocking up on the parts they needed.

The moral of this is not that the new machines are bad, it is that they increase the risks facing the farmer. A decision made over a collective bargaining table between a manufacturer and a union, a wildcat strike by truckers or railroad workers, or a bankruptcy in a major equipment manufacturer could put today's farmer in a real bind. He hasn't got a year or two to convert over to new machines, or to wait out a bad situation. He has to plant the crop when the time is right. When it ripens, it must be harvested or it will be lost. There may be a month or two to get those jobs done—or there may be only a few days. He doesn't want to be shut down during those critical times by a decision made far away, for reasons that have nothing whatsoever to do with agriculture and its needs.

The result is often a yard full of equipment, old as well as new, that can be pressed into service if absolutely necessary. Ancient machines, often two or three of the same model, sit rusting but ready to be "cannibalized" into one working version should the need arise. A fully equipped farm shop, often with very sophisticated tools, could be used for that if needed. But the Achilles heel is petroleum. No gas, no go. So farmers have prepared a strong case for preferential treatment should petroleum rationing ever become necessary. They will argue strongly for "first rights" to available supplies, and, no doubt, get them. But that shoots a hole in the myth of self-reliance.

Most farms today grow little, if any, of the food the farm family eats. Instead, they produce only one, two, or three commodities for sale. Some are edible, some are not; none are a full diet in themselves. Every bushel produced is sold and leaves the farm in many cases. If there are chickens, there are 50,000 of them in confinement; if there are milk cows, there are 100. As for keeping a small chicken flock or one milk cow for the family's use, that is a rarity. Farm wives shop in the same supermarkets as city dwellers and buy the same things. What would happen if "hard times" came again? Those farm families would have to scramble just like everyone else. It may be possible to keep from starving with a 10,000-bushel bin of wheat on a farm that has no livestock or other crops, but it isn't going to be all that pleasant.

These are the kinds of uncertainties that can be lessened by an ecologically sound approach to agriculture. When a farmer has decided to farm his land in a manner that maintains the productivity of the soil, he generally discovers that there must be more than a one-crop system on the land. Crops and livestock in combination offer chances to recycle nutrients and rotate crops to build soil strength. Few farms are so well blessed with high-quality farmland that there are no rough areas that should be in grass or trees for soil protection. Those become the pastures or woodlots that can produce products needed to make the farm more self-reliant. Odd areas become wildlife habitats, growing the game birds or animals that provide not only sport, but occasional variety to the family diet.

Such a farm may not be a production factory, but it need give up very little in the way of financial soundness and can, in many cases, be as profitable as any. The high-input/high-output commercial farms make more gross dollars per acre in most cases, but profit still lies in the difference between what is taken in and what it cost to get it. A farming system that needs less of the purchased inputs can survive lowered outputs and still maintain a profitable margin of difference between the two.

The signs that many farmers are searching for such options are plentiful. The growing move toward organic farming methods is one. So is the interest in producing gasohol or other farm products that can be used in an automobile or farm machine. Many farmers not only want to grow the crop, they want to process the fuel. The reason is, often as not, more rooted in the desire to be self-reliant than in the feeling that the processing would be highly profitable or competitive against industrial-size operations.

There are also signs that farmers are aware of the need to do a better job with their soil. The Harris Poll commissioned by USDA showed that farmers were more conscious of the need for soil conservation than the general public, and this finding has been supported by policy positions taken in farm and conservation organizations. Part of this feeling may be rooted in a general sense of responsibility for stewardship, but a significant part also comes from the knowledge on the part of farmers that naturally fertile soils are the best insurance policy against any kind of outside pressures.

Thus, it seems that the stage is set for a significant conservation revival in American agriculture. The pressures against such changes are great, however. Those pressures won't just vanish overnight; some may get more intractable if petroleum costs continue to climb without a corresponding increase in farm prices. The private marketplace will not create the economic incentives needed for soil conservation to become an integral part of agriculture. As has been repeatedly demonstrated, the incentives in the marketplace all encourage maximum production at whatever loss of soil quality, since there are no short-term or cash costs associated with wasting the soil.

That leaves public programs. Non-market incentives, if they are to exist at all, must be a product of public policy. The United States has a long and distinguished history of soil conservation programs. Those programs must not be working too well, however, judging from the soil waste, prime farmland conversion, and other resource losses that are occurring. If they are intended to serve as the "offsetting" incentives that help farmers carry out their conservation responsibilities in spite of anti-conservation incentives from the private marketplace, they must be falling short.

To some people that means that the current conservation programs must be ill-designed or improperly operated. My judgment is that they have been neglected far too long. The market incentives and other pressures on farmers have changed dramatically in the past decade; the conservation programs have not. Anti-conservation incentives have increased; conservation incentives have decreased. Agriculture has entered the Space Age; soil conservation programs still ride a horse.

Before they can be changed, however, we need to understand the wide array of public programs in existence today. No matter how good or bad they may seem, the fact remains that they are the basis from which new approaches must spring. In polling the public about its program alternatives in the Resources Conservation Act study, USDA received overwhelming opinion that the existing programs were, in the main, what Americans feel are the right kind of approaches. The weaknesses cited were those of conflict, duplication, and gaps; along with a long history of underfunding, made worse in recent years by soaring inflation. In sum, the public told USDA that they had a fairly good set of soil conservation programs, if they could be strengthened and modernized. In order to accomplish that, more people need to understand where those programs are today, and how they got to that point.

If it is feasible to protect our agricultural land and reserve the wasteful patterns of today's agriculture, then it is important to understand why the previous conservation policies and programs of government have not been up to the task. There has been a significant national soil conservation effort, but the problem still persists. Why?

13.

Public Programs, Past to Present

Even though people have expressed concern for the fate of America's topsoil since early Colonial days, it took the Dust Bowl of the 1930's to elicit action from the federal government. The history of that action is replete with bureaucratic battles between agencies over control of different aspects of the program, interspersed by periods of uneasy truce. Much of the story is not terribly exciting, being mainly the recitation of gradual changes in government programs. Understanding that history, however, is the only path to understanding why the programs are—and are not—working today, and it is that starting point from which constructive improvements must begin.

In 1928, Hugh Hammond Bennett, a scientist with USDA's Bureau of Chemistry and Soils, published a circular entitled *Soil Erosion, A National Menace*, that led, in 1929, to the first federal appropriations for soil conservation work. The money—$160,000— was used to set up ten soil erosion measurement stations, and the information that began flowing from those efforts was widely publicized by Bennett and his co-workers.

In 1933, President Franklin D. Roosevelt named Bennett director of the Soil Erosion Service in the U.S. Department of the Interior, and gave him $5 million to employ people on erosion control projects. The Civilian Conservation Corps (CCC) eventually put three million idle young men to work on America's farms, forests, and streambanks.[1]

256

That early work was largely done in demonstration projects, although the word "demonstration" was somewhat a misnomer, since it implies some knowledge about what is to be "demonstrated." Often those first workers didn't really know what to do or what would result.[2] With no precedent to guide it, the trial and error process paid off, not only in terms of learning more about erosion control, but also in terms of on-the-land training for people who would become national leaders in the new soil conservation effort.

Early work to stop soil erosion was done mainly by hand. Here, a system of grassed waterways constructed with picks, shovels, and carefully placed seed and straw mulch protected this Nebraska field. SCS photo.

An early project was undertaken near Poplar Springs, in Spartanburg County, South Carolina, to heal the Berry Gully. This growing scar was 800 feet long and 35 feet deep, with many laterals. It had washed out in only eight years from the uncontrolled water leaving field terraces, destroying many acres of cropland and creating serious sediment damage below. The story is told by South Carolina conservation pioneers:

On December 18, 1933, about 75 local relief workers provided by the Civil Works Administration assembled

around the edges of the gully. Their only equipment was
hand tools — shovels and axes. The technical knowledge of
those who directed the work was equally limited. However,
by diverting the water away from the gully, sloping and
planting the steep slopes to grass, trees, and shrubs, and
building log dams across the main channel, they brought the
menace under control. By early spring, signs of success
were noted. Consequently, many directors and staff
members from other early projects visited the South Tyger
River project as part of their training.[3]

Through these projects came the confidence that soil erosion
could be stopped and the beginning knowledge of how to do it.
People who had worked on the early efforts became articulate
advocates of soil conservation, and effective teachers who could
show others the needed skills. A new science began to emerge — soil
and water conservation — blending the skills of the agronomist,
ecologist, engineer, forester, and hydrologist into a new scientific
mixture that specialized in curing land and water ailments.

In 1934, the first national soil erosion survey undertaken by
the new agency estimated that 50 million acres of land had already
been destroyed for cultivated crops by gully erosion. Another 125
million acres had lost most of its original topsoil. To Hugh Bennett
and his staff, the challenge was clear:

It will be impossible to maintain permanent prosperity over
large areas of the United States if the present rapid
destruction and impoverishment of our most valuable
agricultural lands by accelerated erosion are permitted to
continue. The remedial step that inevitably must be taken is
the application of a coordinated land-use and land-protection
program in accordance with the specific needs and
adaptabilities of all types of valuable land needing
treatment. Whatever may be the wishes or inclinations of the
people of this country, this task of protecting the land
against increased impairment and destruction must be
fought from now on.[4]

Bennett's call for a "coordinated program" resulted in 1935
in the Soil and Water Conservation Act (Public Law 74-46) which
created the Soil Conservation Service (SCS) as a permanent agency
within the USDA to develop and execute a continuing program of

soil and water conservation. For the first time, there was a national policy statement on soil erosion and the threat it posed:

> The wastage of soil and moisture resources on farmlands, grazing lands, and forestlands of the nation, resulting from soil erosion, is a menace to the national welfare and it is. . . the policy of Congress to provide permanently for the control and prevention of soil erosion. . .

That statement of Congressional intent still stands today, since the law has never been amended in its 46-year history. Carrying out the new mandate proved a considerable challenge for the department, however.

USDA, while it had actively sought the transfer of the soil conservation program from the Department of the Interior, immediately ran into internal problems in trying to establish the new agency. Concerns about administrative duplication came from USDA bureaucrats, since soil conservation had traditionally been part of the responsibility of several agencies within the department. To deal with the problem of consolidating responsibilities, Secretary of Agriculture Henry Wallace named an interagency committee to prepare policy guidelines for the new effort.

The committee's most important recommendation, apparently made to help solve the problem of who within USDA would control federal soil conservation efforts, was that soil conservation districts should be established as independent units of special-purpose government at the local level to take major responsibility for planning and operating the program.[5] These districts would handle soil and water conservation as their only mission, much like a school district operates only schools or a sewer district sewers.

The adoption of this recommendation by Secretary Wallace marked a significant new direction in the public administration of agricultural programs. Instead of trying to utilize the land grant colleges working through general-purpose county governments, as had historically been done with USDA programs, an entirely new type of governmental approach was being proposed.

Conservation Districts

Following the secretary's decision to recommend formation of soil conservation districts, USDA lawyers drafted a suggested state law that contained the legal details worked out in the department.

After much review within the administration, President Franklin D. Roosevelt sent a letter in February, 1937, to all the state governors, enclosing the "standard" enabling act and suggesting that each state adopt such a law as part of an effective national effort to conserve the soil.

Arkansas and Oklahoma were the first to act, but by the end of 1937 a total of 22 states had enacted the district enabling law. Within a decade, every state, Puerto Rico, and the Virgin Islands had established the legal framework for landowners to create soil conservation districts. The final state laws varied somewhat, but generally included provisions for establishing a state soil conservation agency (called a committee, commission, or board in the various states) to oversee the formation and operation of districts; a petition and referendum procedure for creating local districts; and a statement of the authorities that would be granted to the new units of special-purpose government.

The formation of local districts began immediately, starting usually with a local petition signed by 25 residents and a referendum election that established the district's boundaries. The first district in the nation, the Brown Creek Soil Conservation District, was established in Hugh Bennett's home county, Anson County, North Carolina, in August of 1937. By 1945, SCS Chief Bennett could report to Congress that 1,415 soil conservation districts had been organized. Today, some 2,930 districts cover more than 99 percent of the nation.[6]

At the local level, the districts entered into a memorandum of understanding with the Secretary of Agriculture, and the SCS agreed to provide technical assistance to the district and its cooperating land users. This assistance was in the form of federal technicians who would live in the district and help farmers solve soil and water problems.

Thus, the national goal of soil and water conservation was translated into local action programs designed to fit local conditions and guided by locally elected leaders. Through conservation districts, the SCS could bring a national research and testing effort and the knowledge of skilled technicians to the farmers of America. A touchy political issue—federal intervention in private land use—had been averted. The federal agent did not have any authority over the landowner; the district provided a local agency to establish a cooperative, voluntary arrangement. If any harsh measures such as sanctions or regulations were ever needed, it would be state law and local enforcement that would carry out the task, not the federal government.

No new idea in government, however, is born without travail, and districts were no exception. This innovation in public administration threatened the turf of the Extension Service and the land grant colleges. Instead of delivering soil conservation programs through the states and their university systems, the new approach would allow more national direction through the SCS. A long and bitter battle ensued, with the formation of conservation districts held up in some states for many years as a result of resistance from the Extension Service.

As late as 1952, an attempt to reorganize SCS and transfer virtually all its functions to the Extension Service was fought out in Congress, and only strong support by the National Association of Conservation Districts (NACD) prevented major restructuring of the national soil conservation program. What did occur was the reorganization of SCS by the closing of the agency's regional offices and the transfer of all conservation research activities to the Agricultural Research Service (later to be merged with Extension Service into the Science and Education Administration within USDA). Thus, despite the failure of its opponents to disband SCS, the effect of the effort has been a serious decline in USDA emphasis on soil conservation research, a subject to which I will return.

Lay Leadership Emerges

Under most of the state laws, the referendum election that created the district also resulted in the election of local leaders to be part of the first governing body. The district officials were called supervisors in most states, "commissioners" and "directors" in some instances.

These new conservation leaders — mostly farmers and ranchers — were embarked on a new adventure for, as George Bagley, NACD past president, points out, "at the time of their beginning [districts] represented a totally new concept in the relationships between federal, state, and local units of government."[7] There was no precedent to rely upon as those first laymen began to construct an effective local program for reaching out to farmers and promoting the soil stewardship ethic.

The state law gave them authority, but not guidance. They were empowered to conduct local surveys, investigations, and research relating to soil erosion and control, carry out preventative and control measures, develop comprehensive plans for soil conservation and prevention of erosion, cooperate with and enter into agreements with landowners, as well as with any governmental agency, and, in most states, adopt land-use regulations. That all sounded fine, until it needed to be converted into action among their friends and neighbors. Then, the real challenges began.

They were unpaid, so the new duties and responsibilities were a major test, not only of their abilities, but also of their time, patience, resourcefulness, ingenuity, dedication, and perseverance. One of their most vexing duties was to try to make peace between the local representatives of the SCS, Extension Service, Forest Service, and the USDA's farm programs agency.* Representatives of all those agencies could help get the local conservation program into action, but only if they would stop fighting among themselves over who had responsibility for what and how the glory and credit would be shared. It fell to the district's lay leaders to try and negotiate those issues and get each agency to contribute.

J. B. Douthit, one of the first district supervisors in South Carolina, tells of the early efforts:

> I recall how amazed I was when Ernest Turner, representing the State Soil Conservation Committee, stopped by my farm to tell me of my appointment. I still am amazed at how little I knew then. My first reaction to Ernest's request was: "Well, this is another of those bothersome love and affection jobs." I thought of the new assignment as one which would require nominal participation, occasional signing of a paper agreeing to something that someone higher up had devised, meanwhile hoping that as a farmer I might be one of the recipients of his planned benevolence. But, as time passed, I learned just how mistaken a fellow can be.[8]

Before long, the new district leaders realized that the issues they faced were not restricted to the boundaries of one district, and they began banding together in regional and state associations to discuss mutual problems and share ideas. In those early statewide meetings, they provided each other with support and inspiration. A problem solved in one district became a case study to spark discussion in a meeting, and shared ideas were taken home and tested. Those that worked gained widespread and rapid adoption. That continues today, with 52 state associations representing all states, Puerto Rico, and the Virgin Islands, as well as a national association that provides opportunities for information-sharing and service to all districts.

*Called by various names over the years, including the Production Marketing Administration (PMA), Agricultural Adjustment Administration (AAA), and, today, the Agricultural Stabilization and Conservation Service (ASCS).

The results were soon apparent to even the most casual observer. Hundreds of thousands of farmers joined the districts, and farming practices began to change rapidly as people installed the new conservation systems. In some communities it was terraces, in others strip-cropping, or the cessation of stubble burning in the fall. The exact methods varied from place to place, but the distinctive patterns of soil conservation began to be seen on the land all over America, and a new breed of farmer—the conservation farmer—emerged. Progress, while too slow to suit the professional conservationists who longed to see all of America's farmland operated without waste, was impressive.

A Second Program —and a Fight

At about the same time that SCS and its programs were being established in USDA, and people were being encouraged to form conservation districts to help carry out those programs at the local level, another soil conservation effort was initiated in the department as a result of a rather unusual chain of events.

The mid-1930's were a time when many farmers were being forced off the land and others needed major help to adjust to the new circumstances brought on by the drought, the Depression, and the mechanization of farming. Congress sought a way to help farmers make adjustments to the new economic realities through a system of payments. This was seen as one way to use the federal system to transfer wealth from those who had it to those who needed it. (Food stamps and a myriad of social programs continue that trend today, although the payers and receivers in society tend to change over time.) The Agricultural Adjustment Act of 1933 made such payments to farmers, from a fund created through a tax on processors. That tax, and consequently the act, was ruled unconstitutional by the Supreme Court.[9]

In searching for a new way to support farm income legally, a farmers' conference recommended that soil conservation payments, financed by general appropriations from the Treasury, should be the vehicle. Congress agreed, and with the passage of the Soil Conservation and Domestic Allotment Act of 1936, USDA was authorized to make federal payments to farmers who shifted cropland from soil-depleting crops to soil-conserving crops. In addition, the department was given the authority to share in the cost of soil-building practices such as the application of fertilizer and lime or the construction of soil conservation practices such as terraces or tile drains.[10]

Thus was the Agricultural Conservation Program (ACP) born, as much an attempt to transfer money from urban America to rural America as a soil conservation program, and, most important, as a dual federal approach to soil conservation incentives and program administration which persists to this day.

The ACP is administered by the Agricultural Stabilization and Conservation Service (ASCS), through a system of offices and committees at the national, state, and local level. State committees of three people each are appointed by the president, from recommendations furnished by the state's senior senator of the party that holds the presidency. The appointees are usually agricultural leaders within the state who have been politically active in the party. County committees (also three people) are elected by farmers in a local USDA-operated referendum. These committees are charged with carrying out the administration of all ASCS programs, including the cost-sharing programs.

The dual—and often conflicting—programs of SCS and ASCS have been the source of a great deal of friction over the years. Every Secretary of Agriculture, from Henry Wallace to Bob Bergland, has been concerned with, but failed to find a solution for, the battling within USDA over the nature of the soil conservation effort.[11]

The SCS has criticized the administration of the ACP, feeling that it has been used as much for farm income support and maintenance of a strong political lobby for ASCS as it has been to address soil erosion. SCS feels that all cost sharing should be based on a complete farm conservation plan, arguing that in many instances the construction of one specific practice might waste federal funds if not supported by all the elements needed in a complete conservation system.

The ASCS has resented SCS technical criteria, often openly suggesting that SCS was "gold-plating" conservation work beyond what was needed. They pointed out that SCS did not have the manpower to do all the individual conservation planning that would be needed to allow all farmers to participate in the ACP, and suggested that the plans were, in any event, not needed.

Politically, the forces arrayed behind both points of view are formidable, and Congress has found itself caught in the middle many times. In 1947 and 1948, such bitter battling broke out between the agencies that Congress designed a new working arrangement between the agencies. Primarily through the actions of Representative Jamie Whitten of Mississippi and the House Agricultural

Appropriations Subcommittee, a compromise was finally struck in 1951. Five percent of the total funds from the ACP program were made available for transfer to SCS for hiring technicians to service the ACP workload. In return, SCS agreed to work with ACP participants whether or not they were cooperators with the local conservation district, a move SCS had earlier resisted.[12]

The new method of operating forced the USDA agencies, and the conservation districts, to work much more closely at the local level, but it did not solve all the problems. There were still major disagreements over the best use of the limited program funds, over how the cost-shared conservation practices should relate to a total conservation plan on a farm, and the quality that should be required when the farmer installed a practice.

In the early 1960's, while I was working in the Palouse region of northern Idaho, it was common for the ACP to cost-share on the cost of weed-killing chemicals and summerfallow as a "conservation" measure. This was a popular practice, in heavy demand by the farmers, and represented a way for the program to reach many farmers that would not otherwise participate. But it brought widespread criticism and opposition from conservation district leaders and SCS conservationists.

Summerfallow, the practice of clean-cultivating a field all summer long to help rid it of weeds, has long been recognized as one of the single greatest contributors to excessive soil erosion in the Palouse region. This area, with its steep slopes and silt loam soils, experiences some of the most severe soil losses of anywhere in America. After the soil has been cultivated—and pulverized—all summer long, then seeded to a winter wheat crop, there is virtually no resistance to soil erosion during the winter and spring months. It was very frustrating for USDA's soil conservationists to preach that summerfallow should be stopped while USDA's farm program was subsidizing it in the name of "conservation," and there was constant friction over the issue.

In other parts of the country, much of the available money was spent to share the cost of applying agricultural limestone, a practice necessary to offset soil acidity, increase production and, at times, establish healthy stands of grasses and legumes for soil-conserving. The soil-saving virtues of limestone were open to question, however, in comparison to its value as a yield-raiser.

In 1953, Secretary of Agriculture Ezra Taft Benson tried to delete some of the production-oriented practices from the program, but Congress would have none of that, and the Agricultural Appropriations Bills from 1959 through 1962 contained language that prohibited USDA from changing the list of approved practices without the approval of ASC state and county committees.[13] In the early 1970's the department again tried to limit the practices, but again the Congress inserted language in the Appropriations Bills to give local ASC committees authority to select from the pre-1970 list of practices.

In 1972, the ACP was renamed the Rural Environmental Assistance Program, in recognition of the new awareness on Capital Hill for environmental matters. In 1973, it was called the Rural Environmental Conservation Program, then the name was changed back to ACP in 1974. Throughout this entire period, every president from Dwight Eisenhower to Gerald Ford recommended that the program funding be deleted from the federal budget. Each year, Congressional action, led by Representative Jamie Whitten of Mississippi, would assure that funding was continued. To say that there have been significant differences of opinion about the need to fund soil conservation efforts, and the ACP in particular, would be a monument of understatement. The tensions remain largely intact today.

On the one hand, there is the resentment at the local level over a restricted list of conservation practices approved at the Washington level of USDA. These national practices, selected ostensibly because they are more effective at conserving soil, often miss the major concerns of a locality. Farmers, ASC committee members, conservation district leaders, and farm organizations all testify regularly to Congress that conservation practices ought to be selected at the local level because farmers know best what their land needs.[14]

The other side of the argument was supported by the continued findings of Congressional investigations that the majority of ACP funds were spent at the local level for practices that were in popular demand and "were oriented more toward stimulating agricultural production than toward conserving cropland topsoil."[15] ASCS, caught between these two points of view, has actively attempted to convince state and county committees of the need to target the program funds toward significant erosion-reducing practices. If that could be consistently done, the benefits of local control over the details of the program would no doubt lead to a far greater degree of autonomy for the local committees.

Reaching Farmers

Although most rural people are aware of the existence of the conservation district, the SCS, and the ASCS, few that are not directly involved with one or the other of the programs can tell the difference between the agencies. Many farmers receive conservation assistance for years without being able to distinguish who is helping them, or why.

A farmer who begins searching for technical assistance usually starts in the local conservation district office, where the confusion is compounded by the fact that it is normally the office of both the conservation district and SCS. He makes application to the conservation district for assistance, signing a non-binding agreement where he agrees to apply a complete conservation plan to his land in return for the technical assistance provided without cost through the district.

After approval of this application by the district, a professional conservationist (usually an SCS employee working with the district under the memorandum of understanding) visits him on his land to view the situation; make the necessary inventories, surveys, and investigations; and propose some alternative solutions for the land user to consider. The farmer is free, of course, to do anything he wants, including nothing, but the conservationist tries to help him select an approach that will solve the soil and water problem and be compatible with what the farmer wishes to grow on the land.

Once a plan has been chosen by the farmer, there may be more surveys and investigations, engineering designs, or other technical work needed to assure that he can install the conservation practices properly.

Farmers who wish to participate in the ACP cost-sharing program must go to the local ASCS office and apply for that program, stating what they want to do and certifying that they would not be able to carry it out if cost-sharing assistance from USDA were not available. If the practice involves engineering, the application is forwarded to the local office of the SCS for the necessary technical assistance.

In this case also, the SCS technician visits the site with the farmer, determines that the proposed practice is needed and feasible, and does whatever surveys and designs are needed. The farmer is given a copy of the standards and specifications that will need to be met in order for the completed practice to qualify for the government payment. Following construction, the finished job is checked to see that the standards have been met and, if they have, the ASCS pays the appropriate amount under the rules of the county program.

The differences created when a farmer utilizes the ACP are subtle, but significant. Once the farmer has applied for government

cost-sharing funds, the technical determination of the SCS technician becomes an important factor in whether or not he will receive
his payment. His relationship is no longer the voluntary one he
enjoyed with the conservation district. If he does not construct the
job according to specifications, he may be rejected and denied the
cost-sharing money. Sometimes this is the result of negligence or
willful attempts at fraud on the part of the farmer, but most of the
time it stems from disagreement between the farmer and the SCS
technician over what is needed to do the job correctly.

*A soil conservation plan, developed by this Massachusetts farmer
with assistance from a professional soil conservationist, identifies
the types of soil on the farm and shows how each field will be
managed to prevent soil erosion. Similar services are available to
virtually every farmer in America through a local soil conservation
district that exists to provide soil conservation assistance. SCS
photo by Konieczny.*

Where the farmer feels the SCS specifications are overly strict or will result in a job that is unnecessarily expensive, he may try to cut corners, or substitute his own judgment for that of the technician. If the SCS technician rejects the final product, the farmer can appeal to the ASC County Committee. If the issue is not resolved there, it can be appealed to the ASC State Committee or perhaps even find its way into USDA headquarters in Washington. The conservation district is not part of this battle, which lies strictly between the farmer and USDA. By seeking the federal cost-sharing support, the farmer gives up a measure of the independence he enjoys in his relationship with the local conservation district.

The Federal Program Broadens

Shortly after the soil conservation programs began operating, it became obvious that soil erosion was not the only resource concern facing local people. In many instances, it was water that was the villain; too much in some places, causing serious flood damages, and not enough for maximum crop production in others. On some farms, it was soil quality and productivity that was declining, even though erosion was not the primary cause. Pastures needed renovation to provide adequate feed for livestock enterprises and farm woodlots needed better management to contribute to farm income. Both pasture and woodland were effective soil-conserving uses in many cases, but if they did not contribute sufficiently to farm income, the farmer might be tempted to plow them up for crops, which would exacerbate soil problems. Thus, encouraging proper land use meant helping farmers use their lands in a financially rewarding way.

In many instances, helping individual farmers either failed to solve the soil and water problems in an area, or proceeded far too slowly to satisfy local conservation advocates. One farmer could seal his irrigation ditches to prevent water loss, but if the major problem was in the community ditch leading to the farm, his individual effort was of little avail. This led to programs for encouraging farmers to work together in informal groups, legal associations, or even community-wide sponsoring organizations that could tackle larger jobs.

Different regions of the nation faced far different situations. The terraces that were adapted in the Corn Belt didn't work in the Palouse, so there was constant effort to develop different approaches and technologies to meet local conditions. Two trends emerged: the gradual evolution and broadening of the basic soil conservation programs of USDA, and requests to Congress for new programs that were specifically targeted to new concerns. Three examples of the new programs that were enacted by Congress as a result of these requests are the Small Watershed Program, the Great Plains Conservation Program, and the Rural Clean Water Program. A brief review of these initiatives may shed some light on the complex forces that have shaped the federal programs in existence today.*

The Watershed Approach

By the end of World War II, the soil conservation program had fairly well established its ability to work with individual farmers, but there were serious situations that could not be addressed one farm at a time. Central among these were the damages being sustained, to both farms and towns, from the lack of flood control. Many small watersheds, often comprised mainly of agricultural land without adequate conservation treatment, periodically created serious flood damages. No one farmer could cure the situation himself, but if every farm in the watershed was given conservation treatment and appropriate flood control dams or structures were placed in the channel, most damages could be avoided.

Once again, however, the soil conservation program became involved in an agency "turf" battle. The Army Corps of Engineers had historically carried out a national flood control mission, and they resented the entry of a new agency on the scene.

The Corps efforts consisted primarily of major dam and levy projects on the larger rivers and their tributaries. The SCS felt this was an after-the-fact effort that did not address the real source of the excess runoff but merely treated the major symptoms. The Corps pointed out that treatment of upstream watershed areas with soil conservation measures was an almost impossibly large task to address, and still did not guarantee that the river would not cause major flood damage. With major metropolitan areas constructed in the natural floodplains of rivers, this argument had much in its favor, since the river would have flooded the town periodically even if the watershed had remained in pristine condition. But on many small watersheds, proper treatment could both improve the farmland's productivity and reduce the flood damages.

In the face of heated political arguments between supporters

*See Appendix F for the major federal programs in existence in 1981, their legislative authority, and the year they were initiated.

of the two approaches, Congress enacted the Watershed Protection and Flood Prevention Act of 1954 (commonly known as Public Law [P. L.] 566). This gave SCS leadership for a nationwide program of technical and financial assistance to communities for watershed protection and flood prevention on watersheds of 250,000 acres or less. The Corps, as the principal dam-building agency of the federal government, retained jurisdiction over the larger projects, but still bitterly opposed the new program. Not only did they express doubts about the effectiveness of upstream watershed work in controlling floods, they also questioned the competence of SCS engineers to make the flood surveys and design the type of dams needed. Men whose only experience was in the design of farm ponds, they asserted, should not be trusted to design major water structures.[16]

Despite the opposition from the Corps (which has largely disappeared with time), the P.L. 566 program was an immediate success. The cost-sharing formula for flood control under the P.L. 566 program (the federal government pays all of the construction cost for structures dedicated to flood control) and the obvious value of flood protection to downstream landowners have kept the number of applicants for assistance far above the funds available for construction. The addition of cost-sharing and technical assistance for fish and wildlife projects, as well as for recreation, municipal and industrial water programs, has also broadened sources of local support, as has the ability of the program to help the economy of rural communities. Local sponsors pay, on the average, around one-third of the total costs of a small watershed project, with average total costs per project running about $3 million.[17]

In the mid-1960's, environmental opposition to the watershed program began to develop, largely around the issue of stream channelization. Environmentalists had intense criticism for projects that eliminated the flooding caused by meandering, lazy streams by converting them into straight, sterile ditches that flushed out the water rapidly but had little biological life or aesthetic quality. The word *channelization* was a battle cry as SCS was accused of having abandoned its conservation mission and joining the Corps of Engineers in an engineering assault on the natural environment.

It took a decade of strife, and a great deal of compromise on both sides, to put the channelization issue to rest. Under new channel guidelines, SCS now works carefully with local sponsors and wildlife agencies to minimize damage to channel values while still improving flood flows. Not all the wounds are healed yet, but the main battle over channel alteration seems to be over. The program may be (although this would still be argued) more capable of providing lasting benefits in both an economic and environmental sense as a result of the new sensitivities that have been forced upon it.

The Great Plains Conservation Program

While the small watershed program was getting started in the mid-1950's to address flooding, the Great Plains were once again in serious trouble from drought. This vast region, always subject to wide swings in climatic conditions, suffered as much damage from drought and high winds in the mid-1950's as had been suffered in the Dust Bowl days of the 1930's.[18] The major climatic problem in the Great Plains is not just drought, but the fact that the climate changes around a point that is critical for crop production.[19] If the region was consistently too dry for farming, people wouldn't be tempted to farm there, but it is not. In the good years, bumper crops are possible, as the rainfall goes well above the lower limit for crop growth. In the bad years, however, the rainfall drops so far below minimum needs that complete crop failures can occur. It was apparent in the mid-1950's that a regional approach was needed to address the problems unique in this fragile, high-risk area.

An unprotected Montana field suffers severe wind erosion while adjacent fields, protected by strip-cropping, are untouched. The Great Plains Conservation Program has been in existence since 1956 to aid farmers in planning and installing effective soil conservation programs to protect the farmland in this vulnerable region. USDA photo.

In enacting the Great Plains Conservation Program, Congress not only addressed the problems of the Plains, but also created another new program approach—cost sharing under a multi-year contract, based on a complete conservation plan—with the entire program, including the cost-sharing aspects, administered by SCS.

Farmers and ranchers sign up for assistance and develop a plan to make the land-use adjustments and conservation treatments needed on their land. In return, they receive cost-sharing payments of up to 80 percent of the cost of installing approved conservation practices. An SCS technician visits the farm each year and, if the farmer is not keeping the agreements contained in the original contract, can compel either the proper application of the conservation plan or the repayment of the federal cost-shares received by the farmer under the program.

By 1980, more than 50,000 farmers had taken advantage of the Great Plains Program, and there were over 5,000 waiting for funds to become available so that they could enter a contract. Over 9 million acres of cropland had been converted to grass for varying periods of time, over 22 million acres had been strip-cropped, and 110,000 miles of field windbreaks had been planted.[20]

The long-term contracting features of the program have been widely praised, and copied by later legislation. But there are critics of the program, as well. Because the program has been voluntary, success at reaching seriously eroding lands where the owners refuse to cooperate has been limited. The contracts between the owner and USDA are binding only for the length of the contract, so many acres that have been seeded to grass under the program have also been plowed out when the contract had expired and the price of wheat made it worth another gamble on the weather. A 1977 study by Congress found that 26 percent of the farmers in the program had plowed out their newly established grasslands after their contracts had expired and that more were planning to do so.[21] There simply were no economic incentives, farmers asserted, to keep those lands in grass, particularly when wheat prices went high enough.

Other critics argued that farmers did not really need to enter into long-term contracts when they could just as easily get cost-sharing from the ACP on an annual basis, without having to go through the planning process or enter into a formal contract. That criticism seemed more a part of the ASCS/SCS feud over leadership of the program than a valid criticism of the long-term contracting technique. The number of farmers waiting for funds to become available so that they could sign contracts seemed to belie any significant reticence on their part.

The Rural Clean Water Program

In the debate over the 1972 amendments to the Federal Water Pollution Control Act, national attention was drawn to the need to control soil erosion as a way of preventing water pollution. Sediment, and the associated nutrients, pesticides, herbicides, organic matter, and bacteria that often were carried with it, was identified as the top contributor (in volume) to the dirtying of the nation's waters. Section 208 of the 1972 law (P. L. 92-500) required state and regional plans for the prevention of this "non-point source" pollution.

The challenge of stopping such dispersed, uncontrollable pollution completely baffled the water-quality specialists in the Environmental Protection Agency (EPA) and their state agency counterparts. They had experience in controlling pollution from point sources (those that come from the end of a pipe, such as municipal or industrial pollutants). Those could either be treated in an appropriate sewage treatment plant, or the offending industry or city could be forced to stop releasing the pollution into the water. But non-point pollution came from everywhere and could neither be outlawed by regulations nor captured for purification in a treatment plant.

Obviously, non-point pollution had to be prevented at its source, as much as possible, and that meant the installation of soil conservation systems on the land. Stop soil erosion, and non-point pollution, at least from agricultural lands, would be dramatically reduced. Working in cooperation with the NACD and USDA, EPA began to encourage state and regional water-quality planning agencies to work with local conservation districts and state soil conservation agencies, in a joint effort to plan programs for preventing non-point source pollution.

In 1977, as amendments to the Clean Water Act were again being considered, Senator John Culver of Iowa introduced an amendment to authorize a new program in USDA that would allow the Secretary of Agriculture to cost-share in the installation of needed water pollution control practices on farms and ranches. The program, called the Rural Clean Water Program (RCWP), would utilize long-term contracts patterned after the Great Plains Conservation Program. It was enacted into law, but efforts to get funds appropriated failed in the Agricultural Appropriations Subcommittee of the House. Supporters of ASCS convinced Chairman Jamie Whitten that the new program, to be administered by SCS, would compete with the ACP program and duplicate the administrative capabilities of ASCS. The old interagency battle had flared up once again, only this time with a new twist.

The Clean Water legislation contained a new provision authorizing SCS to transfer portions of the federal program funds to state and local governments to carry out portions of the RCWP projects. This was done, in large measure, because state and local government had the responsibility for controlling water pollution, and districts had been named in most states to administer the local efforts to prevent agricultural sources of pollution.

But such a departure also meant that, for the first time in the 45-year history of soil conservation programs, the conservation districts would administer portions of a federal program. This proposal won the support of water-quality agencies, environmental groups, and conservation districts, but was bitterly contested by ASCS, who viewed it as direct competition with their ability to administer USDA program funds at the county level.

Representative Whitten, aware of the widespread support for a financial assistance program to help farmers meet water-quality goals, but unwilling to allow the ASCS-SCS power balance to be altered from the pattern he had helped put together in the early 1950's, was hard-pressed for a politically acceptable solution. Finally, in the 1980 Agricultural Appropriations Act, he provided for an Experimental Rural Clean Water Program to be administered by ASCS. The "experimental" program was not based on the authorization provided in the Clean Water Act, but authorized solely in the appropriations bill. ASCS was directed, however, to run the program much as had been envisioned by the Clean Water Act, only without allowing a state or local role in program administration.

The program received $50 million in 1980, and 13 project areas were selected and funded. In 1981, the program was given another $20 million. But the final shape of USDA programs for encouraging water pollution prevention on the nation's farms and ranches is still far from settled, and the uneasy balance of power within the department over conservation programs is still very much a consideration in what may finally emerge.

Districts
Face the Future

The winds of change that rippled through the federal conservation programs did not go unheeded in the soil conservation districts, either. While many district leaders were content to continue the methods that had marked the early efforts, others recognized that America was changing, agriculture was changing, and the institutions of government would have to change as well.

Many districts were running into serious soil and water concerns that had little to do with farming. In the early 1960's, rapid shifts in land use from agriculture to urban-type uses, excessive soil erosion and sedimentation from construction of housing and other developments, and growing public concern for environmental quality presented new challenges that conservation districts were ill-equipped to meet.[22]

At USDA, Secretary of Agriculture Orville Freemen had decided that the SCS must serve as a national soil and water assistance agency, not just a farm agency, and that meant that local districts must broaden their concerns if the historic relationship between SCS and districts was to be maintained. At Freeman's urging, more than 2,700 conservation districts updated their long-range programs and entered into new, broader agreements with USDA.

But district leaders were concerned about the capacity of districts to meet these broader challenges. The basic question arose: What could be done with districts that could not be done without them? Were districts really needed? In 1965, a special study by NACD, *The Future of Districts*, concluded that districts filled a necessary role by providing an appropriate legal body to represent all the people and all community interests in resource matters, but only if district leaders were willing to guide their local programs in new directions.[23]

In many cases, this meant entering into activities that districts were not authorized to do in their state enabling legislation. To help meet that need, NACD prepared sample legislative provisions to broaden district powers. As a result of strong efforts by district officials, over 200 amendments to state soil and water conservation district laws were adopted between the late 1960's and 1975.

Some of the needed changes included broadening the scope of the law in 30 states to include issues such as flood prevention, drainage, irrigation, water pollution and storm water runoff; including urban areas within district boundaries; providing for urban or non-farm representation on district governing bodies; authorizing the levying of taxes or assessments; and the exercise of eminent domain, allowing districts to receive funds from counties.[24] A major transformation in conservation districts, largely unnoticed by the public, was in full swing.

In the early 1970's a new sense of urgency was added to the conservation effort by the passage of Section 208 of the Federal

Water Pollution Control Act Amendments of 1972. This brought the threat of federal regulation over farm activities if, as the law proposed, EPA enforced its authority to stop pollution at whatever source it originated. Farmers and their organizations were adamantly opposed to such federal regulation and the conservation districts were no exception. Within NACD, a major debate raged over whether to fight the new law, ignore it, or work with it to see if an acceptable approach could not be designed.

The latter course of action was chosen, and in a cooperative effort with the Council of State Governments, USDA, and EPA, NACD worked out a model state erosion and sediment control act. The Council of State Governments included the model act in its suggested state legislation for 1973, and NACD carried out 42 state sediment control conferences to increase awareness of erosion and sediment control, explain the provisions of the model act, and encourage state action to meet the challenges posed by water pollution.[25]

As a result of these efforts, 20 states, the District of Columbia, and the Virgin Islands adopted legislation to establish sediment and erosion control programs. Most of these laws prohibit local governments from issuing subdivision approvals or building permits without an erosion and sediment control plan approved by the conservation district. That lets the natural resource expertise of districts be effectively backed by the police powers of the local governments.

In the process of accepting broader responsibilities, districts often changed their political nature, as well. In Nebraska, a sweeping reorganization resulted in 24 Natural Resource Districts, with broad powers of taxation and eminent domain to help carry out programs for soil and water conservation, watershed development, irrigation, drainage, and other resource-related programs. With that added financial and political mandate, Nebraska's districts have developed active, well-financed local programs that often equal the size and capabilities offered by the federal agencies.

In several states, the election of district officials was moved from a special election to the general election ballot, and local political interest in running for district seats picked up. The addition of urban areas brought large non-agricultural constituencies and new political constituencies into districts in many states. Today, over one-half of the nation's districts are working with local government officials, agencies, developers, builders, consultants, engineers, and the public to develop effective urban conservation programs. Coastal zone management, surface mine reclamation, prime farmland preservation, and wetlands management have emerged on the agenda of many districts.

With districts taking on new responsibilities based on local or state laws, it quickly became impossible for SCS to provide all of the technical assistance that would be needed. It was not only a matter of manpower (which SCS was rapidly losing under federal budget constraints) but also authority. Many districts were empowered to carry out functions, such as the enforcement of local ordinances, in which federal employees could not become involved. As a result, districts sought state and local funds to hire district technicians to supplement the personnel available from the SCS. Today, over 6,000 district employees provide engineering, conservation planning, and conservation education services as part of the local programs. These employees receive technical training from SCS technicians, which allows the benefits of national research and technology development to flow to each district. Policy and program guidance, however, is provided by the elected district leaders, which keeps the program local in nature.

Within the federal establishment, leaders in the SCS realized that the rapidly changing circumstances demanded a fresh new look at their objectives and programs. This led to the development of a new "framework plan" for SCS entitled *Soil and Water Conservation for a Better America*, which set out three new long-range mission objectives for SCS. The agency's mission, that document proposed, would be to help the American people enhance the quality of the natural resource base, the environment, and their standard of living.

A long list of goals was established, but the major changes wrought in SCS programs, according to then-Associate Administrator Norman A. Berg, were two-fold: a significant expansion of effort in inventorying and monitoring natural resource data; and a broadening of SCS planning assistance beyond farms and ranches to try to meet as much as possible of the workload developing with urban and suburban landowners and local governments with soil- and water-related problems.[26]

The effort to accelerate the gathering of natural resource data initiated by the SCS framework plan was set into motion by the Rural Development Act of 1972 and resulted in studies such as the Potential Cropland Study of 1975 and the 1977 National Resource Inventories. Expansion of the snow survey and streamflow forecasting program in the western United States, monitoring of wind erosion conditions in the Great Plains, and a newly established capability to monitor the effects of short-term natural phenomena such as floods, earthquakes, and droughts have also risen from that effort. It is interesting to note that the ability of the American people to evaluate the condition of their natural resources in 1981 relies so heavily on a decision made within SCS a decade earlier. The fact that it took that long to establish the programs, carry out the inventories, and interpret the data holds a lesson for policymakers today. What will we need to know in 1991 and 2000? We had better decide and set about the task of gathering that information now.

New State Ventures

For most of the 1940-1960 period, the role of the state soil conservation agencies was largely limited to helping form local districts and providing some guidance in their operation. After the districts were formed, however, many of the state agencies began to take a more active role in conservation programs. In many states, cooperative ventures between the state and the SCS produced acceleration to soil surveys, watershed plans, and other conservation efforts.

During the early 1970's, most states were caught up in a widespread effort to reorganize and streamline state government, and the soil conservation agencies underwent significant changes. While many of them expanded in size and the scope of their operations, it is doubtful that they maintained their stature in most states due to the rapidly expanding role of state agencies for virtually every governmental function from transportation to welfare. During most of the decade, however, state soil conservation agencies were slowly and steadily extending their capabilities and efforts.

As district programs became more independent and active, state appropriations to support their operations became more common. These funds were used to hire district employees and pay for other district programs, so the state soil conservation agency became more active in monitoring district programs, understanding their needs, and telling state legislators what those state funds were buying and why new appropriations were needed.

In some states, such as Iowa, the state agency hires technicians on the state's payroll and stations them in district field offices to assist with the local program, but the more common pattern is for the state agency to transfer funds to the district, where district officials are responsible for hiring and managing the staff. These employees, no matter who pays their salary, are integrated with the federal staff provided by SCS into one local office that provides a broad range of technical services to the public. In most instances, few people at the local level know which of the district staff is paid locally, which by the state, and which by the federal government. To the recipients of the district's services, it really doesn't matter, as long as the service they receive is adequate. To the states, however, the changes have been significant, as more and more governors and state legislatures now consider budgets and program elements of an active state and local soil conservation program that formerly did not exist.

The concept of a long-range program to guide these state investments and coordinate them with all the other conservation activities taking place was not needed for many years, but as states took a more active part in funding district efforts, a need grew for the state government to develop its own soil and water program. When the Soil and Water Resources Conservation Act of 1977 (RCA) (discussed in more detail later) required SCS to develop a national report on conservation program needs, SCS in turn encouraged each state to conduct an inventory and appraisal of its soil and water resources. From this base, most of the state soil conservation agencies are developing a state long-range program to set out a strategy for solving resource problems. Included in these programs are not only the actions that the state itself will take, but also opportunities for USDA programs to orient their programs toward the highest priorities identified by the state.

In general, the states use two methods to develop their programs: One is to develop a statewide summary from the programs of the local districts; the second relies on citizen meetings where statewide concerns are identified, priorities established, and actions developed. Two states, Oregon and Iowa, were early examples of these different processes, with Oregon developing a state program consisting largely of the sum of local programs, while Iowa started with a statewide approach and has developed a program called "Iowa 2000" that has been very successful in calling soil and water conservation needs to the attention of the people of that state.[27]

Even with many variations in the processes being used, the states seem to have a common goal of bringing about a closer working relationship between landowners, conservation districts, the state soil conservation agency, the Soil Conservation Service, other state and federal agencies, and the general public. One result has been the elevation of public concern over the waste of agricultural land.

In Iowa, for example, a 6-month public opinion survey by the *Des Moines Register* identified an eight-point "Iowa in the 80's: A People's Agenda," based on what the people of Iowa want their state to be like ten years from now. Item number 3 on the agenda:

"Erosion of Iowa's topsoil must be stopped. We need a massive program; more money for permanent land improvements, more personal counseling of farmers by conservation specialists, and use of the property tax system to spur conservation."[28] There is no way definitely to link this strong expression of public opinion with the intensive efforts carried on for the past two years by the Iowa Department of Soil Conservation, the local soil conservation districts, and the SCS to develop a statewide soil conservation program, but the conclusion seems fairly obvious.

Another type of expansion in state and local activity has come in their financial contribution to conservation work. In the early years of the conservation effort in the United States, practically all of the financing for soil and water conservation programs came from the federal government. Today substantial contributions are being made by state and local governments and private interests. Table 13.1 demonstrates the rapid growth in state and local funds for soil and water conservation programs. (By comparison, the federal soil conservation effort in 1980 totaled about $900 million.)

Of the approximately 6,000 employees working for the 2,925 conservation districts in 1980, over two-thirds are paid by state and local funds. From 1979 to 1980 the total number of district employees hired with state and local funds increased by 8 percent and included a 15 percent increase in full-time positions and a 31 percent increase in full-time technicians. This indicates the tremendous movement on the part of districts in the past few years to strengthen their contribution to the on-the-land assistance available to farmers—a service formerly provided almost entirely by federally paid technicians.

Table 13.1. **Growth of State and Local Governmental Funds in Soil Conservation District Programs, 1957-1979, in Actual Dollars**

Source of Funds	1957	1963	1968	1973	1979
	(millions of dollars)				
State	$4	$14	$30	$42	$65
County and Local	9	17	33	44	87
Total	$13	$31	$63	$86	$152

SOURCE: National Association of Conservation Districts, *RCA Note* #7, July 25, 1980.

The early state funding efforts were aimed almost entirely at helping districts provide technical assistance or develop resource information, but recently there have been several new programs that provide funds to help farmers actually apply conservation measures. Iowa, Wisconsin, Illinois, Kansas, Missouri, and Minnesota all have cost-sharing programs that supplement the funds available through similar programs in USDA; Montana and Utah have revolving loan funds to help ranchers improve grasslands. These programs are administered by the state soil conservation agencies and the conservation districts, giving them an active assistance program to offer farmers.

In addition, some of the new state laws permit cost-sharing in urban areas, a feature not available in USDA programs. That may indicate that state legislatures, being closer to the problems and not necessarily developing these programs in their agricultural committees, may take broader approaches in the future than the federal programs. While that is not certain, what does seem most likely is that the states have generated considerable new interest in soil conservation and a great deal of momentum in developing independent programs of their own.

Federal Funding Drops

In contrast to the state and local governments, the federal government's financial support for the soil conservation program has been falling off in recent years. Part of that, no doubt, has been due to the conflicts inherent in the various programs and the continued question as to how effectively the program funds were addressing the soil erosion problem.

During the decade of the 1970's, for example, while state and local support for soil and water conservation was doubling in actual dollars and even growing in real dollar terms (budgets with the effect of inflation removed), the federal effort was shrinking. This was occurring despite the efforts of influential congressmen to keep programs at higher levels, as evidenced by the fact that the amounts appropriated have, in every year for the past 20, exceeded the requests contained in the budget submitted by the president. Program dollars in most of the budget items gradually climbed, as Table 13.2 illustrates, except for the ACP, where a level around $200 million stayed fairly static over the years. In addition, Table 13.2 shows that new programs have been steadily added to the total program mix over the years. The total amount of dollars allotted to the federal soil and water conservation effort has grown more from new programs than from additional resources in the old ones.

Table 13.2 tells only part of the story, however, because in terms of real purchasing power, the older federal programs have not fared well over the past few years. Figure 13.2 shows how some of the major USDA conservation programs were supported during the decade when their budgets were measured in terms of constant 1982 dollars. As NACD President Lyle Bauer of Kansas pointed out in 1980, "In a decade when they all talked about the quality of the environment, our national leaders have allowed programs to help protect the most vital elements of that environment—our productive soil and waters—to shrink away."[29]

Table 13.2. **Federal Funds Appropriated to Selected USDA Soil Conservation Programs, 1950-1980**

Program	Administering Agency	Fiscal Year			
		1950	1960	1970	1980
		(millions of dollars)			
Conservation Operations	SCS	51	82	128	274[a]
Watershed Planning	SCS	*	6	7	10
Watershed Construction	SCS	*	41	91	139
Great Plains Conservation Program	SCS	*	10	15	19
Resource Conservation & Development	SCS	*	*	11	32
Agricultural Conservation Program[b]	ASCS	252[c]	213[c]	185	190
Forestry Incentives Program	ASCS	*	*	*	15
Water Bank Program	ASCS	*	*	*	10
Experimental Rural Clean Water Program	ASCS	*	*	*	50

[a] Several new programs such as inventory and monitoring and resource appraisal were added between 1970 and 1980.
[b] Funds for this program are cost-shares only. Salaries and expenses for administering the program are budgeted in a different line item in the ASCS budget.
[c] Cost-shares earned by farmers. (May differ from actual appropriations somewhat, since ACP funds are carried over from year to year if not used in the year for which they are appropriated.)
*Program did not exist.

As the lines on Figure 13.1 indicate, it would take consider-
ably higher budgets in 1982 to provide the same incentives to
farmers that earlier budgets provided. In ACP, for example, it would
have taken around $400 million in 1981 to equal the purchasing
power of the $185 million of the 1970 program. SCS technical
assistance programs fared somewhat better, virtually holding their
own during the period.

Figure 13.1. The Effects of Inflation on Selected Federal Conservation Programs

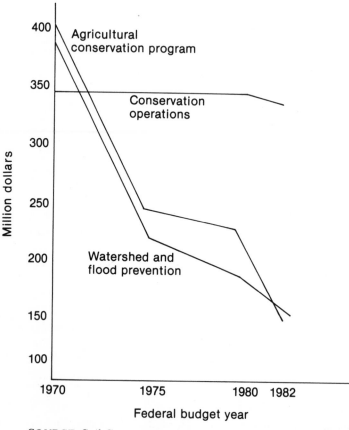

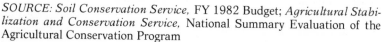

SOURCE: *Soil Conservation Service,* FY 1982 Budget; *Agricultural Stabilization and Conservation Service,* National Summary Evaluation of the
Agricultural Conservation Program

But each appropriation was a hard political fight for the Congressional proponents of soil conservation. As the total federal budget climbed rapidly toward the $1 trillion mark, the percentage allocated to soil conservation dropped steadily. By the time President Reagan proposed his cut-back federal budget proposals for fiscal year 1982, all 30-plus programs that helped provide incentives to farmers for soil conservation totaled less than $1 billion out of a total federal budget of nearly $700 billion.

Why Doesn't It Work Better?

The logical question, after reviewing all the programs and all the effort that has gone into improving them over their history, is "why haven't the problems been solved?" That question has resulted in dozens of Congressional inquiries, USDA studies, and private research efforts that have generated piles of paper and hundreds of speeches, but little satisfaction.

One of the reasons why Congress has found it hard to improve soil conservation programs may lie as much in the complexity of the solution as in the fact that it is not known. Congress needs a simple evil that can be adequately explained in somewhat under two minutes, followed by a clear, concise solution that can be explained fully, understood readily, and measured in terms of costs and benefits in proven dollar terms. Complex problems, with multifaceted solutions involving significant expenditures of public money but only vague estimates of the dollar benefits to the public, are difficult, if not impossible, to handle.

The soil damage being suffered by American farmland is evidence enough that whatever the soil conservation program is doing today, it is not nearly adequate. Most competent observers agree that, if the programs were not in place, the situation would be far, far worse. But that does not cure the deficiencies. It is clear that it is time for a major transformation.

I am convinced that the changes that have taken place in the federal, state, and local programs during the 1970's were needed and demonstrated that the system has great flexibility. But while the programs were expanding to meet a wide variety of non-agricultural needs, U.S. agriculture simply passed them by. In a period when farm mechanization, use of capital inputs like fertilizer and fuel, debt loads, farm size, and commodity demands were increasing at a fantastic rate, the soil conservation programs were concentrating on issues that were often not associated with the conservation problems of production agriculture. Even if they had concentrated fully on farm problems, however, it is doubtful that they could have kept pace, since they were stagnating under no-growth budgets which exacerbated the fighting between USDA agencies as each tried to achieve some measure of dominance, if for no other reason than to attract a larger share of the diminishing soil conservation budget.

The Resources Conservation Act

By the late 1970's, a growing frustration with the national soil conservation effort was keenly felt in Congress. USDA reports told of growing soil erosion problems, and an increasingly worrisome rate of prime farmland loss. A report to the Congress by the Comptroller General of the United States in 1977 was titled, "To Protect Tomorrow's Food Supply, Soil Conservation Needs *Priority* Attention." In the report, all of the federal soil conservation programs were criticized and the existence of a continuing soil erosion problem *despite* the 40-year history of USDA efforts was viewed with alarm.[30]

In Congress, supporters of the soil conservation effort were getting weary of long, tough, often politically divisive fights each year just to salvage a budget that was admittedly inadequate. Worse than that were the criticisms that the program was ineffective. Was it true, as some argued, that the programs were simply starving to death for lack of money and thus had no chance to prove themselves? Or was the real truth that the programs were ineffective, poorly administered, and hopelessly outmoded? If that were the case, as often charged, providing more money was simply to waste more. Farm-state members of Congress and those on the agriculture committees knew from their own experience and the testimony of their constituents that major changes needed to be made, but they were having little luck convincing their non-farm colleagues. They simply lacked the facts to make a politically convincing argument.

Even more frustrating was the jousting that went on each year with the administration's Office of Management and Budget (OMB). The economists in OMB, convinced that the nation was spending money at cross-purposes when some farm programs held down production while others enhanced productivity, could see little value in soil conservation or farmland protection programs.

Those drove up productive capacity, which was, in times of farm surpluses, not only unneeded but unwelcome as well.

As a result, soil conservation advocates in Congress felt they were presented soil conservation budget proposals that were totally unrelated to the concerns and needs of the nation's farmers and farmlands. Short-term political considerations were, they felt, pushing aside longer-term resource management considerations. That point of view was reinforced by resource professionals as well as conservation interest groups and farm organizations.

In 1977, after an effort spanning two years of Congressional work and one veto by President Ford, Congress enacted the Soil and Water Resources Conservation Act of 1977. President Jimmy Carter signed the bill, soon to be nicknamed the Resources Conservation Act and shortened to the acronym *RCA* in the alphabet-soup world of Washington, on November 18, 1977. The act directs the Secretary of Agriculture to make an appraisal of the soil, water, and related resources of the nation. This investigation of the quantity and quality of the nation's soils and waters is to identify important trends in either their amount or productivity and to aid people at all levels of private and public effort to make decisions on resource issues. The act also directs USDA to propose a program to meet the needs that have been identified.

Washington lawyer Philip Glick, one of the original authors of the 1937 model soil conservation act and a long-time observer of the conservation scene, feels that two factors in the RCA law will strongly affect the nation's resource conservation effort. First, he points out that the term *resource base* is given a very broad legal definition, including "the conservation and use of soils; plants; woodlands; watershed protection and flood prevention; the conservation development, utilization, and disposal of water; animal husbandry; fish and wildlife management; recreation; community development; and related resource uses."[31] As a result, the mission of the department's soil conservation programs is no longer limited to more narrow considerations such as the prevention of soil erosion but is expected to address the entire broad range of resource issues. That is consistent with the way the programs had actually developed, but the RCA provides, for the first time, a single statutory base for the combination of all these resource concerns into one set of programs.

The second major virtue of the RCA, says Glick, is the fact that it tied the entire nationwide conservation effort directly into the budgeting and appropriations process.[32] By requiring the continuing appraisal of the resource base, the preparation of a periodically updated program for conserving, protecting, and enhancing those resources, and requiring the president to submit each annual budget with an assessment of how well the budget would meet the needs identified in the appraisal, the RCA could, Glick argues, dramatically change the way in which federal budgets would be viewed.

But there are difficulties in the RCA, as well. The fact that the law required a report and set a deadline did not guarantee that USDA could produce a good appraisal in the short time between November of 1977 and January of 1980, when the first report to Congress was due. Much important data was lacking, and there were few precedents for such a far-reaching review of the nation's resources. Then there was the budget problem. The fact that SCS and USDA would argue for higher conservation program budgets was a foregone conclusion with OMB budget examiners. They had heard all those stories before about how soil erosion and the loss of farmland threatened U.S. farm productivity, but they didn't believe them. The grain bins were bursting; that was the "farm problem." No amount of "proof" from USDA was going to change that opinion. Federal budget policy was to be set in the office of the president, and the idea of having to explain why the president's budget was not in accord with some resource study prepared by an action agency was not welcomed.

A chronicle of the process undertaken by USDA in trying to fill the mandate of the RCA would fill a book in itself and can be better done after a few years have elapsed. By the end of 1980 and the Carter administration, there had been no report to Congress, and it was not certain that one was in the offing. The appraisal, based on the 1977 National Resource Inventories, had been completed, and most of the information made available to the public. (It is the source of most of the data contained in this book.) The crisis that it revealed was beginning to be recognized by many people, both in Congress and around the nation. But a program, and particularly one that sets a budget level based on soil and water needs, is a different matter, particularly when a new administration is taking office with the avowed goal of cutting federal spending.

The real value of the RCA, I believe, is not in its effect upon the budget process, as its sponsors had hoped. Instead, the real value will come with the resource appraisals and the new information they provide the public. The 1980 appraisal, if properly used by USDA and others, should greatly raise the level of public knowledge about soil and water resources. It will provide the basis for significant Congressional action on new conservation programs, I feel confident.

By 1985, when the next RCA appraisal is due, several additional things could be possible. First, people will have a better
feeling for the accuracy of the 1980 trends and projections. Second,
the resource information should be better, as SCS learns from the
trial-and-error process that marked some of the first hurried efforts
in 1977-1979. Third, USDA should understand more about the
effectiveness of the various federal conservation programs and incentives by 1985, based, in part, at least, on attempts to respond to the
1980 program. If one thing was learned in the 1980 effort, it was
that physical facts alone were inadequate. It is one thing to know
that soil erosion is occurring; it is quite another to demonstrate that
a certain new type of federal program or a different funding level in
an old program would result in the needed improvements to that
situation. Finally, Congress will have had some time to study the
whole direction of the nation's conservation effort, as they consider
new programs suggested by the USDA staff work and public
response in the 1980 effort.

The RCA is landmark resource legislation, of that there is
little doubt. USDA's commitment to natural resource programs,
particularly soil conservation programs, has waxed and waned over
the years, with the level of policy and budget support at any one time
being in inverse proportion to the size of the commodity surplus at
the moment. The RCA has helped illustrate the short-sightedness of
such a policy, and that may help build the public support for a
longer-term view.

Basic Public
Policy Issues Remain

Despite 40-plus years of debate over soil conservation, there
are still some important issues that remain unsettled. Some of the
most troublesome "turf" issues over which agency will administer
USDA conservation programs may never go away. The Golden Rule
of Bureaucracy, it must be remembered, is *They That Manage the
Gold Make the Rules*, so these are not unimportant issues, particularly to federal bureaucrats. Any new program, no matter what its
legislative language or history, will be molded by the agency that is
charged with its administration.

It is likely, for example, that any new soil conservation or
farmland protection program assigned to SCS will have a strong
role for state and local governments. The same program, administered by ASCS, will use their politically appointed committees at

the state level and the farmer-elected committees at the local level to inject the necessary local management. If assigned to the Extension Service, the land grant colleges in the states will have a major policy-setting influence. The difference in the amount of control that can be exerted from Washington, D.C., over the program will be significant, so the political decision as to which USDA agency will run a new conservation program is really a decision on the balance of power between federal, state, and local governments. That is not a decision that will be made lightly, by either politicians or bureaucrats.

There is the troublesome issue of voluntary versus regulatory programs that has been cropping up in soil conservation legislation since 1936. The original soil conservation district law contained provisions for states to give districts the authority to enact land-use ordinances to regulate proper land use. Only about half the states would accept such a provision, and, in those where the authority was included in the law, it was virtually ignored by the conservation districts. This gave ammunition to a host of critics, with Aldo Leopold being among the most articulate: "We, the public, will furnish you free technical assistance and loan you specialized machinery, if you will write your own rules for land-use.... But, after a decade of operation, no county has yet written a single rule."[33]

Leopold knew, no doubt, that the political climate in his Wisconsin county was such that no land-use regulation had a chance of survival, as was true across the nation. There are still many areas today where farmers respond with indignation at any mention of land-use regulation. But there are many other areas where this is not the case. Some 20 states now regulate certain aspects of erosion and sediment control, and conservation districts are an important part of most of those programs. Iowa's program has a backdrop of regulatory authority, even though it relies heavily on the voluntary approach through education and persuasion. An Iowa farmer who refuses to control his soil erosion, however, can be forced to do so upon a complaint filed by his neighbor or the conservation district, providing public funds are available to help share in the cost of the necessary preventative work.

It matters a great deal *who* is doing the regulating and *how* it is done. In a series of field hearings held across Iowa in mid-1980, Senator Roger Jepsen asked virtually every witness the same question: "Do you feel that conservation practices should be voluntary or mandatory?" The answers came back virtually 100 percent: "voluntary." But several of the witnesses also talked about the value of the Iowa law. The implications are clear: Given the choice, everyone favors voluntary programs, except when someone else's negligence is causing them serious harm. If regulation is needed, most farmers prefer that it be administered locally and as a last resort.

The value of a regulatory program, ironically, lies not in how much it is used to prosecute wrong-doers, but in how seldom it must be invoked. As one Iowa witness pointed out, there is no way of measuring the effectiveness of the Iowa regulations in encouraging more people to control erosion voluntarily, although, from his observations around Iowa, he felt it might have "had quite a bit of effect."[34]

But the regulatory/mandatory issue will be debated at every turn in any new soil conservation program design. "If we don't do it voluntarily, 'they' will cram regulations down our throats," will be a major battle cry. In my view, there are far better reasons to conserve soil and water than to evade regulations, and there has been far too little effort to make the voluntary program coupled with incentives work to warrant abandoning it for lack of effectiveness. I see the threat of federal regulation of soil conservation as an idle and impractical one that inflames passions but does little to add to the search for effective federal efforts.

Research is another area that needs review. The original charter of the SCS included research—in fact, the first money appropriated to the activity in 1929 went largely for the establishment of ten erosion research stations. In 1952, however, in a massive reorganization, all soil and water conservation research was transferred to the Agricultural Research Service.

Today, soil and water conservation research is carried out by the State Agricultural Experiment Stations and the Agricultural Research Service. Economic and statistical research relating to soil and water conservation programs and their effectiveness is carried out by the Natural Resources Division of the Economic Research Service, largely with funds transferred from SCS for that purpose. USDA is charged with providing coordination to the whole spectrum of agricultural research activities, including the setting of priorities and insuring that high priority research concerns are properly funded.

Total expenditures for soil and water conservation research are difficult to identify with any certainty, since much of the research is done as part of broader research programs in agencies that are not basically oriented to soil and water conservation concerns. Because the SCS, as the action agency for soil and water conservation, has no authority for carrying out its own research program, there is little research done to determine the most effective ways of carrying out the federal soil and water conservation mandate. That is, I feel, a serious weakness.

Since SCS can only be effective by convincing private land-owners to voluntarily carry out conservation, this lack of a "tactical" research capability, plus the need to carry out more on-the-land testing of different conservation systems, are shortcomings that badly need resolution.

The most difficult issue to deal with over the past few decades is still alive and well, but if the facts assembled in this book are accurate, it will be less of a problem in the future. That is the issue of whether or not the nation should spend money to help farmers conserve soil when all that does is maintain or increase the productive capacity of agriculture. Perhaps one of the severest criticisms of the soil conservation program in the mid-1960's was Held and Clawson's book *Soil Conservation in Perspective.* They did not fault the programs for being ineffective in saving soil; they questioned the wisdom in spending federal funds to save soil at all! They argued that, since it was clear to one and all that there was not then, nor would there be in the foreseeable future, any hint of a resource shortage in agriculture, and since the major problem for farm policy was to *slow down production*, it made little sense to fund soil conservation efforts. It was, they said, "stepping on the gas and stepping on the brake simultaneously."[35]

This point of view is indicative of the basic economic theory that has dominated agricultural policymaking for three decades and still finds articulate expression in 1981. It has been used, success-fully, to hold down agricultural research budgets and soil conservation program budgets, and to kill attempts to protect prime farmland from conversion to other uses. It is a "surplus syndrome" that will die hard in a generation of economists, scientists, and agricultural policymakers who have spent their entire careers trying to cope with the vexing issues of overproduction and low prices in agriculture. But it will, in fact, die as the demand for agricultural products rises while the production-raising technologies and investments of the past begin to dwindle in effect, as it appears they must.

14.
A New Land Ethic

Americans need a new prescription for using and caring for their land. There is no longer any doubt that the methods now being used are too wasteful. Yet the knowledge and skills are available to help us to do better, and most farmers are closely attuned to a sophisticated information system that gets new knowledge to them almost instantly.

In short, if the concern for wasting farmland reaches the point where a wide segment of the American public feels that the current rate of land loss needs to be slowed; voices the opinion that it should be slowed; and provides the moral, financial, and program support to slow it; there is no reason why significant improvements cannot be made, and in a fairly short time.

The first step in achieving that public consensus for action is a major turnaround in the way that the public *thinks* about farmland.[1] From the time of the nation's founding, America has been land-rich. There was always another frontier, another technical breakthrough, another type of land that could be put to productive use with new techniques. Crops were in surplus supply more often than they were scarce, and the capacity of American farmers to overproduce their markets was assumed as fact by agricultural experts and ordinary citizens alike. Land was not limited, so why be concerned with saving it or protecting its productivity?

Compare that situation with what has existed for centuries in Holland or the Scandinavian countries. Those nations have never had an endless supply of land. But have they collapsed, in hopeless inability to feed themselves? Of course not. What they have done, and done well, is husband their limited land and cherish its productivity. Much of that has been due to the certain knowledge, on the part of every citizen of those nations, that land was a valuable resource, to be treated with care.

America's land is not limited to the same degree as the land of many other nations. Some countries hit their limits long ago, while we are just now approaching ours. We have millions of acres of potential cropland to be used when needed. But that does not mean that our land supply is endless, and if we continue to treat it as if it were, we will approach the outer limit of our productive capacity at such a rapid rate that we will find it difficult, if not impossible, to turn our society around and begin to treat our land with the care and stewardship it must have.

What is needed in America is a significant change in our *attitude* toward land. Much of what has been explored in the preceding chapters has told of the physical ills being inflicted on the land today, some of it due to negligence, but much of it due to misguided ideas about progress and growth. Many farmers today are seriously abusing their land, not because they want to, but because they are either unaware of the damage they are causing, or feel it is insignificant compared to the losses they might suffer if they changed their farming system. Often, they are being assured by scientists from agricultural industries, agencies, and colleges that what they are doing is real progress. The criteria, too often, is that if it is profitable, it must be right.

Part of this is due to the rapidity of agricultural change over the last decades. When a new technique is introduced, it only takes a few years to tell whether or not it will be profitable. It may take decades, however, to know the full implications to the soil. Intensive fertilization is a good example. Only a year's testing was needed to prove that it greatly increased yields. Only a few years were needed to breed plant varieties that thrived on higher and higher levels of plant nutrients. But what of the soil? Were the major nutrients being withdrawn at unusual or uneven rates? Were the natural biological processes being altered? Was the soil becoming "hooked" on continuous high rates of fertilizer application? These are questions that take many years to answer, and the answers may not always be the same on different soils.

American agriculture is not experimenting in test plots, however, it is experimenting on its land base. If the high-intensity fertilization practices of the past 30 years prove to have damaged the soil, virtually every acre of farmland in America will be affected. We adopted a new technology and applied it; now all we can do is hope that there are no adverse effects over the longer term.

The farmland losses we are suffering today will worsen other land concerns such as the loss of wetlands, the clearing of bottomland hardwoods, the deforestation of vast areas, and the increasing competition of energy-related production for land and water.[2] Wasting the good farmland we have today forces us to look to these other lands to meet our cropland needs, and the many values realized from them today will be lost as we convert them to food-producing uses.

In addition to these concerns about the physical capability of the land, we can point to other, metaphysical reasons why people are uneasy about what is happening to our land. Wendell Berry puts it eloquently:

> . . .we and our country create one another;. . .our land passes in and out of our bodies just as our bodies pass in and out of our land;. . .therefore, our culture must be our response to our place, our culture and our place are images of each other and inseparable from each other, and so neither can be better than the other.[3]

In short, what we do to the land, we do to ourselves.

When we speak of the land in these terms, which take us far beyond simple economics, we are aware that land means more to us than just acres, or tons of topsoil, or dollars. It is inextricably woven into the fabric of our hopes, our aspirations, our very lives. For example, ownership and management of land is part of the American dream; part of our national common heritage; our culture. There are few families in this nation who have not, at some time, attempted to take part in the common vision of land ownership. We share the view that "as many as possible should share in the ownership of land and thus be bound to it by economic interest, by the investment of love and work, by family loyalty, by memory and tradition."[4]

But we are also aware that this goal of land ownership is merely a dream for more and more Americans. The price of land has been rising at about 14 percent per year lately, roughly two and a half times as fast as the rate of inflation. It has become difficult, if not impossible, for new farmers to buy land, and the temptation to sell prime farmland for non-agricultural uses—at non-agricultural prices—is strong. As Charles Little has pointed out, "The capability of land to be sold for development *somewhere,* suggests to each and all that land can be sold for development *anywhere.* Therefore the prices go up *everywhere,* making the marketing of land for non-agricultural use almost mandatory."[5]

So we have a growing recognition that the accelerating conflict between the ever-increasing needs of our society and our shrinking land resources demands a new set of ideas as to how land should be used. These new ideas, in order to eventually evolve into a new land ethic, must be broadly shared by the people of the nation. It requires, as Christopher Derrick has said, "an extensive and effective change of heart and outlook."[6] That will require strong leadership, as well as skill in explaining the new land ethic so that people can understand and follow it.

Social Pressures to Conserve

The need to conserve land and use it wisely has been widely advocated by national leaders, particularly since the soil conservation movement in America began in the 1930's. Agriculture Secretary John Block picked up the theme immediately upon entering office in 1981. He pointed out to farmers that he is fully committed to protecting their freedom, their "right to farm," their property rights in their land, and their liberty. "But liberty," Block told a Chicago audience, "doesn't mean the freedom to squander resources and evade responsibility. It means taking on responsibility so that we have the freedom of making deliberate, conscious choices about our future."[7]

Thus, while Block and virtually all national politicians have no intention of trying to reduce farmers' freedom, they do intend to try to convince them of their responsibility for proper stewardship of the land. Stewardship is, however, more of a social responsibility than a government requirement. During his 1979 visit to the United States, Pope John Paul II told an Iowa audience:

> The land must be conserved with care since it is intended to be fruitful for generation upon generation.... You are stewards of some of the most important resources God has given to the world. Therefore conserve the land well, so that your children's children and generations after them will inherit an even richer land than was entrusted to you.[8]

With those kinds of stewardship admonitions, it would not matter whether or not there was a single public law encouraging soil conservation—farmers would still face the sanction of their neighbors for wantonly wasting land, and many farmers feel, I am convinced, a deep connection with the land and fully believe in the need to conserve it. What is needed is a set of goals that express the expectations of all of us—farmer and non-farmer alike—about how the land should be used and treated. This would be the new "land ethic."

As Aldo Leopold said:

> An ethic, ecologically, is a limitation on freedom of action in the struggle for existence. An ethic, philosophically, is a differentiation of social from antisocial conduct.[9]

The Piedmont Environmental Council of Virginia adopted the following definition:

> An ethic is ". . .a set of prescriptions and proscriptions grounded in fundamental judgements about what constitutes a good society. It expresses ideas of goodness, rightness, and obligation. . .An effective new ethic not only defines desirable ways of acting, but also persuades us to adopt an appropriate sense of responsibility and more adequate values.[10]

What could we hope to accomplish by the adoption of any new ethic? What should the ethic do for us? What are the major problems it should address, and what are the current prevailing attitudes it should change?

A new ethic should attack the notion that the value of land rests solely in its price on the market. We confuse price with value, and money with wealth, and therein lies much of our inability to think straight about the value of land. Land is more than a store of basic wealth—a quarry to be mined for the products needed for survival, comfort, or pleasure. It is the living, dynamic bridge whereon plants convert solar energy to human food. It spans the gap between death and life, between what was. . .and what is to be.[11]

When we measure the value of land solely by economic criteria, we find ourselves measuring the amount of profit—in dollar terms—that the land can return to its current tenant. If it is cheaper this year to import food from abroad than to grow it at home, or more profitable to build houses than to plant corn, short-term economics tells us to convert or abandon farmland. But we are not even sure that is a good economic decision if we are forced to face any kind of long-range economic analysis, and we are fairly certain it is a costly social decision, at least in most cases.

What *is* certain is that we cannot follow such a theory of land value to its logical conclusion. We can't make the greatest profit by destroying all the land. We can't ignore the needs for future people to harvest crops and enjoy a benign and productive environment by focusing entirely on this year's balance sheet.

It is clear that farmland values are not measured simply by price, and that there is a fundamental difference between money and wealth. But economic analysis does not measure—or even recognize—these differences.

One noted economist, the late E. F. Schumacher, was fairly harsh in his assessment of the state of economic theory in dealing with the value of land:

> Economics, as currently constituted and practiced, acts as a most effective barrier against the understanding of these problems, owing to its addiction to purely quantitative analysis and its timorous refusal to look into the real nature of things.[12]

A new ethic needs to overcome the idea that the rights and privileges of ownership include the right to destroy the land for immediate profit, or use it in such a way that its destruction is hastened or assured. Although farmers, in particular, are loath to think about it, the United States is in no short supply of social restrictions on the ability of a person to maximize profit from land.

If you are a farmer, *Cannabis,* not corn, may be the most profitable crop you could grow today, but you'd better not try it. Potential profit notwithstanding, society still frowns on people smoking marijuana, so makes it illegal for farmers to grow it. If society's views on pot smoking were different, the land-use options open to farmers would change.

Do farmers rebel at such restrictions on their right to make a profit? No, not really. They, too, subscribe to the common morality that underlies the limitations. But where is the moral principle that tells us that no one has the right to rape the landscape for a fast buck? We know Wayne Davis is right when he says, "The destruction of the land for personal financial profit is a behaviour that is simply not ethical and cannot long be tolerated," but when will we get up the gumption to say so?[13]

A new ethic needs to kill the notion that we owe the speculator—or the entrepreneur—his profits. Not that profits are bad, not at all. Our free enterprise system encourages them, but it does not guarantee them; and when the prevailing land-use game has become clearly tilted so that some people are reaping windfall profits while others are effectively prevented from profiting at all, there will be strong political pressures to change the rules of the game. "Scratch a farmer, and you'll find a speculator," says Hector McPherson, an Oregon farmer who knows the land-use game from all sides.[14] The reason is simple: The game is set up these days so that speculators make money; farmers don't.

A new ethic must contain some responsibilities for all of us, and the governments that represent us, as well as for the individuals who own land, and are thus able to exercise management control. Individuals use the land, and land-use decisions are essentially private decisions, but it is seldom that those private decisions are made in the absence of conditions imposed by the public.

Take the example of a farmer who knows that his land needs a year or two in clover or alfalfa to build up the soil and make it more resistant to erosion. He may forego that option, not because he is insensitive to the needs of the soil, but because he knows that the government may establish crop acreage limitations on his cash crops that are based on the cash crop history of his farm. If he happens to get "caught," as thousands of conservation-minded farmers have been, with a history based on years when a portion of his cropland was in soil-building crops, his ability to grow cash crops—and, perhaps, stay in business—will be limited so long as the government is in the business of controlling agricultural production.

So he makes a decision balanced between his desire to keep the land healthy and his need to avoid opening himself and his family up to economic penalties. His disinclination to make proper soil stewardship decisions is understandable; the government's policy of penalizing him for that stewardship is unconscionable. The new land ethic must be clear enough, and bold enough, to move those who represent us in Washington to the action needed to eliminate such policies.

The new ethic must state—in clear terms—that antisocial conduct in relation to the land is becoming a very real—and serious—threat to the continued existence of our society. Whether that antisocial conduct stems from private greed or public stupidity makes little, if any, difference.

The capability to produce ample supplies of food—far more than enough to feed our own people, has long been assumed by most Americans. But is it only Americans that we must consider?

The loss of 6 million acres of productive capacity a year equates to over 115,000 acres each week; over 16,000 acres every day! An acre of prime farmland in this country produces enough food to equal the caloric intake of 16 people at the accepted minimum of 2,000 calories per day.[15] If you multiply those numbers out, you get a truly alarming statistic. We may be losing enough potential productivity each day to produce a minimum diet for nearly a quarter of a million people. Many nations, where people starve when the weather turns bad, now look to us as a reliable source of food when they need one. That may be, in the near future, an unfounded hope.

How much suffering and economic stress will be needed before we move toward a new ethical basis for treating our land? How long will we, as noted environmental author Robert Cahn has said, "take for granted that the waters, the soils, the trees, and the wildlife can be manipulated at will in the name of progress, even if that progress is short-lived and far outweighed by the loss of the land's long-term ability to benefit mankind?"[16]

What is needed, clearly, is a new land ethic—an ethic forged of our twin concerns for the land's proper use and its proper care. We must begin to treasure the prime farmlands that have made us the world's richest nation, keep them available for agricultural use, help farmers survive economically and environmentally so that they can profitably produce from them, and insist that they be used in such a manner that soil depletion is minimized.

But the emergence of a land ethic strong enough to offset our current proclivity to measure in narrow economic terms will not come easily. As Rich Collins has pointed out, "An ethic powerful enough to alter patterns of land use must necessarily challenge other accepted habits and vested interests."[17] There are important values which must be confronted, including individualism, liberty, property, the pursuit of happiness, and the proper role of government.

It may be, as Collins goes on to say, that "the land, because it is so intrinsically important as a source of nuture for life and pastoral values, as well as being the basis for deeply rooted Jeffersonian ideals of individualism and human freedom, is the battleground for the nation's future."[18]

If land and land use are to be a battleground, then facts, trends, and ideas must be the major weapons. The new land ethic must be a product of education and social evolution. It can't be enacted as law, to be imposed on people against their will. Human behavior, in the final analysis, is more effectively governed by ideas and attitudes than by rules and regulations. Only when we change the way Americans *think* about land will we successfully alter the ways in which we *use* this vital resource.

A Land User's Ethic

In the work that has been done to set out a new ethic, it appears that most ethical statements are couched in terms of the private land owner's responsibility for stewardship. The Peidmont Environmental Council, in their outstanding work entitled "Toward a New Land Use Ethic," have decided that a new ethic should define the principles that "ought" to govern man's relationship to the land.[19] In a set of ten land ethic prescriptions that the author, Professor Graham Ashworth of Salford University in England, calls "you oughts," they have summarized what they feel a new land ethic should contain:

1. **You ought** to consider land as a resource that may be yours for a time but is also held in trust for the future. Land is not a commodity in the sense that any of us can own it in the ordinary sense of the word.

2. You may be a trustee of the land and that will often confer private benefits on you, but **you ought** not to seek benefits that incur disbenefits on the community or other individuals.

3. If you are presently trusted with the management of a piece of land, **you ought** to use it in a manner that benefits the land and does not damage it. Some land uses are abuses that have irreversible consequences, and **you ought** to avoid such abuses.

4. **You ought** to accept that the use of land should be subject to public scrutiny and control and to exercise your responsibility, with others, in ensuring that no use is permitted that damages society as a whole.

5. **You ought** to ensure that the land-use controls developed in your area prevent irreversible damage, avoid waste, protect your natural and cultural heritage, stimulate visual order, regulate and control the unsightly, and safeguard individual liberties (such as mobility and choice in housing and schooling) so long as those liberties do not curtail the liberties of others.

6. **You ought** to recognize that the exercise of land-use controls in the interest of the community can result in costs and benefits to individuals and be willing to see those costs and benefits equitably adjusted.

7. **You ought** to recognize that these controls can only be exercised democratically through governmental operations. Hence, **you ought** to expect an extension of government to give proper expression to this new land-use ethic.

8. **You ought** to accept that the administration of the ethic must reflect local circumstances and needs, so it will vary from place to place.

9. **You ought** to be ready to give time and talents to fight for this land-use control that is vital for your continued freedom.

10. **You ought** to recognize that you may have to make some sacrifices, along with everyone else, for this control to be effective.

But social desires for land use are expressed in many other ways besides ethical statements or, for that matter, land-use laws or regulations. In fact, governments (local, state, and federal) may do more to instruct people on the manner in which land should be used through tax laws, agricultural policies and programs, and national economic policies than through any set of instructions that relate directly to the land.

Continuous conservation treatment of private lands involves time and expense on the part of private land users. Part of this ought to be within the inherent obligations that attend the rights of private land ownership. But much of the expense borne by the land user produces benefits that aid only the general public and generates little or no revenue to help the land user install and maintain the conservation system.

The people of the United States must think carefully about the manner in which proper land use and conservation will actually be achieved. The job, if it is to be done, will be done by people, on land they own, rent, or manage, using information and technical skills at hand and the money, machines, and labor that they command, for reasons that make sense in their own private situations. Public policies and programs do *not* conserve land and water, but they can, if they are well designed and administered, help those people who will do so.[20]

This help involves many aspects. Public programs and public policy must help people understand the reasons for conservation so they believe it is necessary. Public policy must also create a climate of "permanence" so that individuals have confidence in the future of the land, of agriculture, and of themselves and their children. Only when there is such an aura of confidence in the future will people see the logic in protecting and conserving resources. Another major need is for technical assistance—on the land—so that specific problems can be identified and appropriate solutions developed.

Individuals must see that their private economic interests will not be threatened as a result of their conservation efforts. This requires agricultural policies that protect conservation managers from damage wrought by farm programs, as well as provide economic incentives for carrying out those conservation measures that cost the private producer money but provide mainly public benefits. People who feel that conservation will result in economic losses are understandably reluctant. Any situation that creates added economic stress in agriculture makes that situation worse.

Finally, individuals must see any public program as being "fair" and "acceptable" in light of their sense of proper public/private relationships. They will not accept programs which might cause an unacceptable loss of freedom or personal rights. Means and ends are inextricably interwoven, and any conservation program, to be effective in saving soil, conserving water, or preserving prime farmland, must also be effective in meeting current public expectations of fair play and proper governmental functions.

A Public Land Ethic

From the viewpoint of the private landowner, there are some "ethics" that should be observed by governments when they affect the use and management of private land. Using the same general framework of "you oughts," it may be possible to express some general terms of fairness that landowners might request. As we look further at specific public policies and programs that affect land use, we can test them by these measures of fairness.

1. **You ought** to recognize that land is an important part of our life, and that we have strong feelings about the rights and responsibilities that attend land ownership.

2. **You ought** to help us obtain the information, skills, and techniques we need in order to produce economically from this land while using it in such a way that it will not be damaged.

3. **You ought** to develop economic policies that inspire public confidence in the long-term future of agriculture and forestry as an economic activity as well as a way of life.

4. **You ought** to recognize that we can't change our land management plans from year to year as short-term economic situations change, and develop a set of public policies that will allow us to plan and carry out a proper land-use plan that does not have to be changed each year due to changing national production targets and economic policies.

5. **You ought** to administer public programs affecting land use in

such a way that my opportunity to profit from land value is the same as everyone else's opportunity.

6. **You ought** to eliminate economic penalties that will fall on us as a result of our good land stewardship efforts, and instead see that good stewards of the land are rewarded by the economic and political system.

7. **You ought** to create a climate of laws and administrative methods that help us feel secure in our tenure on the land so that we can realistically plan on staying here as long as we want and passing ownership to our children if they wish to follow in our footsteps.

8. **You ought** to help us protect our property, along with its plants and animals, from harm imposed either intentionally or unintentionally by other people. Whether it is vandals dumping trash, shooting livestock, or tearing down fences; or cities and industries polluting our water supply, allowing hazardous wastes to escape, or sending clouds of pollutants aloft to rain down acid on our land and crops, we need your help in protecting our land and ourselves from harm.

This list of "you oughts" focuses the ethical discussion back at the 97 percent of all Americans who are not active farmers and who might be otherwise tempted to talk about a land ethic as a simple statement of the farmer's responsibility. There is a public responsibility as well, and it must be expressed through governmental action, as part of any new ethical dimension on the use of land.

There must be a new agreement, or social contract, between the society at large and the farmer. Only then will the public be able to encourage farmers to begin taking the individual, private, everyday actions that, taken together, become the solution to our basic problems.

Society gives the landowner the right and privilege of owning land. As part of that right, we grant the owner the unfettered use of that land within moral limits that meet the test of social and political agreement. Those moral limits should not include the right to destroy the land or steal its value from current or future generations for the sole reason of immediate financial profit.

Most landowners accept these limitations as being reasonable, and in so doing, implicitly accept the fact that society has a continuing interest in how they use their land and that government, as the representative of the people, has the right to develop and administer rules to assure that society's interests are maintained.

In return, society must grant the landowner security in his land tenure, assist him in preventing damage to his land caused by other people, create a political and economic climate that encourages and rewards stewardship, and develop and distribute information that helps him conduct his business and lifestyle in a manner that is both economically rewarding and environmentally sound.

This kind of a new statement of ethical behaviour and responsibility on the part of *both* the private land user and the public policies and actions that affect his ability to use the land is a vitally important and timely national need.

The nation has time for action, but not time to waste. The data on land and water resources indicates that the growing pressure on the land of the United States is a serious concern, not an imminent crisis. But the clock is running. The situations that concern us about the land are not going to go away or cure themselves. The forces cutting into America's land and its productivity have a great deal of vitality and inertial force. They are serious now and are certain to get more so. Solutions open to us today will be closed soon, and we have little time to waste. Americans are a people who fear rash or harsh governmental action, particularly in regard to the way government regulates or guides land use, but, by their very inaction, they seriously increase the odds that modest options are foreclosed and heavy-handed methods made more necessary.[21]

No longer is the concern for proper land use and conservation limited to those few dedicated people who have championed it primarily as a moral, ethical, or environmental cause. In the new situations facing our nation, the protection and continued productivity of agricultural lands is an economic imperative vital to the well being of all. As that recognition strikes home, we must have the kind of broad public attention that is essential to any debate on this subject.

The level of the conservation ethic demonstrated by America's land users is too low and getting lower. Too many people are using the land for short-term profit. Too few are using it as though they and their children and grandchildren will need its bountiful production in the future. But this loss of the conservation ethic is not simply the result of a reduction in personal pride or morality. It is also the result of three decades of national policies that have encouraged people to leave farms and rural areas, encouraged the wasting of land, and created an "impermanence syndrome" among farmers, ranchers, and other land users that encourages them to deal with the land as a short-term asset rather than a timeless resource.

In the coming decade, this is one of the major challenges facing America. For it is not just farmers—or ranchers—or soil conservationists—that are damaged when the land is wasted. Every American suffers in that process. We must halt that damage by giving conservation a far higher priority, both in our private and public actions.

15.

New Approaches to Action

The United States needs to devise a system that encourages farmers to protect their topsoil from erosion and save their prime farmlands for future cropping. That action is not only vital, but urgent, in terms of the time remaining to begin the process. It must be done, and soon. It will take a new commitment to private, individual actions as well as a new approach to public policy. Both will demand the effort and involvement of a broad array of citizens.

Outlining what such a new system must encompass is difficult and also a little presumptuous. No one knows what future pressures on farmers will be, and any system of incentives must be relevant to those future pressures, not just the ones of today. When we design a new public program today, it is with certain knowledge that the world in which it functions will be a different world from the immediate past which we used as the pattern for design. It is also a virtual certainty that it will not work the way its designers intended. By the time it passes through many compromises in the Congress or state legislatures, then gets handed to an agency where a bevy of bureaucrats mold it into what they wanted to do all along, and finally is made available to farmers who respond according to what they think is most advantageous for them, the final result is seldom recognizable in terms of the original design.

Those factors beg for programs that are designed with flexibility and latitude for change as time passes. If they are to be helpful in the wide variety of land and agricultural situations found in

America, they must also be highly flexible for local adaptation. With those caveats, we can look at some general directions that new programs should take.

Stability for Farmers

The keystone of farm policy is that somehow, through a combination of market and public policy forces, farmers must be assured more stability in their lifestyle and economic future. If that is not done, the growing sense of insecurity and impermanence on the farm will destroy any chance for development of a new land ethic that results in better care and stewardship of the land. People who feel they have little future see little value in investing for the future, and that is the most deadly ailment that could possibly affect us.

This is a hard issue to address, however, because it is perhaps the key challenge for public policy today. The Reagan administration's economic recovery program is designed to help *all* Americans, farmers included, feel more positive about the future. Efforts to stop inflation are the key to a future that we can afford; one where our economic system has the chance to work.

Most of the standard farm policy issues that deal with commodities, overseas sales, and management of domestic grain stocks address this matter of stability in agriculture through attempts to hold farm prices relatively stable and provide farmers with supports that enable them to ride out periods of low prices. Therefore, if we want a stable farm future, it is not just land policy or conservation policy that must be considered, but the entire spectrum of national economic policy.

Yet to farmers, security is most deeply involved with the land. Many things can go wrong, but if the land is safe and the farmer has full control over it, he will feel secure. This is a major reason why farmers have generally rejected government land-use policies and plans. It is not that such plans would not help them— often they would, and the farmers know it. But if, in the process, the plan restricts the farmer's freedom to use his land as he sees fit (which it must nearly always do if it is to accomplish anything), many farmers question whether the benefits are worth the price. Given the choice, most farmers would choose to retain full and unfettered use of their land rather than accept any government-designed land-use plan, regardless of the level of government involved or the clear benefits to the farmers themselves of cooperating with the plan.

Those feelings affect how government must design policies and programs to guide how farmers will use and treat land. It does not make these policies impossible to design, but it makes them different from urban land-use policies.

Farmers must be convinced that government land policies are designed to increase the individual's control over his own land rather than take it away. Agricultural districts, formed and operated by farmers to assure that their right to farm their land is not unreasonably restricted in the future, are one good example of a supportive program that finds acceptance. Districts like that have been formed in New York, Virginia, and other states, all supported by the farmers themselves.

Farmers resent regulation but respond readily to social pressure. They would vigorously oppose any law that would stop soil erosion by making it a crime to allow excessive soil to leave the farm, but if their neighbors are convinced that it is wrong for a person to abuse his land by letting it erode excessively, most farmers will readily do what is needed to comply with the community morality. That makes information and education a powerful tool for accomplishing public policy, if it is properly directed.

If offered an economic bonus for practicing soil conservation, farmers will weigh the benefits and costs and take the bonus if they deem it fair. To obtain that bonus, they may sign a contract that limits what they can do with the land for at least a few years, as they have gladly done in programs like the Great Plains Conservation Program and the Water Bank. The key to those programs was that they were voluntary—the farmer was not forced to enter the program unless he wanted to do so. Once he entered, following the rules was not a problem. The number of defaulted contracts under the Great Plains Program, for instance, have been very few.

Reaching Serious Problems

What the past programs have not been able to accomplish, however, is to reach the farmer who has really serious soil erosion. Under the Agricultural Conservation Program (ACP), a 1980 evaluation found that more than 52 percent of the cost-shared erosion control practices installed under the program had been placed on lands eroding at rates of less than 5 tons per acre per year.[1] That is not a surprising statistic in light of the program's incentives.

Erosion control on a Class II soil that needs very little

treatment is going to be less expensive than on a Class IV soil that is steep and subject to severe erosion. The farmer will also get higher returns from the better lands. Since the ACP limits the amount of federal cost-share any farmer can receive in a year to $3,500, he can get many more acres treated with the available money on Class II land than on Class IV. In addition, farmers whose land is in good condition are usually more committed to conservation, leading them to take more steps to save soil than their neighbors who care less.

Thus, a totally voluntary program of economic incentives will always find more takers among better farmers and always be addressed to the better lands.

That brings two challenges to conservation programs: how to target efforts toward the soils that are eroding the fastest and need the most assistance; and what to do with the lands that should be converted completely out of cropland into grass or trees.

Targeting is a challenge that has been studied in USDA intensively over recent years. Several suggestions have emerged as a result. Dr. Charles Benbrook of the Council on Environmental Quality has suggested that Soil Conservation Service (SCS) inventory data is now adequate and accurate enough to be used as the basis for locating the soils within a county that are more erosion-prone.[2] That is no doubt true. What is more difficult, however, is to convince the landowners who farm the eroding soils to change their ways and turn to conservation management.

One way might be to have a sliding scale of cost-sharing for conservation practices. Under such a scheme, a farmer who entered into a long-term contract to install and maintain a conservation system on his land would receive, for example, 50 percent cost-sharing on the land that has an erosion potential of under 5 tons per acre per year (t/a/y), but 75 percent on the land that may erode at rates of 5-14 t/a/y.

Such a system of incentives would have to be carefully structured, however, so that the public did not simply end up subsidizing the cultivation of marginal lands through the guarantee of higher cost-sharing rates. One possible way would be to disallow cost-sharing of any kind on lands that suffer excessively high erosion rates and require that those lands be retired to a more conserving use as part of the overall plan that would qualify the remainder of the farm for participation in the cost-sharing program.

Where entire farms, or regions, are so erosion- or drought-prone that cropping is not really a wise idea, the options become much more difficult. In southeastern Colorado, for example, large portions of some counties are so marginal for cropping that any requirement for proper land use would drastically affect the local economy. That would require a difficult adjustment as area farmers shift from grain to livestock, but it looks like an adjustment that must somehow be made. We might make it with a government program, one that tries to ease the adjustment problems, or we can let the wind and weather force the necessary changes. If those soils keep blowing away, change is only a matter of time. My view is that the public should step in now, while some of the more serious adjustments can be mitigated and while the topsoils are still worth saving. If the soils turn to desert, the battle will be lost.

From 1935 to 1946, a federal Land Utilization Program bought up over 11 million acres of worn-out cropland, rehabilitated the land with grass, trees, and other conservation treatment, and converted it into recreation areas, parks, forests, or grazing lands.[3] The program was carried out by five federal agencies, with the SCS responsible for some 7 million acres. Those lands, in a 1953 reorganization, were turned over to the Forest Service for continued management. Today, most of them are managed by the Forest Service as the National Grasslands.

It is doubtful that such a program is possible today, at least not without major changes. People no longer support major transfers of private land into public ownership. One-third of the nation is public land already, and with local government so dependent on property tax revenues, there is little support for removing more land from the local tax rolls. In addition, public support for resettlement of marginal farm families at public expense would not seem easy to obtain.

There still seems to be a potential, however, for using government as an intermediate land-holder to rehabilitate eroded and wasted soils. One way would be for the federal government to fund state and local programs that would purchase the land and assemble large enough tracts to make the effort worthwhile. The land purchased would be limited to land that was marginal or unsuited for cropland by virtue of soil conditions.

The land would be seeded to grass or trees, held until the vegetation was established, then resold at auction to private owners. At the time of its resale, the state or local government could restrict the deed to the property so that the buyer would not have the right to use the land for cultivated cropland, develop it for urban use, or convert it out of range or forest use. The purchaser, fully aware that he was buying property with a restricted deed, would bid his price accordingly and establish his expectations for the land on the basis of what was allowable. Such a scheme is little different from buying and selling land with the mineral rights withheld. People do it all the time, knowing that mining is not one of the options they can consider on that land.

The amount that the public paid for the property, plus the cost of restoring proper cover, minus the receipts from the eventual sale, would constitute the public's cost of getting the land back into productive, healthy condition. Such a cost might be significantly lower than the ultimate costs of letting the land continue to erode until it is ruined, then paying the social and environmental penalties involved.

If the public stepped in with such a program, farmers on marginal lands could realize a fair price for their land that might help them make the transition to better farmlands or other vocations. The transition to a ranching economy would be aided by making established grasslands available for sale at prices commensurate with their income potential, which would help ranchers become established or allow them to expand their units to an economically feasible size. Such a program would not be easy, and there has been little or no discussion of such an effort at the national level to date. But if the marginal and unsuitable croplands of the nation are ever to be retired from cultivation permanently, I think we will need to consider such an approach seriously.

Finding New Incentives

The major productivity loss, however, is not from the marginal or unsuitable croplands. They may be experiencing the bulk of the soil erosion, but they are not losing the most productivity. The most harmful losses are from the Class II and Class III croplands that are suffering erosion rates in the 3-15 t/a/y range. These small losses, affecting millions and millions of our best crop acres, are slowly robbing the productivity from the nation's breadbasket soils. Finding a set of incentives that will apply to farmers on those croplands may be more important, in the long-term national interest, than stopping the more visible and heavy erosion damage on a few million acres of steep soils.

Reducing the amount of topsoil being wasted from the nation's best croplands will require that new programs—no matter where they are designed or implemented—address the real causes behind those losses. There is too little time to waste on ineffective efforts, and the American public is out of patience with government programs that cost too much for what they accomplish.

As a first step, the nation as a whole must be convinced that the current waste of farmland is a serious concern that must be addressed. Farmers may not realize the severity of the losses they are suffering on their own land, and that calls for a more intensive information and education program. Such a program should also educate the non-farmers of the nation, because it is *everybody's* future that is washing and blowing away when soil erosion steals topsoil.

While this is the same kind of public awareness campaign that marked the early days of the soil conservation program, there are significant differences. The waste of farmland today is not happening because farmers do not know how to stop it. If that were the case—as it was in much of the nation during the 1930's—research and education would be a larger part of the answer. If a modern farmer wants to slow the soil erosion on his land, even to the point where virtually no topsoil ever leaves his property, it can be done in nearly every instance. There are management methods, plant varieties, and soil conservation practices that have been adapted to virtually every soil situation. The methods for conserving soil are well known to soil conservationists who are stationed in virtually every county and who will come to the farmer's land on request and help design a workable conservation system.

That assistance would not be readily available if every farmer suddenly decided to install a conservation system, because there is only about one SCS technician (or technician hired by a state or conservation district) per 200 farmers on the average. But so long as there is no mad rush to the conservation district's doorstep, it is still possible for a farmer to get the information he needs to plan and install soil conservation methods on his land.

If soil conservation *can* be done, but *isn't*, then it seems safe to assume that farmers are not finding it feasible to do it from an economic standpoint. Several reasons for this have been explored in earlier chapters, and a different type of solution may need to be developed to meet each of the disincentives that now discourage farmers from practicing conservation.

Removing Government Penalties

One of the most logical ways to increase the willingness of farmers to practice soil conservation is for USDA to stop punishing those who voluntarily do so. Despite the apparent simplicity of such a step, however, it has eluded conservationists for decades. It is time for Congress to *insist* that it be done, because it isn't really that hard.

Commodity stabilization programs that regulate cropland acreage should establish a category of cropland called "voluntarily diverted cropland" that would recognize any land that the farmer had converted from crops to grass or trees within the past five years. From thenceforth, that land should be a recognized part of the farm's cropland base in the event that commodity circumstances called for a reduced planting of certain crops. So long as the farmer does not convert it back into cultivated crops, he should be allowed to count it against any new acreage diversion requirement imposed by USDA.

The effect of such a move would be to lower the actual acres diverted in any one year in relation to USDA's targets. If, for example, USDA estimated that wheat acreage needed to be reduced by 10 percent, the existence of some land that could be credited as already diverted would mean that farmers could qualify acres that had not actually been in wheat the year before. In order to offset that, USDA would have to target a slightly higher diversion. If they need to lower plantings by 10 percent, they might require wheat farmers to carry out a 12 percent diversion.

The impact would be to make farmers who had never voluntarily diverted cropland to carry out a full 12 percent reduction, while a conservation farmer who had already diverted 5 percent of his cropland would only need to make an additional 7 percent cutback. The conservation farmer would be rewarded for having already removed his most erosion-prone land from cultivation instead of being punished, a historic reversal.

To keep such a program from being abused by farmers who might plow out land for a year or two in order to qualify it as cropland, then reseed it to grass to claim the diversion credit, USDA should, from the time the new policy went into effect, only recognize soils that are Class IV or better as cropland.

A closely related reform—one that would no doubt be controversial—would be for USDA henceforth not to recognize as cropland *any* new lands that do not meet the criteria for Capability Classes I through IV. In addition, programs such as crop insurance and disaster relief should not be available on lands that do not qualify as Class IV or better, unless the farmer can demonstrate that he has applied an effective conservation system on the land.

That such changes would anger farmers who have such lands, and who have used these programs in the past, is certain. What must be considered, however, is how many other farmers are facing unfair competition sponsored by the government when federal insurance and drought programs make it feasible to risk growing crops on land that really does not have the soil and climate conditions that would warrant cropping without such backup support. It is clearly essential that federal programs stop promoting the cultivation of unsuited soils, for the public is paying two prices in that process: one in terms of the cost of the subsidy and the second in terms of topsoil loss, water and air pollution, and all the other costs of soil erosion.

Sharing the Costs

Soil conservation management often results in significant costs to the farmer but seldom returns cash benefits to him in the short run. The costs he incurs are of several types, and each may need to be addressed differently in public programs.

First, he may need to construct specific soil conservation practices to enable him to reduce soil erosion to tolerable limits. Such practices might be expensive to install and pay back little or nothing in terms of crop production. In these circumstances, public programs that share a portion of the cost, based on the ratio of public versus private benefits, are needed. Such programs have a long history—the Agricultural Conservation Program (ACP) is the oldest, largest, and best-known—and there are several other USDA and state programs that provide similar benefits. All together, however, they are not funded at levels that effectively encourage conservation. ACP, for example, is $190 million in 1981, and President Reagan's budget proposal would drop it to $150 million in 1982. It has been near $200 million for over a decade, during which time the effect of inflation has cut the amount of conservation work it can buy in half.

There are some who feel that ACP funding has sunk so low that the program deters soil conservation as much as it encourages it. In most counties, the funds are limited and used up rapidly. Farmers who need to install conservation practices decide that, with the local cost-sharing fund depleted for the year, they will wait until next year when cost-sharing money is available. Unfortunately, next year may be no better, and in the meantime, the soil is wasting away and costs are rising. The expectation that there *will* be public cost-sharing money for conservation work leaves many farmers an "out" when that money is not available, which is more and more often as the federal appropriations shrink.

If this problem is significant, two solutions exist: Either abandon the program and kill the expectations that public cost-sharing will be available, or fund it properly. Farmers, farm groups, and conservationists would generally support the latter course of action, largely on the basis that the ratio between public and private benefits from soil conservation measures requires that the public share in the costs in order for farmers to justify the investments.

The second kind of expense facing the farmer is the additional operating costs he may face with certain types of conservation systems. Contour farming, for example, may cause more machine turning and wasted time as the farmer winds around a complex topographic pattern rather than keeping his rows straight. Terraces on the land may cause additional machine time to be needed on a field, in addition to periodic maintenance work. Some farmers estimate that their additional operating expenses may be in the range of 10-20 percent as a result of the conservation system on their land.

The third type of costs facing a conservation farmer may be lost income. If a steep hillside is producing wheat, but losing soil, seeding it down to grass will reduce both the soil erosion and, often, the income from the land. If organic matter or other topsoil quality is going down because a field has been in row crop for too many years without rest, seeding it to grass or legumes for two-three years may improve the topsoil, but it will also cut income drastically. Too many farmers are operating on financial margins far too narrow to consider such income loss, so the steep hillsides stay in wheat, and the row crop field gets still another year of intensive cultivation.

These last two kinds of costs are not easily paid by public cost-sharing programs. In general, the only way a public program can fairly share a cost with the farmer is for the farmer to present proof that the cost was actually incurred. The easiest example occurs when a farmer builds a conservation practice by hiring a contractor to do the work. The contractor's bill is paid, then presented to USDA as proof of the cost incurred. With additional machine time or lost income, however, there is little way to demonstrate the extent of the cost incurred. Farmers can use up too much machine time just by wasting it, and to predict that yields—or income—would have been at a certain level on a certain field if it had remained in crops is highly speculative.

For these reasons, I feel there must be another type of economic incentive program available to farmers who are willing to install and maintain a conservation system on their land. USDA could, I believe, offer conservation incentives in a wide variety of programs that affect a farmer's finances. These could be in the form of either added benefits or reduced costs to the farmer.

In recent months, there have been at least two proposals as

to how such a system of conservation incentives might be structured. One of these is the cross-compliance concept. The other is the Green Ticket (or conservation incentives) concept. Both of these ideas are based on the same general philosophy—farmers who carry out conservation programs should be rewarded in some way by the public programs that are available to them. That way, a whole array of public incentives could be aimed at encouraging farmers to carry out soil and water conservation.

Even though the basic idea is the same, the mechanisms would be totally opposite. The cross-compliance method would essentially penalize farmers who did not cooperate, while the conservation incentives program would reward those who did. The economic impact would be roughly the same, but the political impact would be far different. It is arguable which would be better; my feeling is that the incentives approach would be superior because it would be more acceptable to farmers and ranchers from a philosophical point of view. Since it is farmers, not USDA programs, who put conservation on the land, the first requirement of any program is that the farmers should feel positive about it and willing to cooperate in it. If you don't have that, any conservation program is doomed. But there are people who argue stongly for a cross-compliance program, and it could become part of some future farm program.

Cross-Compliance

Several authors over recent years have proposed that farmers should be required to have a conservation management system on their land to be eligible to participate in federal agricultural programs.[4] Under this approach, a farmer would not be eligible for crop insurance, commodity payments, farmland tax assessment benefits, and so forth unless he had received his annual conservation certification. It would make soil conservation mandatory for participation in USDA programs.

There was a loud and sustained howl of protest when this idea was included in the Resources Conservation Act (RCA) program proposals circulated for public comment in 1980. The opposition was based on several arguments. First, such a scheme would not necessarily assure that conservation systems would be required on the farms with the worst soil erosion problems. The only farms that would be clearly affected would be those that, for one reason or another, participate in federal farm programs. Those are not always the farms with conservation problems—or with owners who don't care about the land.

Second, farm programs such as commodity programs are not in effect every year, and in those years when there were no acreage allotments or price supports, the conservation cross-compliance would be meaningless. Such a situation seems to be facing us in the next few years, and if there are no income-support programs in effect, it is of little value to talk about requiring farmers to comply with anything in order to qualify for them.

There is also the matter of practicality. If national farm support programs were to be needed in the future and the law required all farmers who wished to participate to install conservation systems, the capability of conservation districts, SCS, and others to help people install conservation work would be swamped. The time between imposition of the requirement and the first year of the program would be far too brief. Pleas for technical assistance would go unanswered, and farmers and ranchers would have legitimate reasons to call the requirements unreasonable and get them removed. Little would be gained, except the further alienation of rural people against government and its ability to cope with problems.

But there are arguments in favor of the cross-compliance idea, as well. Tom Barlow of the Natural Resources Defense Council has pointed out that farmers are not forced to participate in any USDA program, but when they do, they should be required to agree to protect their soil. "If a farmer abuses the land, the federal government should not be providing the assistance year after year to help continue this wasting," he wrote in 1980.[5]

Barlow also noted that farmers who install conservation measures incur a cost level, while those who abuse the soil avoid conservation costs. That makes unfair competition, and every bushel of grain that enters the market as the result of land abuse tends to drag prices down. If all farmers were required to prevent soil erosion, competition would be more fair, Barlow argued.[6]

Finally, there is some evidence that farmers themselves are beginning to accept the idea of cross-compliance. A recent study conducted by agricultural economists in the land grant colleges in ten states recently asked almost 5,000 farmers about their views on the matter. The statement they were given was: "To help achieve national and state soil erosion control goals, each farmer should be required to follow recommended soil conservation measures for his farm to qualify for price and income support programs."[7]

In three states, a majority of the farmers either strongly agreed or agreed with the statement, while in four more, the percentage who either strongly agreed or agreed was larger than the percentage who either disagreed or strongly disagreed. In one state, the percentages came out equal, while in two states, the percent in opposition was larger than that in support.

Given the rather blunt and brief nature of the question, the findings are somewhat surprising. Farmers generally criticized the cross-compliance notion in the review of the RCA, and farm organizations have been almost unanimous in their rejection. If there is now beginning to be support for the idea, that support may be relatively new and could mean that the farm community itself is beginning to accept the idea of tying federal program benefits to proper soil conservation on the farm.

A Green Ticket Program for Conservation Incentives

There is one basic question with any kind of program that would provide public benefits to private citizens: Who deserves it? Who is spending the time and effort to accomplish what the public wishes to see done and, therefore, deserves special treatment? The Green Ticket program was originally named because it would be a way to certify those farmers who are installing, managing, and maintaining conservation systems on their land. Each farmer that carried out a satisfactory conservation program would earn a Green Ticket that could be used to apply for special benefits under USDA's farm programs. Thus, qualifying farmers would be identified locally, in a logical and time-proven method of operation between farmers and local governments.

The district would enter into an agreement with each farmer, based on a plan developed for his farm, ranch, or forest management unit. The plan would set forth the conservation practices to be installed, the schedule of installation, the regular management practices to be carried out, and any harmful practices that would be avoided. By voluntarily signing the agreement, the farmer would certify his belief that the conservation system proposed was both feasible and reasonable, as well as his willingness to carry it out and maintain it.

The district could seek the technical certification of SCS (or other agencies for forest management, wildlife habitat, or factors calling for other skills) to assure that the agreement would result in a technically sound conservation system. Once accepted by the district, the agreement would form the basis for issuing an annual conservation certificate, or Green Ticket, to the land user. Periodic reviews on the land would assure that the farmer was keeping his end of the bargain.

Under such a system, the district would base its decision to issue or not to issue a Green Ticket on the single question: Is the farmer keeping the conservation agreement or isn't he? This would help reduce the chance for arbitrary or capricious decisions. Since districts operate under state law, there would always be an appeal recourse to the state soil conservation agency for farmers who felt that the decision made in their case was unfair.

Basically, the Green Ticket idea calls for a more active relationship between the people who work the land on a daily basis and the local conservation district. It would involve more service to people on the land, so would require the replacement of local conservation technicians that have been lost as federal conservation program budgets have shrunk in recent years.

While this may sound simple, it is not. Many conservation districts serve over a thousand farmers, and working out agreements with any significant percentage of them will take a great deal of time and technical manpower. Most of the agreements will need to be based on a new conservation plan, developed with the aid of a trained soil conservationist. Many plans are already in existence, but most would have to be updated to be adequate. Developing these plans takes more than money or people—it takes skilled people, and those are not developed overnight.

There are many possible economic and social benefits that could be made available to Green Ticket holders, depending on the willingness of local, state, and federal governments to provide incentives. At the very least, public recognition within their community would help develop individual pride in those conservation farmers who were Green Ticket holders. By issuing an annual Green Ticket, conservation districts would be reinforcing the land user's commitment to conservation, as well as keeping him aware that the local district maintains a continuing interest in his personal conservation progress.

Economic incentives through programs other than USDA farm programs could also strengthen the Green Ticket program. Local and state legislation could give Green Ticket holders a break on their real estate taxes on the basis that the conservation farmer is less of a tax burden upon the community than those who are not following such a system. Likewise, there could be state and federal income tax credits to Green Ticket holders similar to the credit that is now available to people who install energy conservation measures in their homes.

Crop insurance premiums could be reduced, and Farmer's

Home Administration and private lending agencies could lower interest rates on loans to Green Ticket borrowers, since the conservation system provides assurance of continued levels of productivity.

The Green Ticket could also certify that the farmer is meeting the requirements of certain state and federal laws. The Federal Water Pollution Control Law, Section 208, is one example. By certifying that a farmer was meeting the water pollution prevention requirements, it would guarantee him freedom from regulatory hassles that might otherwise develop, and this would be a significant benefit to his peace of mind, as well as his pocketbook.

All of these benefits added together could be the total package offered the conservation farmer. None of them alone might be enough incentive to encourage the necessary soil conservation investments, but taken together they make soil conservation an attractive alternative from the farmer's point of view. For those farmers who have a sense of ethical responsibility to the land, but who are prevented from applying the needed level of soil conservation management by financial pressures, such a program would certainly make a difference. For farmers who don't care about the land, a more coercive approach may be necessary. It is my opinion, however, that the number of farmers who don't care about their land is relatively insignificant. In my contacts with farmers all over America, I remain convinced that the major problem is economic pressure, coupled with a lack of consistent technical advice about the probable result of their current management system on the land.

Better Technical Advice

Farmers get more information, about more subjects, than most non-farmers can possibly imagine. Information flows in a steady stream from land grant colleges through Extension Service programs, brochures, and stories in the farm press. Results from federal and state research stations and from the more than 1,000 economists employed by USDA reach farmers in weekly bulletins as well as in farm press reports.

Added to that are private information sources including equipment manufacturers, fertilizer and pesticide manufacturers and dealers, seed companies, and a whole host of similar agribusinesses. Then there are the farm organizations, cooperatives, farm management services, and the financial institutions that provide credit to farmers. Finally, there are a host of onlookers, from environmentalists to consumerists and public policy analysts that are eager to tell the farmer what he ought to be doing differently.

In the midst of all that noise, soil conservation gets mentioned once in a while. But in relationship to the total information load, soil conservation doesn't get much attention. Instead, many of the recommendations to a farmer encourage him to establish soil-depleting practices. He is sold machines, fertilizers, and irrigation equipment without regard to whether his land is capable of sustaining intensive production or not. He is tempted by advertising that shows how some farmers make a profit doing things a certain way, yet often does not have enough information to know for sure that it will work under his conditions. When dealing with a salesman, he is almost always assured that the salesman's products work perfectly and profitably.

This confusion can't be avoided, and many of the conflicting ideas a farmer must weigh are an essential part of the free enterprise system which has many strengths, not the least of which is the encouragement of innovation and competition. There are some steps that could be taken in the public sector, however, to give the farmer a better balance of information so that he can make his choices with better knowledge.

First, public research and extension should be aimed almost exclusively at determining the long-term effects of agricultural methods. Private research does a fine job of evaluating the potential profitability of various methods and materials and developing products for farmers to use. Public research should provide the countering information to help farmers spot the weaknesses and pitfalls to avoid.

As an example, there has been a great deal of innovation in earth-moving equipment and methods to cut the time and cost of constructing terraces. Most of that research and development has been done by contractors and equipment manufacturers, as it should be. Their discoveries have improved competition and lowered prices to farmers. Those private businesses have little incentive, however, to develop improved terrace designs which lengthen the life of a terrace, make it function more effectively, or keep maintenance costs down. Any improvements along those lines would be difficult to sell commercially and would not be likely to result in either lower installation costs to the farmers or higher profits to the contractors. Those kinds of innovations, which have little commercial promise but great social value, should be the target of public research, which can afford to spend time looking for new ideas that benefit all terrace users and make their conservation systems more effective.

In order to redirect their priorities to such needs, however, researchers must first convince Congress and state legislatures of the absolute need to be financially independent of commercial funding in their soil and water research programs. In the past, much of our agricultural research capability has been purchased at bargain prices by commercial and commodity interests. By funding a small share of the total research effort, private interests were able to influence the priorities on many of the research stations. Those priorities were nearly always aimed at short-term production goals, such as higher-yielding or better-milling varieties of grain. What was forgotten, too often, was the implications of the new technology on long-term soil productivity. It is easy to understand why the companies and commodity groups want this type of research — their customers and members demand it. And it is equally easy to see why the research stations are eager to cooperate with them — their constituents, the farmers, stand to benefit, and research budgets are so tight that outside contributions are sorely needed.

But soil and water conservation research priorities are not going to get attention when the research agenda is controlled by corporations and commodity groups. That means the budgets must be solid enough to have that portion of the research program funded by adequate public appropriations. There simply are not enough private economic incentives to assure proper priorities unless this is done.

It is also important to find ways to shorten the time between the discovery of new methods in the laboratory and their practical application in the field. There is a tremendous reservoir of technical and innovative skill in America's farmers, but often they are inadequately involved in actual research, and their practical experience is not well enough recognized by researchers. We must harness the tremendous capability of farmers themselves in adapting systems to the soil and modifying them to make them more practical. Much of what is needed is not so much *research* as it is *technology transfer*, and that can mean transfer of good ideas between farmers and from farmers to scientists as well as from scientists to farmers. I have urged the SCS to establish an on-farm testing program for new conservation methods that would be similar to their current program that tests new plant varieties. This would allow leading farmers actually to help mold and develop SCS recommendations through their experience and expertise.

There is one whole area of research that is getting little attention, and it may often be the most important in terms of reducing soil erosion rates. This is what might be called "tactical research." Nobody is studying questions such as, "Which of the current USDA programs work best, and under what conditions?" "What is the best combination of incentives to offer today's farmer to encourage him to practice soil conservation?" "How can the programs be made more appealing to farm people?" Those questions have not been studied in nearly the depth they deserve, largely because the agency that is charged with the task of selling conservation to farmers (the SCS) has no authority to conduct research, while the research agencies of both the federal and state governments, lacking the responsibility for actually seeing that anything happens on the land, feel little urgency for the matter.

In addition to focusing public research on the long-term effects of different farming and conservation systems on the soil, public information and extension programs must also concentrate on those issues. There are ample sources of information from the private sector about the most effective and profitable ways to apply commercial products such as fertilizers and pesticides. If extension agents concentrated on teaching people how to use those products safely, and in a way that contributed to soil and ecosystem stability, they would provide a public service that cannot be realistically expected from the private sector. Some of this is already being done, but a fair percentage of the information provided by extension agents is a straight promotion for various commercial products and soil-depleting cropping systems. "If the farmer can't make a profit, he won't be here to care for the soil," I've had professional farm advisers tell me. "Yes," I reply, "but if he wrecks the soil making that profit, no one will be here before long."

Middleground Approaches to Farmland Preservation

How to keep development from using up millions of acres of America's prime farmlands is a problem equally as vexing as soil erosion for agricultural policymakers today. That the loss of these lands is a direct contradiction to our need to increase farm productivity is a certainty, but stemming the flow of prime farmlands to other uses is no easy task.

Most of the state and local programs in effect today use some form of zoning to prevent undue development and conversion of farmlands. While zoning is widely used, it is not widely popular, nor is it thought to be the final answer by many students of the subject.

Charles E. Little of the American Land Forum (ALF), for example, feels that agricultural zoning will become more and more difficult to maintain as rural growth and development continue. If developers continue to be willing to bid for farmland at high prices, there will always be farmers wanting to sell and get out of farming. Local regulations preventing such sales would be under constant attack and soon would probably be changed.

One way to overcome that problem would be for government to purchase and hold the "development rights" to farmland. These would be in the form of an easement that would be sold by the farmer to the government, thus restricting the deed of the property from that day forward. Several programs such as this have been tested; one in Suffolk County, New York, and the other in Burlington Township, New Jersey. Both stalled on the same issue: money. In those highly urbanized areas, the cost of the development rights was somewhere in the range of two-thirds to three-quarters of the total land cost. Purchasing development rights to all the farmland in New Jersey might cost somewhere in the range of $1 billion.[8] For the nation as a whole, my calculations of the total cost of this option come out somewhere close to $1 trillion. Obviously, this is not likely to be a feasible solution in the near future and, if land costs continue to rise, it will get no better.

The general criticism of zoning as a cumbersome tool to protect farmland, coupled with the realization that public purchase of the development rights on all farmland was financially impractical, has led to a study of other methods. That search has focused on farmland conservancies, which ALF's Little calls "middleground" approaches, because they lie somewhere between the use of government's police powers such as zoning and outright purchase. The heart of the idea is that there would be a local organization that could, as a last resort, purchase farms to prevent their sale for development, then resell them to farmers with a restricted deed insuring continued agricultural use.

The local organization (called a farmland conservancy, it might be a soil conservation district, the local government itself, or a private organization with a public charter) would be empowered by state law to buy and sell land or rights in land for the purpose of maintaining prime, unique, and locally important farmland in farm use. It could combine or split properties to strengthen the abilities of families to purchase and operate economically feasible farm units and carry out soil conservation projects as needed on the land.

The farmland conservancy would have the right to intervene in any sale of land which it had previously designated as prime, unique, or locally important farmland. It could meet a bona fide price being offered or go to arbitration to establish a fair price. It could use a wide variety of real estate techniques to facilitate purchasing the land and spreading out the payments for the benefit of the seller.

In addition to restricting the deed to prevent sale of the land for development purposes, the conservancy could, if it felt it desirable, require the buyer to maintain soil conservation systems on the land under penalty of reversion of the land to the conservancy if soil conservation were not practiced.

This approach sounds much like the "land conversion" idea I proposed for marginal croplands, and it is. The same organization could be chartered to do both jobs. The underlying key is that, if the public wishes to restrict the manner in which private individuals use the land, there must be a way to compensate the private citizen for that restriction. A capability to enter the real estate market, hold some of the land rights in the public domain, then resell on the private market, may help achieve some of the public's goals without trampling on private rights and values.

This is not to suggest that such a program could be easily instituted. Any kind of public intervention in private land use runs into difficulty in this country, no matter how fairly it is posed.

It also does not mean that farmland conservancies are the only way to save farmland, either. Every study to date seems to point toward the need for many different types of farmland retention programs, even within one county, to meet the various situations facing farmers and local governments. A complex and perhaps difficult program seems the only realistic answer.

What does seem certain is that we have finally decided there is one option we do not have: We can't sit idly by and let the greatest body of prime farmland on earth simply slip out of our hands. The question of whether or not we need to preserve agricultural land is answered. The question of whether or not we want to do it with fairness and equity to landowners seems obvious. What is not so obvious is just how we will accomplish that difficult task.

Becoming a Conservation Farmer

For those who believe, as I do, that improving America's soil and water conservation effort is a task worthy of their personal effort, there exist many ways to be of constructive service. Much of what needs to be done is in the field of public policy, but it would be a mistake to think that there is no role for the average individual who has little or no connection with national politics. One thing is certain—it will be farmers on the land who make the difference in our future productivity.

Every farmer has both the opportunity and obligation to become a conservation farmer. Whether it is ten acres of grass, trees, and gardens or a multi-thousand-acre Texas ranch, the person who assumes the privileges of ownership also undertakes the responsibility for stewardship. Failure to carry out that stewardship is not only a personal failure, but also a public one that threatens the well-being of current and future citizens.

Along with jeopardizing the land and our future economic capability, failure to manage soil and water properly also threatens the freedom and right Americans now enjoy to own and use private land. That right, most fiercely guarded by American farmers, gives them a great deal of freedom to do with their land as they please. It could be jeopardized most, not by people who like to pass new laws, but by those who abuse their privileges. The American public will only allow the needless destruction of their basic food-producing lands to a certain point, then political pressure will grow to take legal steps forcing landowners to stop wasting land. Farmers, who would be the most adversely affected by any such action, are also the only people with the ability to prevent it. By taking the responsibility for stewardship and writing their success boldly on the face of the land, they can strike an effective blow for continued freedom for the nation's landowners. Where there is no problem, there need be no suggested remedy.

Over 99 percent of America's farmers live within the boundaries of a soil conservation district. If you are a farmer you can, without charge or obligation, apply to the district for assistance in developing a soil and water conservation management program for your land. Millions of farmers have already done so, and millions of acres of land have been improved or protected as a result. The responsibility for seeking help lies mainly with you. If you want help, you can usually get it. But because the local offices are normally busy and understaffed, if you don't seek their help, they will seldom have time to seek you out.

Not every farmer who has worked with the district has kept his conservation program up to date, and many may need to rethink their efforts in light of the new demands wrought by fast-changing farm technology and methods. Conservation district files are full of conservation plans that are outdated or which have been abandoned by the farmer over the years. That is partly because the staff of professional conservationists available to follow up with those farmers and encourage them to keep their plans current has dwindled over the years and was never adequate to provide the regular assistance most farmers need to keep their conservation systems working.

Those deficiencies can be remedied more by pressure from farmers than any other way. If you can't get the kind of technical assistance and advice you need to solve your soil and water problems, you should protest. The fact that your area isn't getting the help it needs is largely the result of a perception on the part of people who make public decisions that it doesn't matter. They may have heard from professional conservationists and conservation district officials who have objected to the gradual decay of America's soil and water conservation programs, but they have not heard nearly enough from the recipients of those services—the farmers and ranchers on the land.

Start with your local conservation district officials. They will be five or more local citizens, elected or appointed to their task, serving without pay. Let them know about your needs, and if you have not been able to get the kind or quality of service you need from the district office or the USDA agencies that serve you, give them the details. If they are doing their job, they will be glad to hear from you and try to help solve your problem. If they don't seem to care, you may have uncovered a major reason why the local program isn't working.

If efforts to improve the local service fail, don't hesitate to go further. Conservation districts operate under state law, and if the program in your area is inadequate, let your state legislators and governor know. Many states are now providing significant inputs into the conservation program, and the leaders who try to justify those expenditures need to know when the needs of the people on the land are still being inadequately addressed.

Most of the programs that are available locally are still funded by the federal budget. The SCS is a federal agency, funded through the USDA. There are supposed to be SCS technicians available to assist farmers in every conservation district, and they are supposed to have the kind of training and experience needed to help solve your problems. Where that help is inadequate, and your capability to install and maintain a good conservation system on your land is jeopardized as a result, you should let officials in Washington know about it. Write your congressman and senators, as well as the Secretary of Agriculture. Often, they will be fighting to gain funds for soil conservation programs in the face of other demands on the federal treasury, and they need requests like yours to bolster their argument.

It remains, however, your responsibility, not the government's, to see that your land is properly managed. The fact that there are no federal cost-sharing funds available this year or that the SCS technician who visited your farm may not have satisfied you may make your job more difficult, but certainly not impossible. Information is available from a multitude of sources, starting with the local conservation district and SCS office, but also including the Extension Service, the land grant colleges, other federal and state agencies, and many private firms who have capable agriculturalists on their staff. Publications ranging from USDA pamphlets to full textbooks are readily available, and a copy of the soil survey for your farm is almost certain to be yours free for the asking at the local conservation district office.

Too often in the past, soil conservation promoters have attempted to show farmers that soil conservation can be done easily and profitably. I doubt that those arguments are always helpful. The simpler such promotors try to make conservation seem, the more trivial it appears. Even though most of us dread tackling projects that pose considerable effort, we also know that our ultimate reward will be, in large measure, a reflection of what we invested. As a farmer, you practice soil and water conservation because it is beneficial to you and your family, essential for society's welfare, and the mark of a responsible steward of the land—not because it comes easily. It will often require you to learn some new techniques, make an added investment in the land, or pass up some immediate income so that the soil can be made richer. Those investments will all pay dividends, although some may not come for many years. Investing now will not always seem simple, but that is where the line is drawn between the stewards and the wasters. Stewards know that the land is a resource, and treat it as such; people who feel land is only a commodity tend to waste it.

Promoting Soil Conservation

Whether you own land or not, there is a multitude of ways in which anyone can become directly involved in the soil and water conservation effort. All of us, no matter what our technical background or skills, can become participants, advocates, or supporters for conservation. For some people, it may become a life's work, for others an interesting and challenging way to do volunteer service in their community or even to assume an official position in local conservation organizations or units of government.

Every citizen has the opportunity to be aware of what is happening in his or her community, to try to understand what it means, and to work to improve it. On your next drive through farmland, look for the tell-tale signs of soil erosion. Remember that even a few visible rills mean that many tons of topsoil have been lost forever. If that is common in your community, let people know. Go to the local conservation district, or SCS office, or county agent, and ask for the facts. Find out whether what you have observed is a common occurrence and whether it is widespread or isolated in small areas. Ask what can be done and what is being done. Make your own assessment about the local commitment to reach the farmers who are abusing their land, and satisfy yourself as to whether the problem lies in complacency on the part of local conservation officials, lack of authority on their part, or inadequate funding support from local, state, or federal sources to allow them to do their job. Once you have identified the reason why the local program isn't working effectively enough to prevent the waste of your community's land, you may have discovered what you can do to help.

You may want to carry out a local information campaign, starting with letters to the editor, or discussions in local service clubs, social organizations, or schools. Get local resource professionals from SCS, Extension Service, or other agencies to present the facts, or ask a conservation district official to explain the local program addressing the problems. Often these efforts will bring better community support and understanding, often with good results.

You may need to write a letter to an official in the state capital or Washington, D.C. Tell them about the local situation, why you feel it is occurring, and what they can do to help improve it. If you can recommend specific program needs or other remedies that they should consider, let them know so that they will be better able to address your request.

Writing a letter to Washington is easier, and more effective, than most people think. It only takes a few — perhaps 20 or so — good letters on a topic for a member of Congress to recognize that perhaps the issue warrants a more careful study. That is the beginning of action. Congressmen can be reached by addressing your letter to them by name at the House of Representatives, Washington, DC 20515. Letters to senators should be addressed to the senator by name, United States Senate, Washington, DC 20510. You can write the Secretary of Agriculture by addressing your letter to USDA, Washington, DC 20250.

For those who would prefer to work directly with the conservation program, and who may be seeking a vocation or career, it may be worthwhile to check with local conservationists or school counselors to see what kinds of employment may be open, and what educational background will be needed. For a full career, it is advantageous to have a college degree in some aspect of biology, agriculture, or science, but there are many jobs at the local level that do not require such a degree, and it may be worthwhile to investigate them.

There are also a multitude of volunteer jobs that can be done. In the past, conservation districts have not been aggressive at seeking volunteer help, and legislation is only now being considered in Congress that would allow USDA to use volunteers effectively, but it is an idea that should be explored. There are many jobs, ranging from clerical work to local information and education programs or work to stabilize a gully or an eroding hillside, that could be carried out very effectively by volunteers. If you feel that soil conservation is important, and you have some time on your hands, go to the local district office and challenge them to find a meaningful way for you to contribute.

There are many conservation organizations, and no doubt more will spring up in the future. Membership in them is a good way to stay aware of new developments and opportunities in conservation.*

Finally, there is the opportunity to serve as a conservation official yourself. Most adults who live outside major urban areas are eligible, depending on the laws of the state where they reside, to be elected or appointed to their conservation district governing body. It takes time, demands dedication, can cost money, and will not always be easy. For those who wish to move into the action at the local level, however, it is one of the best opportunities available. As districts get more local and state funding and are responsible for running a more active program of their own, the job becomes both more demanding and more meaningful. District officials report that, as their local program becomes more active, they find their rewards from service far more plentiful, even in the face of added demands on their time.

Some of the things that conservation district officials do are involved in guiding local programs and working with state and national legislators on the need for new programs or funding for efforts that are underway. In addition, they have the opportunity to develop innovative local efforts designed specifically for the people and the circumstances in their community.

I once worked with a conservation district that wanted to

*The National Wildlife Federation publishes an excellent directory of conservation organizations. Copies can be purchased for $6 from the National Wildlife Federation, 1412 16th St., NW, Washington, DC 20036.

stop the local practice of burning off wheat stubble after harvest. This practice, common in many areas of the nation several years back, was terribly damaging to soil organic matter and was the cause of much soil waste. The district made up a supply of signs that carried catchy short verses telling about the evils of stubble burning. They nailed them on roadside fence posts throughout the area, much in the same way that Burma-Shave signs were once displayed. As you went down the road, one or two words at a time would appear on small signs until after a short way, you had read the entire message. The result was that stubble burning became a community topic, and those who persisted in burning off their land were subject to gentle and some not-so-gentle ribbing from their neighbors. In a few years, stubble burning was finished.

My purpose in telling that story is not to promote roadside signs, but to illustrate the power of local solutions, often carried out with little money or manpower but with considerable innovation and skill. Those solutions don't take an act of Congress; they take a dedicated individual or group willing to work together to improve their own community's future. The opportunity to join such a group is universal. If you have decided that soil and water conservation is important to your future, and to the entire nation's future, get involved. What you can do may be vital in affecting the future—a future that could be very different, depending on the actions we all take.

Scenarios for the Future

Summing up all the trends we have been examining into a forecast for the future is not a task that can be taken lightly. The future one builds in imaginary form can take so many different shapes—each one dictating what the appropriate response or strategy might be. My view of the future facing American agriculture is that increasing demand will soon overtake our ability to produce, leading to strong farm commodity prices coupled with strong market incentives that encourage farmers to abuse land resources in order to cash in. At the same time, I see farmers going from the high-risk world they face today to an even higher risk situation because of higher real prices for petroleum and the growing risk of supply interruptions.

Farmers will seek to "hedge" against some of the controllable risks, and that may make them more eager to adopt methods that retain more of the natural productivity in their soils as a "safety net" upon which they can always depend. But that will not be easy, in the face of rising costs that make them reluctant to invest in soil conservation practices or to reduce their gross income by seeding land to soil-building crops instead of cash crops.

At the national level, policy leaders will need to develop a method of balancing agricultural sales abroad with the nation's sustainable capacity to produce, for it makes little sense to ruin our land in order to raise foreign sales by some modest percentage for a few short years. If our maximum possible foreign sales were given a value of 100, and the land abuse caused by that level of production suggests that only 50 years of such high-intensity production is possible, it is more than likely that our sustainable level of production is somewhere around 80 or 85. Holding that level of production indefinitely makes a great deal more sense and probably holds the promise to be far more financially rewarding than pushing for all-out production that can only last a relatively short time.

Agreeing on that sustainable level and finding a way to achieve and maintain it will not be easy, however. Given the budgetary and political pressures facing America, the natural tendency will be to do nothing and hope that agriculture finds a way to work out of the current mess without any more national political attention (and money) than is absolutely necessary.

The options facing us, however, are far too risky for such a do-nothing strategy. We are losing farmland productivity at startling rates—rates that will bring us to the limits of our supply of good land sometime before the year 2000. If rising energy costs and competition for agricultural water continue, if research fails to find a miraculous breakthrough that causes yields to jump dramatically, and if farmland is needed to produce energy as well as food crops, this nation faces a historic reversal in its agricultural situation.

This turnabout is being hastened by national policies designed to sell as much as possible abroad in order to keep farm markets growing and prices high. We are selling more than we can safely continue to produce, creating a treadmill where the more we sell, the more land we will ruin, and the more land we ruin, the harder we must work to produce more product to sell. It is a losing game, where we must continue to run faster each year to stay even. Because such a treadmill must eventually run out of energy for increased speed, we will go, in less than two decades, from surplus to scarcity, and this situation will be made doubly difficult by the fact that our people have had virtually no cultural training or experience in how to deal with such a situation.

It is hard to paint a picture of such a future without sounding apocalyptic in the process. But I see only one reaction to what is on the horizon—the price of food and other products from the land will rise steadily as a result of resource limits and constraints. To many people that may not pose serious harm, but the repercussions in terms of the ability of the United States to trade abroad, feed its own people, and reach out at times to some of the world's truly needy, seem serious indeed.

Most vulnerable of all, I fear, are American farmers. They face, on the one hand, an era of stronger demand for farm commodities which may be reflected in higher farm prices and, hopefully, better profits. Profitability is by no means assured, however, and if farmers continue to grow more and more dependent on petroleum and other purchased inputs, they could find themselves simply generating more gross income in order to pay higher costs. If that occurs, the onset of a bad year or two (which, in farming is not usually a matter of *if* it will come, but when) could cause financial ruin to vast numbers of farmers.

Such a disaster could result in the most tragic resource loss of all—the loss of thousands of enterprising businessmen-farmers who have the skills needed to coax a harvest from the land. This is the kind of future that I feel Americans are willing to work to avoid. It is also avoidable, if people will establish a different vision of the future—one that they want to see happen—and work to make it come to pass.

That other scenario for a far better future will require that the United States establish production levels that it can sustain indefinitely with the land, water, technology, and capital at hand. Such levels will not be static, and should grow over time, as science and technology bring new techniques that raise productivity. But they should not grow at the expense of "mining" our soil and water resources. Only those new increases in productivity that can be accommodated without destroying the soil should be counted as real gains. The others are losers, no matter how appealing they may seem in the short term.

A sustainable agriculture is one that produces the food, fiber, energy, and other crops that we need as a nation, including a marketable surplus that can be sold abroad. It produces this on the average year, not just during times of unusually good weather. It can stand a bad year by drawing on stored fertility and moisture in the soils; stored water in reservoirs; stored wealth in financially secure farms; and stored food products in the granaries of farmers, industries, and government. It can profit from a good year by setting aside extra commodities or making an extra effort to see that they are sold abroad, without driving prices through the floor and creating financial hardship and ruin among producers. In addition to meeting our food needs, a sustainable agriculture would reduce the waste and pollution of water, provide better wildlife habitats, slow the advance of desertification and soil salinization, reduce the loss of prime farmlands and fragile topsoils, and, in general, make rural America a far more healthy, satisfying, and financially rewarding place to live. It is a utopian goal, perhaps, but the essential goal nonetheless.

With a sustainable agriculture as our long-term goal, one that may take from 20 to 50 years to achieve, it is important that we look at the strategies open to us today that will help us proceed from where we are to where we want to be.

It appears that we must develop three different types of strategies in order to address this challenge. Each is essential to help us through a long phase-in process that leads from today's depleting agriculture to tomorrow's sustainable agriculture.

First, we must address the short term: the next 10-20 years. Meeting the rapidly expanding food and fiber demands of this period may be the most difficult of all, for there is no time to develop new yield-increasing technologies and very few are on the horizon for near-term application. In this period, however, we have several excellent opportunities to enhance productivity:

a. Reduce topsoil erosion;
b. Reduce the conversion of prime farmland to other uses;
c. Develop soil and water resources as appropriate; and
d. Guide urban growth into compact, land-conserving patterns.

Each of these actions, taken today, can result in immediate benefits. Topsoil that doesn't wash away in 1981 will add to crop yields in 1982. Prime farmland saved from a shopping center in 1981 can grow crops in 1982—and every year thereafter. A wet cropland soil, drained, is immediately more productive, as is a newly irrigated soil. Urban growth, to the extent it can be kept in more compact patterns, will result in significant savings of both land and energy.

I don't profess that any of these will be easy to do or that they will ultimately solve all the problems facing American agriculture. But they offer immediate results and they *can be done, if we only decide they must be done.*

Next, we must address the mid-term, 20 to 50 years away. By the turn of the century, we will need *significantly* higher levels of crop production than will be available, even if we carry out *all* of the conservation and development efforts outlined as short-term possibilities. This will call for much higher per-acre yields, and those will only be realized through a greatly accelerated research effort. That effort must be started immediately, because the time lag between the research start-up and the field application of significant results must be measured in decades—at least one and perhaps two to three.

A list to consider for the mid-term strategies might include:

a. Intensive research in *sustainable* methods of increasing crop and animal yields and productivity;

b. More efficient methods of food harvest, storage, processing, and preparation;

c. A reduction of the land requirements for food through shifting consumption patterns, aided by the development of more nutritious grains, etc.; and,

d. A shift back toward growing food nearer the areas where it will be consumed, to cut transportation and energy costs and increase regional self-reliance.

Finally, there needs to be an intensive effort to develop the outline for the ultimate goal—a sustainable, healthy agriculture. Some of the challenges involved are:

a. How to keep plentiful supplies of healthful, nutritious food available to the American consumer at prices that include the cost of resource protection, but which stay within affordable ranges;

b. How to keep American farmers free to innovate, compete, and produce in a market economy, while preventing the unnecessary abuse and ultimate ruin of the land; and,

c. How to sell as much product abroad as feasible, at the best possible prices, without destroying soil and water resources in the process.

Those items sound like an agenda for all of agriculture and not just soil conservation, because I think that is where the ultimate answer lies. Soil conservation policy cannot be separated from agricultural policy, it must be an integral part of it. High levels of production accompanied by high levels of resource waste are senseless, in both the social and the long-term economic sense. Only when the American farmer can produce a profitable, balanced output keyed to the sustainable productive capacity of the soil will we have achieved a permanent agricultural policy and a sustainable society. The advantages of such an achievement would be far-reaching, ranging from the obvious economic benefits to the social and cultural virtues of living in an environment that is clean and green where humans are a positive part of nature rather than an adversary.

All of the three kinds of strategies (short-, mid-, and long-range) must be started immediately if they are to be ready when they are needed. In addition, each must be continued once it is begun. We cannot plan to protect topsoil and prime farmland for 20 years, then abandon that effort when some new and wondrous technical "fix" cures all our production problems. Pinning one's hopes on a miracle is fine, but keeping on a solid, sensible course for the long range is also necessary.

We are faced, then, with the need to begin, and the most important decision is to set the general direction we wish to travel. If we are content with the continuation of our current trends—increasing total production at the expense of topsoil, farmland, water, and, ultimately, farmers—we need do nothing. Those forces are so well established, and moving at such speed, that it seems almost trite to say they will continue. Some will say, despite the evidence at hand, that there is no cause for concern, but I know of no logic to indicate that we can build national strength through resource exhaustion, and that is clearly the course upon which we are embarked.

If we wish a different future, one with a sustainable agriculture, we are challenged to do many things and do them soon. Such a future will only come with great effort on the part of governments, businesses, industries, and, most of all, farm and non-farm individuals. Government needs to help us identify where we should aim our combined efforts. Business and industry must be willing to forego the quick profits of farmland waste and design products that will earn the steady long-term profits of a permanent society. And individuals must do their part—whether as farmer, consumer, or policy advocate—to assure that appropriate goals for a sustainable agriculture are quickly set and vigorously pursued. Clearly, for all Americans, the emerging crisis of farmland waste calls for new directions. Will we continue as we are currently heading, or will we begin to define and practice a new land ethic that produces a sustainable American agriculture? Deciding which way to go is up to us, and it is time to choose.

Appendices

Appendix A
Land Capability
Classification*

The capability classification[†] is one of a number of inter-
pretive groupings made primarily for agricultural purposes. The capability classification begins with the individual soil-mapping units.
In this classification the arable soils are grouped according to their
potentialities and their limitations for sustained production of the
common cultivated crops without specialized site conditioning or
site treatment. Non-arable soils (soils unsuitable for long-time sustained use for cultivated crops) are grouped according to their
potentialities and their limitations for producing permanent vegetation and according to the risks of damage if mismanaged.

The capability classification provides three major categories:
capability unit, capability subclass, and capability class. A capability unit is a grouping of soils that are suited to the same kinds of
common cultivated crops and pasture plants and have about the
same responses to systems of management.

The second category, the subclass, is a grouping of capability
units having similar kinds of limitations or hazards. Four kinds of
limitations or hazards are recognized: erosion, wetness, root zone
limitation, and climate.

In the third and broadest category of the capability classifi-

*SOURCE: National Agricultural Lands Study, *Soil Degradation: Effects on
Agricultural Productivity* (Washington, D.C.: NALS, 1980).
†Klingebiel, A. A. and P. H. Montgomery. *Land-Capability Classification.* U.S.
Dept. Agr. Hdbk. 210. 21 pp. 1961.

338

cation all the soils are grouped in eight capability classes. The risks of soil damage or the limitations in use become progressively greater from Class I to Class VIII. Under good management, soils in the first four classes are capable of producing adapted plants, such as forest trees, range plants, and the common cultivated field crops and pasture plants. Soils in Classes V, VI, and VII are capable of producing adapted native plants. Some soils in Classes V and VI are also capable of producing specialized crops, such as certain fruits and ornamentals and even field and vegetable crops under highly intensive management that includes elaborate practices for soil and water conservation. Soils in Class VIII do not return onsite benefits for inputs of management for crops, grasses, or trees without major reclamation.

Capability Classes

Land suited for cultivation and other uses.

Class I. Soils in Class I have few limitations that restrict their use. They are suited to a wide range of plants and may be used safely for cultivated crops, pasture, range, forest, and wildlife. The soils are nearly level, and the erosion hazard (wind and water) is low. They are deep, generally well drained, and easily worked. They hold water well and are either fairly well supplied with plant nutrients or highly responsive to fertilizers.

These soils are not subject to damaging overflow. They are productive and can be cropped intensively. The local climate is favorable for growing many of the common field crops. The soils that are used for crops need only ordinary management practices to maintain productivity.

In irrigated areas, soils may be placed in Class I if the limitation of the arid climate has been removed by relatively permanent irrigation systems. Such soils are nearly level, have deep rooting zones, have favorable permeability and water-holding capacity, and are easily kept in good tilth. Some of the soils may require initial conditioning, including leveling to the desired grade, leaching of a slight accumulation of soluble salts, or lowering of a seasonal high water table. If the limitations of salt accumulation, high water table, overflow, or erosion are likely to recur, the soils are regarded as subject to permanent natural limitations and are not included in Class I.

Class II. Soils in Class II have some limitations that reduce the choice of plants or require moderate conservation practices. They require careful soil management, including conservation practices, to prevent deterioration or to improve air and water relations when the soils are cultivated. The limitations are few and the practices are easy to apply. The soils can be used for cultivated crops, pasture, range, forest, or wildlife habitat.

Limitations of soils in Class II may include singly or in combination: (1) gentle slopes, (2) moderate susceptibility to wind or water erosion or moderately adverse past erosion, (3) less-than-ideal soil depth, (4) somewhat unfavorable soil structure and workability, (5) slight to moderate salinity or alkalinity easily corrected but likely to recur, (6) occasional damaging overflow, (7) wetness that can be corrected by drainage but is a permanent moderate limitation, and (8) slight climatic limitations on soil use and management.

Soils in this class give the farm operator less latitude in the choice of either crops or management practices than soils in Class I. They may also require special soil-conserving cropping systems, soil conservation practices, water-control devices, or tillage methods when used for cultivated crops.

Class III. Soils in Class III have severe limitations that reduce the choice of plants or require special conservation practices, or both. These soils have more restrictions in use than those in Class II. When they are used for cultivated crops, the conservation practices are usually more difficult to apply and to maintain. They can be used for cultivated crops, pasture, forest, range, or wildlife habitat.

These soils have limitations that restrict the amount of clean cultivation; timing of planting, tillage, and harvesting; choice of crops; or a combination of these items. The limitations may result from one or more of the following: (1) moderately steep slopes, (2) high susceptibility to water or wind erosion or severe past erosion, (3) frequent overflows causing some crop damage, (4) very slow permeability of the subsoil, (5) wetness or some continuing waterlogging after drainage, (6) shallow depth to bedrock, hardpan, fragipan, or claypan that limits the rooting zone and water storage, (7) low moisture-holding capacity, (8) low fertility not easily corrected, (9) moderate salinity or alkalinity, or (10) a moderate climatic limitation.

Class IV. Soils in Class IV have very severe limitations that restrict the choice of plants or require very careful management, or both. The restrictions in use for these soils are greater than for those in Class III, and the choice of plants is more limited. If these soils are cultivated, more careful management is required and the needed conservation practices are more difficult to apply and maintain. They can be used for crops, pasture, forest, range, or wildlife habitat.

Soils in Class IV may be well suited to only two or three of the common crops or the yields may be low in relation to the inputs over a long period. Their use for cultivated crops is limited because of one or more permanent features such as: (1) steep slopes, (2) high susceptibility to water or wind erosion, (3) severe past erosion, (4) shallowness, (5) low moisture-holding capacity, (6) frequent overflows causing severe crop damage, (7) excessive wetness with a continuing hazard of waterlogging after drainage, (8) severe salinity or alkalinity, or (9) moderately adverse climate.

In humid regions, many sloping soils in Class IV are suited to occasional but not regular cultivation. Some of the poorly drained, nearly level soils in Class IV are not subject to erosion but are poorly suited to intertilled crops because of the time required for the soil to dry out in the spring and because of their low productivity for cultivated crops.

In subhumid and semi-arid regions soils in Class IV may produce good yields of adapted cultivated crops during years of above-average rainfall, low yield during years of average rainfall, and failures during years of below-average rainfall. During low-rainfall years the soils must be protected even though there is little probability of producing a marketable crop.

Land limited in use, generally not suited for cultivation

Class V. Soils in Class V have little or no erosion hazard but have other limitations that are impractical to remove and that limit their use largely to pasture, range, forest, or wildlife habitat. They have limitations that restrict the kind of plants that can be grown and that prevent normal tillage of cultivated crops. They are nearly level, but some are wet, are frequently overflowed, are stony, have a climatic limitation, or have some combination of these limitations. Examples of soils in Class V are: (1) soils of the bottomlands subject to frequent overflow that prevents the normal production of cultivated crops, (2) nearly level soils in an area in which the growing season prevents the normal production of cultivated crops, (3) level or nearly level stony or rocky soils, and (4) ponded areas where drainage for cultivated crops is not feasible but the soils are suitable for grasses or trees. Cultivation of the common crops is not feasible, but pastures can be improved and benefits from proper management can be expected.

Class VI. Soils in Class VI have severe limitations that make them generally unsuited to cultivation and that limit their use largely to pasture, range, forest, or wildlife habitat. The physical conditions of these soils are such that it is practical to apply range or pasture improvements if needed, such as seeding, liming, fertilizing, and water control by contour furrows, drainage ditches, diversions, or water spreaders. These soils have continuing limitations that cannot be corrected, such as: (1) steep slope, (2) hazard of severe erosion, (3) past erosion, (4) stoniness, (5) shallow rooting zone, (6) excessive wetness or overflow, (7) low moisture-holding capacity, (8) salinity or alkalinity, or (9) severe climate. Because of one or more of these limitations, these soils are not generally suited to growing cultivated crops but they can be used for pasture, range, forest, and wildlife habitat or some combination of these.

Some soils in Class VI can be safely used for the common crops if unusually intensive management is used. Some of the soils are also adapted to special crops, such as sodded orchards, blueberries, and similar crops. Depending on their characteristics and the local climate, the soils may be well or poorly suited to growing trees.

Class VII. Soils in Class VII have very severe limitations that make them unsuited to cultivation and that restrict their use largely to grazing, forest, or wildlife habitat. The physical condition of these soils is such that it is impractical to apply pasture or range improvements, such as seeding, liming, fertilizing, and water control by contour furrows, ditches, diversions, or water spreaders. The restrictions are more severe than those for soils in Class VI because of one or more continuing limitations that cannot be corrected, such as very steep slopes, erosion, shallowness, stoniness, wetness, presence of salts or alkali, unfavorable climate, or other limitations that make them unsuited to common cultivated crops. Under proper management, these can be used safely for grazing, forest, wildlife habitat, or some combination of these. Depending on their characteristics and the local climate, these soils may be well or poorly suited to growing trees.

Class VIII. Soils and landforms in Class VIII have limitations that preclude their use for commercial crop production and that restrict their use to recreation, wildlife habitat, water supply, or aesthetic purposes. They cannot be expected to return significant benefits from management for crops, grasses, or trees, although benefits from their use for wildlife habitat, watershed protection or recreation may be possible.

Limitations that cannot be corrected may be one or more of the following: (1) past erosion or erosion hazard, (2) severe climate,

(3) wetness, (4) stoniness, (5) low moisture-holding capacity, and (6) salinity or alkalinity. Badlands, rock outcrops, sandy beaches, river wash, mine tailings, and other nearly barren lands are included in Class VIII. It may be necessary to protect the soils and landforms and to manage them for plant growth to protect other more valuable soils, to control water, for wildlife habitat, or for aesthetic reasons.

Capability Subclasses

Subclasses are groups of capability units within classes that have the same kinds of dominant limitations to agricultural use as a result of soil and climate. Some soils are subject to erosion if they are not protected; others are naturally wet and must be drained if crops are to be grown. Some soils are shallow or droughty or have other deficiencies. Still other soils occur in areas in which climate limits their use. The four kinds of limitations recognized at the subclass level are: risk of erosion, designated by the symbol e; wetness, w; root-zone limitations, s; and climatic limitations, c. The class and subclass designations provide information about both the degree and the kind of limitation. Capability Class I has no subclasses.

Subclass e (erosion hazard) consists of soils for which susceptibility to erosion or past erosion damage is the dominant problem or hazard in their use.

Subclass w (excess water) consists of soils in which excess water is the dominant hazard or limitation in their use. Poor soil drainage, wetness, high water table, and overflow are the criteria for determining which soils belong to this subclass.

Subclass s (other unfavorable soil conditions) consists of soils in which the soil characteristics of the root zone are the dominant limitations in their use. These limitations are such factors as shallowness, stoniness, low moisture-holding capacity, low fertility difficult to correct, and salinity or sodium.

Subclass c (climatic limitation) consists of soils for which the climate (temperature and lack of moisture) is the major hazard or limitation in their use.

Because limitations imposed by erosion, excess water, shallowness, stoniness, low moisture-holding capacity, salinity or alkalinity can be modified or partially overcome, they take precedence over climate in determining subclasses. The dominant kind of soil limitation or hazard determines the assignment of capability units to the e, w, and s subclasses. Capability units that have no limitation other than climate are assigned to the c subclass.

If two kinds of limitations that can be modified or corrected are nearly equal, assignments to subclasses have the following priority: e, w, and s. For example, in grouping soils of humid regions that have both an erosion hazard and an excess water hazard, the erosion hazard takes precedence over wetness; for soils having both an excess water limitation and a root-zone limitation, wetness takes precedence over the root-zone limitation. In grouping soils of subhumid and semi-arid regions that have both an erosion hazard and a climatic limitation, the erosion hazard takes precedence over the climatic hazard; for soils that have both a root-zone limitation and a climatic limitation, the former takes precedence over the latter.

Appendix B

The National Resource Inventories[*]

The National Resource Inventories (NRI) provided much of the data used in preparing the *1980 RCA Appraisal*. For the NRI, the Statistical Laboratory of Iowa State University selected random sample areas known as primary sample units (PSU's) for each county in each state. Most PSU's in midwestern, western, and southern states were 160 acres; most in eastern states were 100 acres. Some were as small as 40 acres and some as large as 640 acres.

Three points were selected at random within each PSU (only two points were used in PSU's of 40 acres). The Soil Conservation Service (SCS) examined about 200,000 sample points in the field and compiled data for the NRI. SCS field specialists and technicians collected the data. State SCS staffs and the Iowa State University Statistical Laboratory made quality control checks. SCS reexamined more than 6,000 sample points in the field for correctness, and the Statistical Laboratory made other special computer checks for consistency.

County Base Data — Basic data about the gross area of each county and the net area of land and water in each county were obtained from the U.S. Department of Commerce, Bureau of the Census. Estimates from the 1970 Census were provided to SCS field offices. Field personnel reported any changes in land and water

*SOURCE: Soil Conservation Service, *RCA Appraisal 1980, Review Draft, Part II,* (Washington, D.C.: USDA, 1980), p.7-1.

areas between 1970 and 1977. Such changes might have been caused by county boundary changes, construction of large reservoirs, or other activities.

The Forest Service reported land it administered as National Forest System or National Grasslands, and the Bureau of Indian Affairs reported the acreage of land it administered. Field personnel determined from state and local sources the acreage of land administered by other federal agencies.

SCS used existing data to measure roads and railroads that connect rural and urban areas to determine the amount of land used for major rural transportation systems. Transportation categories included:

1. Interstate highways
2. Paved primary federal and state highways
3. Other paved roads
4. Gravel roads
5. Dirt roads
6. Railroads

The number of miles of roadway in each transportation category, the average width of the corridor, and the total acreage occupied were recorded.

PSU Data — Maps were submitted to the Statistical Laboratory showing the location and extent of urban and built-up land of more than 40 acres and the location and extent of irrigated land. The Statistical Laboratory used these maps in selecting the size and location of PSU's and then notified the SCS field offices that were to gather the field data. SCS obtained the following information for each PSU:

• Size — The actual size of each PSU in acres was recorded. For irregularly shaped PSU's the acreage was determined by dot grid or by planimetering the area on a map or photograph.

• Urban and Built-Up Land — The acreage of urban and built-up land in each PSU was determined. This acreage included contiguous areas of more than ten acres used for residences, industrial and commercial sites, institutional sites, railroad yards, small parks, cemeteries, airports, and similar urban facilities.

• Small Built-Up Land — The acreage of small built-up areas was also determined in each PSU. These areas are like "Urban and Built-Up Land" except that they are smaller than 10 acres but larger than 0.25 acre.

• Farmsteads — The acreage of farmsteads in each PSU was

determined. This acreage included land used for dwellings, buildings, barns, pens, corrals, windbreaks, family gardens, and other purposes connected with operating farms or ranches.

• Water Bodies Less Than 40 Acres—All permanent water bodies of less than 40 acres were identified and their use was recorded. This information was recorded for all water bodies even if only part of their total area was within the PSU. SCS field personnel recorded at least one but no more than three of the following uses for each water body:

1. Irrigation
2. Livestock water
3. Water supply (municipal, industrial, household, fire fighting)
4. Recreation, fish, and wildlife
5. Erosion and sediment control
6. Flood prevention and flood control
7. Water quality control (livestock waste lagoons and sewage lagoons)
8. Other (power, navigation, cooling, etc.)

• Perennial Streams Less Than ⅛ Mile Wide—SCS also collected data on the width, length, and acreage of the parts of perennial streams less than ⅛ mile wide that were within each PSU. The field personnel determined that the water in each perennial stream was used for at least one but no more than two of the following purposes:

1. Irrigation
2. Livestock water
3. Water supply (municipal, industrial, household, fire fighting)
4. Recreation, fish, and wildlife
5. Other (power, navigation, cooling, drainage, etc.)
   The most important use was recorded first.

• Perennial Streams More Than ⅛ Mile Wide—Field personnel recorded whether the PSU contained part of a perennial stream wider than ⅛ mile.

• Construction—Data were recorded about any construction activities within the PSU involving an area of more than one acre. Construction areas were defined as land areas where man has modified the land surface, that were bare of vegetation at the time of observation, and that were expected to remain without plant cover for more than 30 days.

• Roads—SCS also recorded any rural road within the PSU.

For the purpose of this inventory, roads included farm lanes, logging roads, woods roads, and other private roads as well as paved or gravel public roads. (Roads included in "Urban and Built-Up Land" were not in this category.)

• Active Gullies—Field personnel recorded the number of active gullies in each PSU. An active gully was defined as an eroding channel through which water flows only during and immediately after heavy rains or during the melting of snow. For the purpose of this inventory, a gully was further defined as a channel one foot or more deep.

Data on construction sites, roads, and gullies were recorded as preliminary information for use in a subsequent phase of the NRI concerning roadside, streambank, construction site, and gully erosion.

Sample Point Data—The PSU data were the main source of information about the total acreage of farmsteads and small urban, built-up, and water areas. SCS obtained more specific information from the point data in the NRI. An SCS representative visited each PSU and made observations at random points in the PSU selected by the Statistical Laboratory. Some information had to be obtained from the owner or operator of a farm. For data points on land that had been in crop production at some time during the previous four years, the kinds of crops and residues were determined for each year. This information was used in the wind erosion and universal soil loss equations.

Other information was gathered at each point. For all land areas, this included the soil name and symbol, the land capability class and subclass and the soil loss tolerance. A determination was made as to whether the point was on prime farmland. For urban areas, SCS gathered information on the density of urban development. For rural lands, the information obtained included the type of irrigation, the kinds of conservation practices being applied, the treatment needs, the type of ownership, and data for the universal soil loss and wind erosion equations. For rural non-cropland, SCS gathered data on potential cropland, and for water, on the size of the stream or water body.

In addition to the soil and water data, other information was collected at each sample point. This information included the land use and whether or not the point was in a flood-prone area or in an area that met the definitions of type 3 to 20 wetlands (Shaw and Fredine, 1956). For urban lands, SCS estimated the amount of impervious cover. For points on irrigated land, the type of irrigation was recorded. Field personnel recorded the type of ownership, the existing conservation practices, and the type of treatment needs.

For undeveloped land not in cropland, the potential for conversion to cropland was determined, the major soil and water problems or other problems that might hinder conversion to cropland were noted, and the type of effort necessary for conversion was determined.

Use of Soil Surveys

For the NRI, states had the option of mapping the entire primary sample unit (PSU) in accordance with the standards and procedures of the National Cooperative Soil Survey or of determining the specific soil map unit at individual sample points. This means that uniform soil survey interpretations were made for each sample point. These interpretations provided such information as the K and T values for the universal soil loss equation, the I factor and the wind erosion equation and the land capability class and subclass. The subclasses define the limitations of the soil, including wetness, erodibility, and such climatic and inherent soil problems as stoniness, droughtiness, and salinity.

Appendix C
The Cropland Resource Pool

In order to evaluate the current and possible future trends in agricultural land availability, it is helpful to start with the concept of a "resource pool." From the standpoint of cropland, that pool consists of the following elements, as estimated by the Soil Conservation Service (SCS) in 1977:

Cropland planted	343 million acres
Cropland slack	70 million acres
Total cropland now in use	413 million acres
High- and medium-potential cropland	127 million acres
Total cropland resource pool	540 million acres

Existing USDA data also gives some insight into the dynamics of land use affecting this resource pool, for it is certainly not a static resource base. It is constantly shifting. Land is constantly going out of the pool, and the size and nature of those shifts are a vital element in evaluating the agricultural productivity and future potential of the nation's land and water resources.

Figures C.1 and C.2 lay out a schematic design that may be useful in understanding the nation's non-federal lands. If, through study of the data on hand, agreement can be reached on the size of the flow arrows in Figure C.2 for the period from 1967-1977, these estimates will provide one basis for making projections into the future.

Figure C.1. The 1977 Cropland Resource Pool, Its Use and Potential

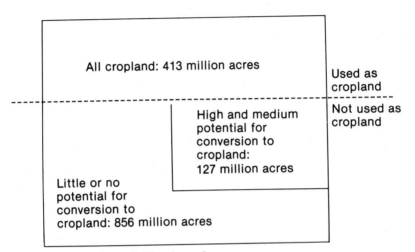

SOURCE: *Compiled by the author*

Figure C.2. Elements of Land-Use Change Affecting the Cropland Resource Pool, 1967-1977

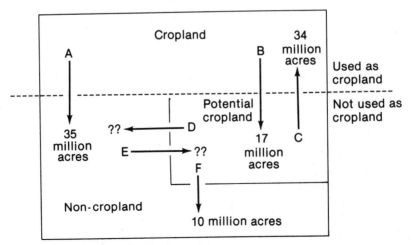

SOURCE: *Compiled by the author*

The flows between the segments of the resource pool are indicated by the arrows on Figure C.2 and identify the following types of land-use change:

A is land that shifts out of cropland use with little or no chance of ever returning to the cropland base. This would include not only conversion to urban and built-up, water, or other uses, but also land-use changes that occur when soil deterioration or other factors preclude future cropping.

B is acreage that shifts out of cropland use but remains in condition to come back into crops fairly readily. This is land shifted to grass, trees, or "other" uses which do not preclude a return to cropland. (Most of the land that is temporarily diverted or rotated from active cropping does not leave the cropland category. It simply shifts from harvested to non-harvested cropland.)

C is high- and medium-potential land that is converted to crops through irrigation, drainage, clearing, or, in the case of high-quality grassland, simply plowing.

D and E indicate land that may become more—or less—suitable for cropping for a variety of reasons, both economic and environmental. The "boundary" between low and medium potential is an economic boundary, based on 1976 costs and prices. As those conditions change, the amount of land that will be feasible for conversion to cropland will shift. Economists use this fact to argue that there will never be a shortage of good farmland, since rising prices will simply call more land into production. To the extent that the good land has already been used, however, all rising prices can do is call more acres of mediocre land into production. If that happens, the costs of growing food will be permanently raised.

F is high- and medium-potential land that is shifted to a new use that precludes its future use as cropland. This includes shifts to urban, water, and "other" uses. In addition it includes land that is compromised by adjoining land uses to the extent that future cropping is unlikely.

Table C.1 estimates the amount of land-use shifts that are projected to occur in the future, using the same types of shifts that were shown in Figure C.2. In order to use the 1977 National Resource Inventories (NRI) data as the authority on the amount of land in each use at the present time, while using the 1975 Potential Cropland Study (PCS) as the source of information on land use shifts, it is necessary to make data adjustments to reconcile the amount of land use change found in 1975 to the 1977 data.

The 1975 PCS estimated the acreage that moved from one use to another. At the same time, it recorded the capability of the land and the potential for future cropland use. The 1977 NRI measured the amount of land in each use as of 1977, but did not identify which land had experienced a use change since 1967. Thus, while both surveys measured the *net* shift in use over a period of years, they were not done with the same sampling methods and did not come out with comparable estimates.

Table C.1. **Land-Use Shifts Identified in the 1975 Potential Cropland Study**

Type of Shift	1967-1975 PCS Estimates	1967-1977 Adjusted Estimates	Land with Low or No Potential for Return to Cropland[a]	
	(million acres)		(percent)	(million acres)
Cropland to Urban	4.8	7*	100	7
Cropland to Water	0.6	1	100	1
Cropland to Other	13.0	8	86	6.9
Cropland to Grass	52.9	31	57	17.7
Cropland to Trees	8.2	5	83	4.1
Total out of Cropland	79.5	52		36.7
Pastureland & Rangeland to Cropland	31.9	22		
Forestland to Cropland	11.0	7		
Other to Cropland	5.8	5		
Total into Cropland	48.7	34		
Total into Urban	16.6	28.8		
Total into Water	6.7	4.5		
Total out of Resource Pool	23.3	33.3		

[a] See Table C.2 for source of percentage estimates.

*Adjusted to give estimates comparable to the 1977 National Resource Inventories totals.

The major value in Table C.1 may be to illustrate the tremendous amount of acreage constantly being shifted to other uses within the land resource pool. If 5.2 million acres are being taken out of crops each year, being partially replaced by a conversion of

3.4 million acres going into crops, this rate of turnover is higher than the previous estimates made by USDA. In 1975, for example, Krause and Hair pointed out that cropland had been abandoned at the rate of 2.7 million acres a year between 1944 and 1964, while new cropland averaged about 1.3 million acres annually during the same period. Further study is needed to see if these 1967-1977 estimates are too high or if they represent a higher rate of shift back and forth in recent years, or whether the data and methodology differences make the two estimates incomparable.

The data at hand leads to the estimate that shifts A and B, as shown in Figure C.2, totaled about 52 million acres between 1967 and 1977. Next, we need to estimate which portion of that shift was, for practical purposes, permanent, and which was temporary or easily reversible.

There has been some controversy about this estimate among USDA analysts. Some people feel that when cropland is taken out of crops but goes into other agricultural uses such as grassland or forest, it remains available for future reconversion to cropland any time the economics of conversion become favorable. Others argue that there are many acres taken out of cropland after soil erosion has completely stripped the topsoil, or when surrounding land uses have become so competitive that the farmer can no longer afford to farm the land. These conversions, it is argued, are permanent and result in the land being lost to agriculture as certainly as if it had been urbanized.

To shed some light on this question, the data from the 1975 PCS was studied to see how much of the land that had been converted out of cropland into another agricultural use between 1967 and 1975 would be rated as having a high or medium potential for return to crop use. This would, it was felt, indicate how much of the converted land could be counted as a continuing part of the "ready reserve," and how much should be counted as lost. Table C.2 shows the results of this analysis, which provides the best data at hand for projections into future decades.

The 1975 PCS data indicated that only about one-third of the land shifted out of crops between 1967 and 1975 would have a high or medium potential for return to cropping. The loss was most direct, of course, where the land went to urban or water uses (shown in Table C.2 as "other"), but it was also true in the other land-use shifts. For instance, only half of the Class I-III cropland that went to pastureland or rangeland shows a high or medium potential for future cropping. Other factors such as parcel size, land-ownership patterns, or adjoining land uses are involved, but we don't know what they are from the data at hand.

Table C.2. **Potential for Return to Cropland of the Lands in Other Uses, 1975**

| 1975 Land Use | Percent with High or Medium Cropland Potential (by Land Capability Class) | | | Total |
	I-III	IV	V-VIII	
Pastureland & Rangeland	51.1	26.7	22.2	42.7
Forestland	21.4	14.3	0	16.9
Other Land	17.4	7.7	0	13.8
Total — All Uses	40.1	20.3	7.3	32.3

This leads to the conclusion that, in Figure C.2, the ratio between shift A and shift B should be somewhere around 2/3:1/3. If A+B = 52 million acres, then A = 35 million and B = 17 million acres. Shift C = 34 million acres, and A+B - C = 18 million acres, which is the net acreage shifted out of cropland over the decade as identified in the 1977 NRI.

The net balance in the set of shifts labeled D and E in Figure C.2 is not known for 1967-1977, since the potential cropland definition was not used in 1967. Forecasting the future is hazardous, as well. As energy prices rise and water becomes a more serious constraint, some acres that were judged to have medium potential under 1976 economic conditions will become less feasible.

On the other hand, if commodity prices rise and political pressure for added production overwhelms environmental concerns for loss of wetlands, drainage could be allowed, encouraged, or even subsidized, and land that was rated as having low potential in 1975 could look very different in 1995. For the analyst today, those issues have so much economic and political volatility that it seems wise to guess that the two kinds of pressure are most likely to balance out in terms of net acreage shift, at least in the near future.

We can find the acreage in the shift labeled F in Figure C.2 by identifying the 1967-1975 acreage shift that was non-cropland, but was in Capability Classes I-III, and use it as a surrogate estimate for the "potential cropland" that was lost from the cropland resource pool during the decade. From the PCS data, this looks like about 10 million acres per decade.

Knowing the magnitude of the 1967-1977 shifts, it is possible to make some projections about the future. One way is to assume that the trends of the 1970's are likely to be repeated in the future. On that basis, the future "resource pool" would be diminished by the sum of the acreage shifts labeled A and F in Figure C.2. (Shifts B and C take place within the existing resource pool and are, therefore, not considered as part of the net change. Shifts D and E are unknown and are assumed to balance each other out.) Using this technique and assuming a straight-line continuation of 1967-1977 trends into the future, the projected losses from the 540-million-acre cropland resource pool between now and the year 2000 would be 103 million acres; by 2030, they would total 238 million acres. Table C.3 lists the acreages that would be involved in these changes if current trends continue. The size of the potential pool would be, therefore, 437 million acres in 2000 and 302 million acres in 2030. These numbers are very startling, in view of the estimates by USDA as to the cropland needed in those years to meet domestic needs plus projected exports.

Table C.3. **Land-Use Shifts within the Cropland Resource Pool***

Type of Shift	1967-1977	1977-1990	1990-2000	2000-2030	1977-2030 Total
	(millions of acres)				
A. Cropland to Non-Cropland	35	45	35	105	185
B. Cropland to Potential Cropland	17	22	17	51	90
C. Potential Cropland to Cropland	34	44	34	102	180
D. Potential Cropland to Non-Cropland	†	‡	‡	‡	‡
E. Non-Cropland to Potential Cropland	†	‡	‡	‡	‡
F. Potential Cropland to Non-Crop Uses	10	13	10	30	53
Total loss from Cropland Resource Pool (A+F)	45	58	45	135	238

*Assuming a continuation of current trends.
†Unknown.
‡Unpredicted.

For example, Table C.3 projects the need to convert 180 million acres of potential cropland to cropland by 2030. Since there are only 127 million acres now rated to have high or medium potential for conversion, this would mean that a great many acres of poor land, or land that is very expensive to convert, would be needed. Either that, or there would not be acres available. The limits of the resource base begin to be visible.

Using the estimates developed in Table C.3, it is possible to make a land base analysis that will reflect the continuation of the current trends. There is little evidence that current trends will—or must—continue, but by the same token, there is no evidence that they have slowed or stopped. Table C.4 shows how the future land base might be affected if these trends are allowed to continue. The "demand" estimates for 2000 and 2030 were taken from the draft documents circulated by USDA for the Resources Conservation Act (RCA) study. In both 2000 and 2030, the RCA estimates the cropland requirements to be somewhere in the range of 450 million acres. That is not too much more cropland than is currently used in the United States. The problem, however, is the diminished resource base that would result if the current trends of land conversion and soil erosion are allowed to continue.

Table C.4. **Projected Cropland Resource Base If Current Trends Continue**

	1977 Actual Cropland	1977 Potential Cropland	Cropland Lost to Other Uses	Net Available Cropland	Cropland Planted	Cropland Slack	Buffer
(millions of acres)							
1977 Actual	413[a]	127[b]	0[c]	540	343[d]	70[e]	127[f]
2000 Projected	413	127	103	437	385	68	-16
2030 Projected	413	127	238	302	389	60	-147

[a] SOURCE: SCS National Resource Inventories 1977
[b] SOURCE: SCS National Resource Inventories 1977
[c] See Table C.3 for estimates of land-use change.
[d] SOURCE: *RCA Appraisal 1980, Part II*, Table 3A-5.
[e] This includes summerfallow, idle cropland, cropland pasture, and other non-harvested cropland uses. It does not include crop failure, which runs 5-10 million acres per year.
[f] This could be estimated as the amount of potential cropland that would remain available for future crop use if needed, under the land use and commodity demand assumptions utilized.

Changing the Assumptions about Future Trends

Projecting the rate of land-use change being experienced today into the future—especially for 50 years—leads to a rate of loss that seems far in excess of what normal economic adjustments, let alone public policy, would allow, so I have prepared another set of estimates that assume a slow-down of the current rates of cropland conversion. This could be caused by many factors, but most likely would be the rise in farm prices relative to other aspects of the economy. Should such a price rise occur, it would dampen agricultural land conversions to some extent.

Another possibility would be effective action at the local, state, and national levels to protect farmland from conversion to other uses. Although there is no indication today that such action is immediately forthcoming, the impending resource shortage indicated by the RCA analysis may hasten action.

The following tables are based on the following assumptions: (1) The rate of cropland loss experienced in 1967-1977 will

Table C.5. **Land-Use Shifts, 1977-2030, If Current Trends Slow**

Type of Shift	1967-1977	1977-1990	1990-2000	2000-2030	1977-2030 Total
	(millions of acres)				
A. Cropland to Non-Cropland	35	45	20	17	82
B. Cropland to Potential Cropland	17	22	9	8	39
C. Potential Cropland to Cropland	34	44	17	14	75
D. Potential Cropland to Non-Cropland	*	†	†	†	†
E. Non-Cropland to Potential Cropland	*	†	†	†	†
F. Potential Cropland to Non-Crop Uses	10	13	5	4	22
Total Loss from Cropland Resource Pool (A+F)	45	58	23	20	104

*Unknown.
†Unpredicted.

remain essentially the same through the decade of the 1980's. (2) At that point, rising pressures will dampen the rate of change so that each decade to follow will see only one-half of the conversion experienced in the previous decade. Such an assumption would mean that the current loss of cropland to non-crop uses would be essentially halted by 2030.

Using these assumptions, Table C.6 compares the predicted availability of cropland with the demand scenarios developed in the *RCA Appraisal*. As it shows, there is still a high probability of cropland shortages affecting the nation's productive capability, despite an assumption that cropland losses can be virtually halted within the next 50 years.

Table C.6. **Projected Cropland Resource Base If Current Trends Slow**

	1977 Actual Cropland	1977 Potential Cropland	Cropland Lost to Other Uses	Net Available Cropland	Cropland Planted	Cropland Slack	Buffer
(millions of acres)							
1977 Actual	413[a]	127[b]	0[c]	540	343[d]	70[e]	127[f]
2000 Projected	413	127	82	458	385	68	5
2030 Projected	413	127	104	436	389	60	-13

[a] SOURCE: SCS National Resource Inventories 1977
[b] SOURCE: SCS National Resource Inventories 1977
[c] See Table C.3 for estimates of land use change.
[d] SOURCE: *RCA Appraisal 1980, Part II*, Table 3A-5.
[e] This includes summerfallow, idle cropland, cropland pasture, and other non-harvested cropland uses. It does not include crop failure, which runs 5-10 million acres per year.
[f] This could be estimated as the amount of potential cropland that would remain available for future crop use if needed, under the land use and commodity demand assumptions utilized.

Appendix D
Measuring Soil Loss

The Universal
Soil Loss Equation

The universal soil loss equation (USLE) is a formula used to predict soil losses caused by water erosion. It was used to estimate sheet and rill erosion for the 1977 National Resource Inventories (NRI).[1]

The use of equations to calculate field soil loss began around 1940 in the Corn Belt. A national committee met in Ohio in 1946 to adapt the Corn Belt equation to cropland in other regions. This committee reappraised the Corn Belt factor values and added a rainfall factor. The resulting formula, generally known as the Musgrave Equation, has been widely used for estimating gross erosion from watersheds in flood abatement programs.

The USLE was developed at the National Runoff and Soil Loss Data Center, which was established in 1954 by the Agricultural Research Service (ARS) in cooperation with Purdue University. Federal-state cooperative research projects at 49 locations contributed more than 10,000 plot-years of basic runoff and soil loss data to this center for summarization and statistical analyses. After 1960, rainfall simulators operating from Indiana, Georgia, Minnesota, and Nebraska were used on field plots in 16 states to fill some of the gaps in the data needed for factor evaluation.

Analyses of this basic data provided several major improvements for the soil loss equation: (a) a rainfall erosion index evaluated from local rainfall characteristics, (b) a quantitative soil erodibility factor that is evaluated directly from soil properties and is independent of topography and rainfall differences, (c) a method of evaluating cropping and management effects in relation to local climatic conditions, and (d) a method of accounting for effects of interactions among cropping systems, productivity levels, tillage practices, and residue management.

Developments since 1965 have expanded that use of the universal soil loss equation by providing techniques for estimating site values of its factors for additional land uses, climatic conditions, and management practices.

The equation is: A = RKLSCP.

A is the average annual soil loss in tons per acre predicted for a given area.

R is the rainfall erosion factor. Soil is eroded from cultivated land in direct proportion to the product of kinetic energy multiplied by the maximum 30-minute intensity of a rainstorm. This product, called the erosion index, shows the erosion potential of the rainfall within a given period. Annual erosion indexes and monthly rainfall distribution curves have been computed for locations throughout the United States where sheet and rill erosion is a problem. These curves were developed using Weather Bureau and ARS data accumulated over more than 20 years.

K is the soil erodibility factor, which expresses soil loss in tons per acre per unit of rainfall erosion index (R) for a slope of specified dimensions, steepness, and length. K factors vary with soil type, series, and degree of erosion. K values have been determined for all soils on the basis of the soil characteristics that determine erodibility.

L is the length of slope factor. This factor is the ratio of soil loss from a specific length of slope to that from the length specified for the K factor of the equation. Slope length is defined as the distance from the point of origin of overland flow to the point where the slope decreases and deposition begins, or to the point where runoff enters a well-defined channel.

S is the steepness of slope factor. It is the ratio of soil loss from the field slope gradient to that from a standard slope under otherwise identical conditions.

C is the cover and management factor. This factor takes into account the combined effect of crops, crop sequence, and various management practices on soil erosion. It is the expected ratio of soil loss from continuously cultivated fallow land with identical soil, slope, and rainfall conditions. The Soil Conservation Service (SCS) has estimated C factors for rangeland and forestland as well as for cropland.

P is the erosion control practice factor. This factor is the ratio of soil loss under a specified conservation practice to that with uphill and downhill farming operations when other conditions, such as soil, slope, and rainfall, are equal.

In simple terms, the equation shows that erosion for any given soil will increase as rainfall and rainfall intensity go up and as slopes become longer and steeper. Erosion will go down as more cover is kept on the land or as conservation practices are applied.

Use of the USLE in the National Resource Inventories (NRI)

SCS gathered data from each NRI sample point to determine each factor in the USLE. The annual soil loss per acre was computed at each point classified as all cropland, cultivated cropland, forestland, rangeland, or pastureland. Computations of sheet and rill erosion did not include points in water; snow and ice fields; farmsteads; other land in farms, quarries, and pits; barren lands; or urban lands where C factors were not available and the USLE did not apply.

The Wind Erosion Equation

Estimating the rate of wind erosion is also done by means of a mathematical equation developed by USDA agricultural researchers. The wind erosion equation (WEQ) combines the primary factors that influence the rate of wind erosion.

The equation is: $E = f(IKCVL)$.

E is the potential annual soil loss in tons per acre per year.

f is a mathematical function.

I is the soil erodibility factor developed by soil scientists.

K is the soil ridge roughness factor.

C is the climatic factor based on average wind velocity and on the precipitation-evaporation index for the location.

V is the vegetative cover index.

L is the unsheltered distance across the field along the prevailing wind direction.[2]

Use of the WEQ in the NRI

For selected points in states where wind erosion is considered a serious problem, SCS tried to determine the value for each factor in the wind erosion equation and compute the expected soil loss per acre under existing conditions. This attempt was unsuccessful, except in the ten Great Plains states. For the *RCA Appraisal 1980*, therefore, SCS confined the use of the wind erosion equation to designated counties in the Great Plains.

Estimating Erosion from Snowmelt

A great deal of soil can be moved in the spring when the winter snows melt, but estimating the amount of this snowmelt-caused erosion through the use of the USLE has proven very difficult. In areas such as the Pacific Northwest, where up to 90 percent of the soil erosion may occur in the spring as the winter snows melt, use of the standard R factor for the region seriously underestimates soil losses. USDA recognizes this problem and has called the need to develop better understanding and predictive capability an "urgent" problem.[3] They currently suggest, as an interim measure, that R be increased by adding a factor equal to 1.5 times the December-March precipitation. The effect of this would be to change a typical R value from 20 to around 38. In the equation, such an adjustment would have the effect of doubling the soil loss predicted by the formula.

Such an adjustment may be grossly inadequate, however. Unpublished field research that I carried out in the late 1960's in Idaho indicated that, based on field measurements of actual soil erosion, R factors of 150-250 would be required to make the USLE realistic as an erosion predictor. In the 1977 NRI, the erosion prediction for Class IVe cropland in the Palouse region of eastern Washington was 3 tons per acre (t/a/y) per year. An SCS River Basin study, based on extensive field measurements, estimated 24 t/a/y for the same type of land. On Class VIe cropland, the NRI still estimated 3 t/a/y, while the field study showed 55.[4] The basic message is that, for those areas where snowmelt causes a significant part of the soil erosion, the current predictive tools need a great deal of additional work.

The major use of both erosion equations (WEQ and USLE) in the past have been as conservation planning tools to help farmers see how a change in crops, crop rotations, or conservation practices would affect the rate of soil erosion that could be anticipated on an individual field. In areas where soil erosion is caused mainly by rainfall, the USLE has proven to be a reliable indicator. The wind erosion equation has been similarly reliable in the Great Plains region. Their use in the 1977 NRI was, however, a new and different application that stretched the capability of both equations somewhat. This has led me to the conclusion that the 1977 NRI does not overstate the amount of soil erosion taking place in the United States and, in fact, understates it rather significantly in some regions such as the Pacific Northwest. People who use the data can feel safe, I think, that they are erring, if at all, on the conservative side.

Appendix E

America's Land Base in 1977

Produced by:
National Agricultural Lands Study

Interim Report No. 2

Data Sources

Land Area:
   U.S. Department of Commerce, Bureau of the Census, 1970. Areas of the United States, E20, No. 1, May 1970, As modified for new water developments and changes in Federal ownership.

Land Use, Cropland Potentials:
   Soil Conservation Service, USDA, *1977 National Resource Inventories,* except data for Alaska from Forest Service, USDA, 1979 *RPA, An Assessment of Forest and Range Land Situation in the United States,* (Review Draft).

Definition of Agricultural Land:
   The National Agricultural Lands Study (NALS) defines "agricultural lands" as lands currently used to produce agricultural commodities or lands that have the potential for such production. These lands have a favorable combination of soil quality, growing season, moisture supply, size, and accessibility.
   Land with at least a 25 percent tree canopy cover or land at least 10 percent stocked by forest trees of any size, including land formerly having such tree cover and suitable for natural or artificial reforestation.

Table E.1 **Land Area**

Farm Production Regions and States	Total	Federally Owned	Non-Federal
	(1,000 Acres)		
Northeast			
Connecticut	3,113	8	3,105
Delaware	1,260	38	1,222
Maine	19,843	134	19,709
Maryland	6,323	152	6,171
Massachusetts	5,051	75	4,976
New Hampshire	5,778	709	5,069
New Jersey	4,809	143	4,666
New York	30,589	229	30,360
Pennsylvania	28,746	643	28,103
Rhode Island	677	7	670
Vermont	5,931	276	5,655
Total Northeast	**112,120**	**2,414**	**109,706**
Appalachian			
Kentucky	25,399	1,095	24,304
North Carolina	31,208	1,924	29,284
Tennessee	26,403	1,210	25,193
Virginia	25,477	2,306	23,171
West Virginia	15,404	1,080	14,324
Total Appalachian	**123,891**	**7,615**	**116,276**
Southeast			
Alabama	32,434	876	31,558
Florida	34,473	2,488	31,985
Georgia	37,160	2,036	35,124
South Carolina	19,301	1,111	18,190
Total Southeast	**123,368**	**6,511**	**116,857**
Lake States			
Michigan	36,364	3,074	33,290
Minnesota	50,698	3,219	47,479
Wisconsin	34,844	1,707	33,137
Total Lake States	**121,906**	**8,000**	**113,906**
Corn Belt			
Illinois	35,661	460	35,201
Indiana	23,087	465	22,622
Iowa	35,828	153	35,675
Missouri	44,126	2,094	42,032
Ohio	26,195	301	25,894
Total Corn Belt	**164,897**	**3,473**	**161,424**

Table E.1 — *Continued*

Farm Production Regions and States	Total	Federally Owned	Non-Federal
		(1,000 Acres)	
Delta States			
Arkansas	33,291	3,087	30,204
Louisiana	28,746	1,108	27,638
Mississippi	30,225	1,551	28,674
Total Delta States	**92,262**	**5,746**	**86,516**
Northern Plains			
Kansas	52,354	638	51,716
Nebraska	48,972	648	48,324
North Dakota	44,303	1,723	42,580
South Dakota	48,568	2,784	45,784
Total No. Plains	**194,197**	**5,793**	**188,404**
Southern Plains			
Oklahoma	43,920	1,099	42,821
Texas	167,782	2,898	164,884
Total So. Plains	**211,702**	**3,997**	**207,705**
Mountain			
Arizona	72,587	31,888	40,699
Colorado	66,407	23,607	42,800
Idaho	52,739	33,287	19,452
Montana	93,131	27,115	66,016
Nevada	70,318	59,682	10,636
New Mexico	77,720	26,416	51,304
Utah	52,537	35,933	16,604
Wyoming	62,140	29,488	32,652
Total Mountain	**547,579**	**267,416**	**280,163**
Pacific			
Alaska	362,516	348,516	14,000
California	100,076	46,369	53,707
Hawaii	4,112	337	3,775
Oregon	61,462	32,313	29,149
Washington	42,595	12,220	30,375
Total Pacific	**570,761**	**439,755**	**131,006**
Total All Regions	**2,262,683**	**750,720**	**1,551,963**

Table E.2 **Use Status of Non-Federal Land Available for Agriculture**

Farm Production Regions and States	Cropland		Pastureland		Rangeland	
	Non-Irrigated	Irrigated	Total	With Cropland Conversion Potential	Total	With Cropland Conversion Potential
			(1,000 Acres)			
Northeast						
Connecticut	193	8	112	30	0	0
Delaware	505	37	23	19	0	0
Maine	885	22	249	83	0	0
Maryland	1,638	39	486	187	0	0
Massachusetts	264	18	91	38	0	0
New Hampshire	273	0	95	53	0	0
New Jersey	622	155	144	81	0	0
New York	5,894	75	2,286	726	0	0
Pennsylvania	5,651	10	1,797	599	0	0
Rhode Island	30	0	18	4	0	0
Vermont	590	7	534	119	0	0
Total Northeast	**16,545**	**371**	**5,835**	**1,939**	**0**	**0**
Appalachian						
Kentucky	5,419	9	5,735	2,424	0	0
North Carolina	5,926	271	2,030	978	0	0
Tennessee	4,902	26	5,474	2,363	0	0
Virginia	3,127	82	3,274	823	0	0
West Virginia	981	10	2,037	299	0	0
Total Appalachian	**20,355**	**398**	**18,550**	**6,887**	**0**	**0**
Southeast						
Alabama	4,462	37	4,122	2,360	0	0
Florida	1,469	1,720	5,483	2,121	3,017	640
Georgia	5,851	636	3,234	1,888	0	0
South Carolina	3,287	44	1,242	700	0	0
Total Southeast	**15,069**	**2,437**	**14,081**	**7,069**	**3,017**	**640**
Lake States						
Michigan	9,256	228	1,230	642	0	0
Minnesota	22,518	398	2,889	1,463	110	15
Wisconsin	11,401	340	2,738	1,116	4	0
Total Lake States	**43,175**	**966**	**6,857**	**3,221**	**114**	**15**
Corn Belt						
Illinois	23,770	66	3,070	1,226	0	0
Indiana	13,180	140	2,147	1,001	0	0
Iowa	26,356	75	4,530	1,803	0	0
Missouri	13,797	776	12,823	5,764	35	23
Ohio	11,719	43	2,615	1,077	0	0
Total Corn Belt	**88,822**	**1,100**	**25,185**	**10,871**	**35**	**23**

| Forestland | | Other Land in Farms | | Farmsteads | Total |
Total	With Cropland Conversion Potential	Total	With Cropland Conversion Potential		
		(1,000 Acres)			
1,416	55	50	15	16	1,795
360	93	4	2	13	942
16,520	161	39	10	45	17,770
2,148	325	74	6	91	4,476
2,756	104	38	6	24	3,191
3,976	160	7	4	12	4,363
1,965	230	81	52	52	3,019
15,445	444	609	240	215	24,524
14,349	540	531	160	257	22,595
301	12	3	0	3	355
3,928	73	22	6	16	5,097
63,164	**2,197**	**1,458**	**501**	**744**	**88,127**
10,648	644	130	28	282	22,223
16,813	3,893	114	92	328	25,482
11,638	1,319	64	60	287	22,391
13,233	1,262	83	32	208	20,007
9,805	113	25	19	76	12,934
62,137	**7,231**	**476**	**231**	**1,181**	**103,037**
19,792	1,763	96	24	231	28,740
12,140	814	238	38	117	24,184
21,566	3,861	41	29	242	31,570
10,770	1,515	33	29	121	15,497
64,268	**7,953**	**408**	**120**	**711**	**99,991**
15,323	894	962	278	315	27,314
13,806	1,919	1,080	314	698	41,499
13,259	1,353	1,275	127	350	29,367
42,388	**4,166**	**3,317**	**719**	**1,363**	**98,180**
3,028	497	527	212	501	30,962
3,534	596	418	197	384	19,803
1,487	183	215	179	708	33,371
10,832	727	216	86	422	38,901
5,865	629	384	162	366	20,992
24,746	**2,632**	**1,760**	**836**	**2,381**	**144,029**

(Continued on next page)

Table E.2—*Continued*

Farm Production Regions and States	Cropland		Pastureland		Rangeland	
	Non-Irrigated	Irrigated	Total	With Cropland Conversion Potential	Total	With Cropland Conversion Potential
			(1,000 Acres)			
Delta States						
Arkansas	5,547	2,443	5,628	1,966	248	47
Louisiana	4,738	1,161	2,945	1,344	326	0
Mississippi	6,948	354	4,041	1,806	30	0
Total Delta States	**17,233**	**3,958**	**12,614**	**5,116**	**604**	**47**
Northern Plains						
Kansas	25,631	3,175	2,701	1,356	16,276	3,992
Nebraska	13,794	6,905	2,899	1,253	22,001	2,499
North Dakota	26,835	78	1,544	816	10,564	1,913
South Dakota	17,684	472	2,413	1,289	22,198	4,101
Total No. Plains	**83,944**	**10,630**	**9,557**	**4,714**	**71,039**	**12,505**
Southern Plains						
Oklahoma	11,073	710	8,713	2,785	14,566	2,709
Texas	22,510	7,929	18,768	4,781	95,401	8,781
Total So. Plains	**33,583**	**8,639**	**27,481**	**7,566**	**109,967**	**11,490**
Mountain						
Arizona	145	1,167	11	9	35,091	313
Colorado	7,699	3,394	1,598	254	23,801	2,408
Idaho	2,743	3,547	1,109	500	6,589	793
Montana	13,294	2,061	2,647	1,136	38,834	4,476
Nevada	4	1,103	298	24	7,351	231
New Mexico	1,203	1,079	382	33	42,096	1,249
Utah	655	1,160	626	145	9,385	363
Wyoming	1,320	1,650	736	254	26,169	1,600
Total Mountain	**27,063**	**15,161**	**7,407**	**2,355**	**189,316**	**11,433**
Pacific						
Alaska	46	0	1	0	6,276	0
California	1,920	8,153	1,127	613	17,554	1,789
Hawaii	139	154	992	74	0	0
Oregon	3,139	2,009	1,767	512	10,110	390
Washington	6,179	1,772	1,252	480	6,041	605
Total Pacific	**11,423**	**12,088**	**5,139**	**1,679**	**39,981**	**2,784**
Total All Regions	**357,212**	**55,758**	**132,706**	**51,417**	**414,073**	**38,937**

Forestland		Other Land in Farms			
Total	With Cropland Conversion Potential	Total	With Cropland Conversion Potential	Farmsteads	Total

(1,000 Acres)

Total	With Cropland Conversion Potential	Total	With Cropland Conversion Potential	Farmsteads	Total
14,072	1,194	99	74	197	28,234
12,595	1,550	58	26	162	21,985
14,412	1,916	92	44	246	26,123
41,079	**4,660**	**249**	**144**	**605**	**76,342**
788	115	328	95	378	49,277
444	36	253	157	478	46,774
366	62	1,010	91	397	40,794
330	20	531	83	477	44,105
1,928	**233**	**2,122**	**426**	**1,730**	**180,950**
4,931	253	82	47	274	40,349
9,240	524	874	140	619	155,341
14,171	**777**	**956**	**187**	**893**	**195,690**
1,803	1	17	14	17	38,251
3,343	30	225	42	183	40,243
4,230	144	51	4	126	18,395
6,341	55	194	32	162	63,533
230	0	19	7	17	9,022
3,426	1	191	4	76	48,453
1,071	0	23	10	33	12,953
1,164	4	9	7	52	31,100
21,608	**235**	**729**	**120**	**666**	**261,950**
6,900	124	0	0	1	13,224
9,855	63	281	112	257	39,147
1,443	11	90	3	3	2,821
10,066	219	127	12	151	27,369
12,382	430	173	33	234	28,033
40,646	**847**	**671**	**160**	**646**	**110,594**
376,135	**30,931**	**12,086**	**3,444**	**10,920**	**1,358,890**

Table E.3 **Non-Federal Land Unavailable for Agriculture**

Farm Production Regions and States	Urban & Built-Up	Rural Transportation	Other Non-Farm	Water	Total
			(1,000 Acres)		
Northeast					
Connecticut	971	56	244	39	1,310
Delaware	151	20	95	14	280
Maine	433	224	1,173	109	1,939
Maryland	1,196	139	310	50	1,695
Massachusetts	1,243	106	385	51	1,785
New Hampshire	337	89	230	50	706
New Jersey	1,176	67	367	37	1,647
New York	2,994	603	1,982	257	5,836
Pennsylvania	3,370	609	1,360	169	5,508
Rhode Island	248	12	46	9	315
Vermont	316	100	114	28	558
Total Northeast	**12,435**	**2,025**	**6,306**	**813**	**21,579**
Appalachian					
Kentucky	1,222	450	255	154	2,081
North Carolina	2,151	651	709	291	3,802
Tennessee	1,698	583	370	151	2,802
Virginia	2,074	321	574	195	3,164
West Virginia	798	211	304	77	1,390
Total Appalachian	**7,943**	**2,216**	**2,212**	**868**	**13,239**
Southeast					
Alabama	1,728	605	257	228	2,818
Florida	4,273	600	2,432	496	7,801
Georgia	2,026	439	693	396	3,554
South Carolina	1,543	439	533	178	2,693
Total Southeast	**9,570**	**2,083**	**3,915**	**1,298**	**16,866**
Lake States					
Michigan	3,287	798	1,680	211	5,976
Minnesota	1,397	1,072	3,105	406	5,980
Wisconsin	1,602	724	1,198	246	3,770
Total Lake States	**6,286**	**2,594**	**5,983**	**863**	**15,726**
Corn Belt					
Illinois	2,662	830	444	264	4,239
Indiana	1,834	518	306	161	2,819
Iowa	857	1,062	163	222	2,304
Missouri	1,375	1,056	328	372	3,131
Ohio	3,532	626	542	202	4,902
Total Corn Belt	**10,260**	**4,092**	**1,822**	**1,221**	**17,395**

Table E.3—*Continued*

Farm Production Regions and States	Urban & Built-Up	Rural Transportation	Other Non-Farm	Water	Total
		(1,000 Acres)			
Delta States					
Arkansas	1,049	516	148	257	1,970
Louisiana	999	434	3,862	358	5,653
Mississippi	1,389	521	229	412	2,551
Total Delta States	**3,437**	**1,471**	**4,239**	**1,027**	**10,174**
Northern Plains					
Kansas	747	1,089	310	293	2,439
Nebraska	509	814	55	172	1,550
North Dakota	269	986	150	381	1,786
South Dakota	256	733	474	216	1,679
Total No. Plains	**1,781**	**3,622**	**989**	**1,062**	**7,454**
Southern Plains					
Oklahoma	1,132	745	283	312	2,472
Texas	5,171	2,088	1,599	685	9,543
Total So. Plains	**6,303**	**2,833**	**1,882**	**997**	**12,015**
Mountain					
Arizona	900	153	1,384	11	2,448
Colorado	827	584	1,010	136	2,557
Idaho	377	232	381	67	1,057
Montana	438	705	1,091	249	2,483
Nevada	235	106	1,254	19	1,614
New Mexico	717	334	1,748	52	2,851
Utah	564	157	2,895	35	3,651
Wyoming	222	320	913	97	1,552
Total Mountain	**4,280**	**2,591**	**10,676**	**666**	**18,213**
Pacific					
Alaska	83	257	405	31	776
California	4,116	850	9,355	239	14,560
Hawaii	139	18	790	7	954
Oregon	788	325	557	110	1,780
Washington	1,368	431	389	154	2,342
Total Pacific	**6,494**	**1,881**	**11,496**	**541**	**20,412**
Total All Regions	**68,789**	**25,408**	**49,520**	**9,356**	**153,073**

Table E.4 **Conversion of Agricultural Land**

Farm Production Regions and States	Agricultural Land Converted to Urban and Built-Up, Transportation and Water 1967-1977	Prime Farmland	
		Total	In Cropland Use
	(1,000 Acres)	**(1,000 Acres)**	
Northeast			
Connecticut	360	394	133
Delaware	60	350	276
Maine	⟨1	853	324
Maryland	780	1,262	814
Massachusetts	300	448	169
New Hampshire	190	144	86
New Jersey	140	1,249	502
New York	810	4,000	2,286
Pennsylvania	1,280	4,448	2,351
Rhode Island	150	84	23
Vermont	220	374	128
Total Northeast	**4,290**	**13,606**	**7,092**
Appalachian			
Kentucky	760	5,994	3,334
North Carolina	1,280	5,606	2,729
Tennessee	770	6,447	3,078
Virginia	1,400	4,324	1,508
West Virginia	480	502	285
Total Appalachian	**4,690**	**22,873**	**10,934**
Southeast			
Alabama	890	7,856	2,913
Florida	3,470	1,417	404
Georgia	1,400	7,767	3,655
South Carolina	920	3,484	1,543
Total Southeast	**6,680**	**20,524**	**8,515**
Lake States			
Michigan	1,220	8,382	5,695
Minnesota	490	19,513	15,302
Wisconsin	190	10,319	6,475
Total Lake States	**1,900**	**38,214**	**27,472**
Corn Belt			
Illinois	1,060	21,400	19,100
Indiana	740	14,162	11,515
Iowa	440	19,127	16,875
Missouri	480	15,067	9,544
Ohio	1,310	11,280	9,216
Total Corn Belt	**4,030**	**81,036**	**66,250**

Table E.4—*Continued*

Farm Production Regions and States	Agricultural Land Converted to Urban and Built-Up, Transportation and Water 1967-1977	Prime Farmland	
		Total	In Cropland Use
	(1,000 Acres)	**(1,000 Acres)**	
Delta States			
Arkansas	370	13,250	6,633
Louisiana	250	9,353	5,267
Mississippi	720	10,227	5,200
Total Delta States	**1,340**	**32,830**	**17,100**
Northern Plains			
Kansas	‹1	27,318	19,520
Nebraska	250	14,203	11,899
North Dakota	320	13,915	12,701
South Dakota	530	5,071	4,312
Total No. Plains	**1,100**	**60,507**	**48,432**
Southern Plains			
Oklahoma	250	15,622	8,390
Texas	2,260	37,498	17,631
Total So. Plains	**2,510**	**53,120**	**26,021**
Mountain			
Arizona	320	1,161	1,086
Colorado	400	1,760	1,613
Idaho	190	3,512	2,998
Montana	350	1,240	889
Nevada	‹1	303	243
New Mexico	290	524	504
Utah	290	650	641
Wyoming	‹1	259	224
Total Mountain	**1,840**	**9,409**	**8,198**
Pacific			
Alaska	‹1	0	0
California	1,500	7,805	6,545
Hawaii	60	227	184
Oregon	270	2,373	1,823
Washington	630	2,016	1,453
Total Pacific	**2,460**	**12,421**	**10,005**
Total All Regions	**30,840**	**344,540**	**230,011**

EXPLANATION OF TABLE COLUMN HEADINGS AND ELEMENTS

FARM PRODUCTION REGIONS
   Grouping of states used by USDA agencies to present natural and related resource data.

LAND AREA
   Based on Bureau of the Census data and adjusted for: (1) new water bodies greater than 40 acres in size; (2) changes in Federal ownership.

USE STATUS OF NON-FEDERAL LAND AVAILABLE
FOR AGRICULTURE
   The definitions of the land use and cover categories are those of the USDA Soil Conservation Service's 1977 National Resource Inventories. Definitions of rangeland and forestland are used by both the USDA Forest Service and the Soil Conservation Service.

Cropland
   Land used to produce adapted crops for harvest, either alone or in rotation with grasses and legumes. Cropland production includes row crops, close grown field crops, hay crops, rotation hay and pasture, nursery crops, orchard crops, and other similar specialty crops, summerfallow, and other cropland not harvested or pastured.

Pastureland

Lands producing forage plants, principally introduced species for animal consumption. In addition to regulating the intensity of grazing management practices typically include such cultural treatments as reseeding, renovation, reestablishment, mowing, weed or brush control, liming or fertilization. Pastureland may be on drained or irrigated lands. Also included in pastureland, regardless of treatment, is land being managed to establish or maintain stands of grasses such as bluegrass, bromegrass, or bermuda grass, either alone or in mixtures with clover or other legumes.

Rangeland

Land on which the potential or natural vegetation climax species are predominantly grasses, grass-like plants, forbs, or shrubs. Included are lands revegetated either naturally or artificially and managed to duplicate native vegetation. Rangelands include natural grasslands, savannahs, shrublands, most deserts, tundra, alpine communities, coastal marshes, and wet meadows. They include land with less than 10 percent stocking with forest trees of any size.

Rangelands in Alaska are dominantly tundra and alpine communities. There is some use of the rangeland by caribou and also some limited reindeer herding.

Forestland (Woodland)

Land with at least a 25 percent tree canopy cover or land at least 10 percent stocked by forest trees of any size, including land formerly having such tree cover and suitable for natural or artificial reforestation.

Other Land in Farms

Land reserved for wildlife and windbreaks, not directly associated with farmsteads. Includes commercial feedlots, greenhouses, and nurseries.

Farmsteads

Land for dwellings, buildings, barns, pens, corrals, farmstead windbreaks, family gardens, and other uses connected with operating farms and ranches.

Potential for Cropland of 1977 Pastureland, Rangeland, Forestland, and Other Land in Farms

Determinations of cropland potential for each data point were made by a group representing a variety of USDA agencies. They were made on the basis of 1976 commodity prices, as well as development and production costs. A "high potential" rating required favorable physical characteristics and also evidence of similar land being converted to cropland during the last three years. A "medium potential" rating required favorable physical characteristics, but generally conversion costs were expected to be higher than those for soils with a high potential rating.

Potential for Cropland
Potential ratings for Alaska were made by the State USDA Land Use Committee. They represent conversions to cropland anticipated by 1985.

Areas with cropland potential include acreages rated as having "high" or "medium" potential. They represent a portion of the pastureland, rangeland, forestland and other land in farm acreages, and thus are *not* a component of the "totals" of non-federal land available for agriculture.

NON-FEDERAL LAND UNAVAILABLE FOR AGRICULTURE
The definition of land use and cover categories are those used by the Soil Conservation Service in its *1977 National Resources Inventories.*

Urban & Built-Up
Land used for residences, industrial sites, commercial sites, construction sites, railroad yards, small parks of less than 10 acres within urban and built-up areas, cemeteries, airports, golf courses, sanitary land fills, sewage treatment plants, water control structures and spillways, shooting ranges and so forth. The rights-of-way of highways, railroads and other transportation facilities are included if they are within urban and built-up areas. In 1977, this category included four million acres of "small" built-up areas of from ¼ to 10 acres in size.

Rural Transportation
Land used for roads and railroads in rural areas. Generally, this includes the entire right-of-way.

Other Non-Farm
Land used for greenbelts, large unwooded parks and other non-farm uses not elsewhere defined. This category also includes land in strip mines, quarries, gravel pits, and borrow pits that have not been reclaimed for other uses. Between 2 and 3 million acres have cropland conversion potential.

Water
Water bodies less than 40 acres in size and streams less than ⅛ mile wide.

AGRICULTURAL LAND CONVERTED TO URBAN, BUILT-UP, RURAL TRANSPORTATION & WATER 1967-1977
Data sources and definitions are those of the USDA 1967 *Conservation Needs Inventory* and the USDA SCS 1977 *National Resources Inven-*

tories. There are some differences in the inventory procedures used to determine the extent of urban, built-up, rural transportation and water in these two inventories. Therefore, the "converted" acreages should be interpreted as estimates rather than precise measures of land use change over the period.

PRIME FARMLAND

Prime farmland is the best land for farming. Prime acres are flat or gently rolling, and susceptible to little or no soil erosion. They are our most energy-efficient acres, producing the most food, feed, fiber, forage and oilseed crops with the least amount of fuel, fertilizer and labor.

Their soil quality, growing season and moisture supply assure continuous, high productivity without degrading the environment.

Prime farmland includes cropland, pastureland, range and forestlands. It does not include land converted to urban, industrial, transportation or water.

Appendix F

USDA Conservation Programs

In addition to the programs discussed in Chapter 13, there are several of the 30-plus different soil conservation programs administered by the USDA that merit some explanation. Many of them do not relate as directly to the prevention of soil erosion or water conservation as do the basic programs of technical assistance and cost-sharing, but they are important just the same.

Table F.1 lists most of the different USDA programs, along with the law that authorized them and the agency that is charged with their administration. Many of the original laws have been amended, and many times the programs are guided by policies set forth in the language of the Agricultural Appropriations Acts, so a reading of the original law may not be adequate as a guide to today's program directions.

What is readily apparent, however, is that Congress has created a wide variety of programs, lodged in several USDA agencies, to carry out a complex effort involving research, data gathering, education, information, technical assistance, cost-sharing, project construction, and loans. Each has been in response to a demonstrated need, and each has been altered as conditions have evolved and national priorities shifted.

In addition, there are many federal programs that relate to public lands, fish and wildlife management, water resource development, and environmental protection that are part of the national effort to manage natural resources. Those programs have not been listed, because they are outside the scope of this book, but many of them have important influences on the way farmers use and manage their farmland. For example, federal water resource projects have greatly affected water use and conservation in the West, while federal land management policies often dictate how a rancher who leases grazing on the public lands must manage his private land in order to have a balanced ranching operation.

Table F.1. **Authorizing Legislation and Type of USDA Conservation Programs***

Agency	Conservation Program	Public Law	Date Enacted	Type of Program
Agricultural Stabilization & Conservation Service (ASCS)	Agricultural Conservation Program	74-461	1936	Cost-Sharing
	Emergency Conservation Program	95-334	1978	Cost-Sharing
	Water Bank Program	91-559	1970	Cost-Sharing
	Forestry Incentives Program	95-313	1978	Cost-Sharing
	Rural Clean Water Program	96-108	1979	Cost-Sharing
Farmers Home Administration (FmHA)	Watershed Loans	83-566	1954	Loans
	Soil & Water Loans to Individuals	87-128	1961	Loans
	Resource Conservation & Development	87-703	1962	Loans
	Irrigation, Drainage & Other	92-419	1972	Loans
Forest Service (FS)	State & Private	68-348	1924	Technical Assistance
	Forestry	81-729	1950	Technical Assistance
	Forestry Research	70-466	1928	Research
Agricultural Research Service (ARS)	Agricultural Research	74-46	1935	Research

(Continued on next page)

Agency	Conservation Program	Public Law	Date Enacted	Type of Program
Cooperative State Research Service (CSRS)	Cooperative Research	79-733	1946	Research
Extension Service	Land and Water Conservation	95-306	1978	Extension/ Education
Soil Conservation Service (SCS)	Technical Assistance	74-46	1935	Technical Assistance
	Snow Survey & Water Forecasting	74-46	1935	Data Collection
	Soil Survey	74-46	1935	Data Collection
	Plant Materials Program	74-46	1935	Testing New Plants
	Flood Prevention	78-534	1944	Technical Assistance/ Cost-Sharing
	Emergency Water- shed Protection	81-516	1950	Technical Assistance/ Cost-Sharing
	Watershed Planning	83-566	1954	Technical Assistance
	Watershed Operations	83-566	1954	Technical Assistance/ Cost-Sharing
	Flood Plain Management Assistance	83-566	1954	Technical Assistance
	River Basin Surveys & Investigations	83-566	1954	Technical Assistance
	Great Plains Con- servation Program	84-1021	1956	Technical Assistance/ Cost-Sharing
	Resource Conservation & Development	87-703	1962	Technical Assistance/ Cost-Sharing
	Land Inventory & Monitoring	92-419	1972	Data Collection
	Rural Abandoned Mine Reclamation	95-87	1977	Technical Assistance/ Cost-Sharing
	Resource Appraisal and Program Development	95-192	1977	Analysis of National Needs

*Public laws cited are basic authorities but may have been amended.

Other USDA
Cost-Sharing Programs

Beginning in 1957, the Agricultural Stabilization and Conservation Service (ASCS) has had authority to cost-share with farmers on measures to correct damage caused by natural disasters. This program, called the Emergency Conservation Program (ECP) allows local ASCS committees to apply through the state committee to USDA for designation as a "disaster" county, then set up special measures designed to help farmers rehabilitate farmland damaged by floods or droughts.

In 1970, Public Law 91-559 established the Water Bank Program to help conserve surface waters; to preserve and improve habitat for migratory waterfowl and other wildlife; and to preserve, restore, and improve wetlands. The program, administered by ASCS, offers ten-year agreements based on a conservation plan developed by SCS and the conservation district. Under the program, farmers in 172 counties in 15 states can receive financial incentives to leave surface wetlands in a wet condition and manage surrounding croplands so that wildlife values in the wetlands are preserved. Most of the activity in the program is in the northern Great Plains, particularly in North Dakota and Minnesota.

In 1974, the Forest Incentives Program was established to allow ASCS to enter into annual or long-term cost-sharing agreements with private forest owners to plant trees or improve an existing stand. To date, 1,450 counties have participated in the program, and more than 1.2 million acres of trees have been planted and timber stands improved.

Section 406 of the Surface Mining Control and Reclamation Act of 1977 authorized USDA to enter into contracts with landowners to reclaim abandoned or inadequately reclaimed coal-mined lands. Participants in this Rural Abandoned Mine Program, administered by SCS, must develop a conservation and development plan that sets forth the conservation treatment required on the land and the post-reclamation uses planned. The contract, based on the plan, provides for federal cost-sharing and technical assistance in carrying out the reclamation activities. Funds for the program come from a per-ton reclamation tax levied on each ton of coal mined in the United States. The tax goes into a special fund, is then appropriated out of that fund by Congress to the Department of the Interior, and then up to 20 percent of the money allotted to Interior can be transferred to SCS for the Rural Abandoned Mine Program.

Loan Programs

Over the years, Congress has authorized a variety of loan programs to meet the needs of U.S. farmers. The basic farm loan program administered by the Farmer's Home Administration (FmHA) is called the Farm Ownership Loan program. In addition to borrowing for the purchase or enlargement of a farm, money can be secured for the improvement of farm facilities, including soil and water conservation practices.

In addition, FmHA administers several loan programs aimed more directly at soil and water conservation. Soil and water loans can be made to assist farmers carry out projects for soil conservation, water development and conservation, drainage, forestation, pasture improvement, or pollution abatement. Recreation loans are available to farmers who wish to establish income-producing recreation enterprises on their farms, and emergency loans are available for repairing damage caused by natural disasters.

Organized associations of farmers can qualify for irrigation and drainage loans to build or renovate systems that serve several farms, and non-profit grazing associations can obtain grazing loans to acquire and develop pastureland and rangelands. In addition, there are separate loan programs for participants in Public Law 566 small watershed projects and Resource Conservation and Development areas.

The Resource Conservation and Development Program

Since 1962, USDA has provided assistance to multi-county areas that develop plans for land conservation and development to benefit rural communities. Under this program, administered by SCS, communities can be provided both technical and financial assistance for installing measures and facilities for water quality management, improving rural community water supplies, controlling agriculturally related pollution, disposal of solid wastes, and development of water-based fish and wildlife and recreational developments.

In addition to technical and financial assistance from SCS, Resource Conservation and Development (RC&D) loans are available from the Farmers Home Administration. Other USDA agencies such as Extension Service, Forest Service, and ASCS provide assistance to project sponsors as well. There are now 194 RC&D

areas in the nation, covering 1,325 counties in 49 states and 1 territory for a total of over 829 million acres. Attempts by the Carter administration to phase out the program as a way of reducing USDA's budget met with stiff political resistance from sponsors and others convinced of the program's effectiveness, and Congress rejected it emphatically. By the end of 1980, several bills had been introduced in the 96th Congress to provide a new legislative authority for the program, but action will need to be taken in the 97th Congress if the program's supporters are to prevail.

Notes and References Cited

Chapter 1

1. Walter C. Lowdermilk, *Conquest of the Land Through 7,000 Years*, USDA-SCS Information Bulletin 99 (Washington, D.C.: USDA, 1953).

2. Ibid., p. 5.

3. Ibid., p. 10.

4. Richard Barnet, *The Lean Years* (New York: Simon & Schuster, 1980), p. 17.

5. Julian L. Simon, "Resources, Population, Environment: An Oversupply of False Bad News," *Science*, June 27, 1980, p. 1,432.

Chapter 2

1. Angus McDonald, *Early American Soil Conservationists*, Misc. Pub. No. 449 (Washington, D.C.: USDA, 1941), p. 3.

2. Ibid., p. 11.

3. D. Harper Simms, *The Soil Conservation Service* (New York: Praeger, 1970), p. 9.

4. Hugh H. Bennett, "The Land We Defend" (Speech given before the 78th Annual Meeting of the National Education Association, Milwaukee, Wis., July 2, 1940).

5. Hugh H. Bennett, quoted in testimony of Peter R. Huessy, The Environmental Fund, Washington, D.C., before the Subcommittee on Environment, Soil Conservation and Forestry of the U.S. Senate Agriculture Committee, August 2, 1977.

6. Walter C. Lowdermilk, *Conquest of the Land Through 7,000 Years*, USDA-SCS Information Bulletin 99 (Washington, D.C.: USDA, 1953), p. 30.

7. Franklin D. Roosevelt, in a letter addressed to all state governors urging the states to set up soil conservation districts to work with the newly created Soil Conservation Service, February 26, 1937.

8. Ann Crittenden, "Soil Erosion Threatens U.S. Farms' Output," *New York Times*, October 26, 1980.

9. A report prepared at the request of concerned members of Congress was: Comptroller General of the United States, *To Protect Tomorrow's Food Supply, Soil Conservation Needs Priority Attention*, Report #CED-77-30 (Washington, D.C.: General Accounting Office, 1977).

10. G. V. Jacks and R. O. White, *Vanishing Lands: A World Survey of Soil Erosion* (New York: Doubleday, 1939).

11. Keith O. Campbell, *Food for the Future* (Lincoln: University of Nebraska Press, 1979), p. 22.

12. Stuart Chase, *Rich Land, Poor Land* (New York: McGraw Hill, 1936).

13. William and Paul Paddock, *Famine: 1975* (Boston: Little, Brown & Co., 1967), p. 8.

14. *Food for the Future*, p. 26.

15. Julian L. Simon, "Resources, Population, Environment: An Oversupply of False Bad News," *Science*, June 27, 1980, p. 1,432.

16. Ibid., p. 1,436.

17. Richard Barnet, *The Lean Years* (New York: Simon & Schuster, 1980), p. 175.

18. Orville Krause and Dwight Hair, "Trends in Land Use and Competition for Land to Produce Food and Fiber" *Perspectives on Prime Lands* (Washington, D.C.: USDA, 1975), p. 12.

19. *The Lean Years*, p. 173.

20. Orville Freeman, *Agriculture/2000* (Washington, D.C.: USDA, 1967), p. 7.

21. John F. Timmons, "Agricultural Land Retention and Conversion Issues: An Introduction," *Farmland, Food and the Future* (Ankeny, Iowa: Soil Conservation Society of America, 1979), p. 1.

22. Lyle Schertz et al., *Another Revolution in U.S. Farming?* (Washington, D.C.: USDA, 1979), p. 48.

23. Bob Bergland, Statement before the Committee on Science and Technology, U.S. House of Representatives, Washington, D.C., May 4, 1979, p. 15.

24. Ibid., p. 26.

25. USDA, *1979 Handbook of Agricultural Charts*, Agriculture Handbook No. 561 (Washington, D.C.: USDA, 1979), p. 75.

26. Ernest Conine, "U.S. Should Consider Two-Tier Pricing System for Grains," *Albuquerque Journal*, October 23, 1980.

27. "Resources, Population, Environment," p. 1,433.

28. *The Lean Years*, p. 165.

29. *Food for the Future*, p. 24.

30. Lester R. Brown, *Resource Trends and Population Policy: A Time for Reassessment*, Worldwatch Paper 29 (Washington, D.C.: Worldwatch Institute, 1979), p. 50.

31. USDA, *RCA Appraisal 1980: Review Draft, Part II* (Washington, D.C.: USDA, 1980), p. 2-2.

32. *The Lean Years*, p. 163.

33. Council on Environmental Quality and Department of State, *The Global 2000 Report to the President* (Washington, D.C.: CEQ, 1980), p. 8.

34. Ibid., p. 9.

35. *The Lean Years*, p. 159.

36. *RCA Appraisal: Part II*, p. 2-6.

37. John F. Timmons, "Solving Natural Resource Problems of Food and Energy," *Journal of Soil and Water Conservation* 34, no. 3 (1979): 122.

Chapter 3

1. Lance Gay, "Block Urges Preservation of Farm Land," *Washington Star*, February 10, 1981.

2. Wendell Berry, *The Unsettling of America: Culture and Agriculture* (New York: Avon, 1977), p. 46.

3. Don Paarlberg, *Farm and Food Policy: Issues of the 1980's* (Lincoln: University of Nebraska Press, 1980), p. 10.

4. John Kramer, "Agriculture's Role in Government Decisions," *Consensus and Conflict in U.S. Agriculture: Perspectives from the National Farm Summit* (College Station: Texas A & M University Press, 1979), p. 208.

5. USDA, *1979 Handbook of Agricultural Charts*, Agriculture Handbook No. 561 (Washington, D.C.: USDA, 1979), p. 77.

6. Ibid., pp. 77, 136, 137.

7. *Chicago Tribune*, October 23, 1980, pp. 1-3. Grains included are wheat, corn, sorghum, rice, barley, millet, rye, oats, and mixed grains.

8. John F. Timmons, "Agricultural Land Retention and Conversion Issues: An Introduction," *Farmland, Food and the Future* (Ankeny, Iowa: Soil Conservation Society of America, 1979), p. 1.

9. Carl C. Campbell, "Comments by a Task Force Member," *Consensus and Conflict in U.S. Agriculture: Perspectives from the National Farm Summit* (College Station: Texas A & M University Press, 1979), p. 148.

10. Steve Huntley, "Why Farmers are Singing the Blues," *U.S. News and World Report*, May 19, 1980, p. 55.

11. Sen. Larry E. Pressler, "End the Grain Embargo," *Washington Star*, October 11, 1980.

12. Ibid.

13. *Denver Post*, October 28, 1980.

14. Lee Lescaze, "Farm Prices are Up, Says Carter Proudly," *Washington Post*, October 26, 1980, p. A-2.

15. Martha M. Hamilton, "New Data Show Tightening of Livestock Feed," *Washington Post*, November 11, 1980.

16. Dan Morgan, *Merchants of Grain* (New York: Viking, 1979).

17. Ray Jergeson, "Economic Realities," *The Future of Agriculture in the Rocky Mountains* (Salt Lake City: Westwater Press, 1980), p. 123.

18. U.S. Department of Housing and Urban Development, *1976 Report on National Growth and Development: The Changing Issues for National Growth*, HUD-386-2-CPD (Washington, D.C.: HUD, 1976), p. 5.

19. Louis Harris and Associates, Inc., "Outline for Press Briefing on a Survey of the Public's Attitudes Toward Soil, Water, and Renewable Resources Conservation Policy," Washington, D.C., January 17, 1980, p. 3.

20. S. E. Durcholz, "How Soil-Saving Programs Look to a Working Farmer," *Catholic Rural Life* 29, no. 3 (March 1980): 13.

21. Charles McLaughlin, "Statement before the Subcommittee on Environment, Soil Conservation and Forestry, Ames, Iowa, August 14, 1980," *Soil Conservation* (Washington, D.C.: Government Printing Office, 1980), pp. 109-10.

22. USDA, *1980 Handbook of Agricultural Charts*, Agriculture Handbook No. 574 (Washington, D.C.: USDA, 1980), p. 16.

23. Ibid., p. 15.

24. "The New American Farmer," *Time*, November 6, 1978, p. 96.

25. Ibid., p. 102.

26. Emery N. Castle, "Resource Allocation and Production Costs," *Consensus and Conflict in U.S. Agriculture: Perspectives from the National Farm Summit* (College Station: Texas A & M University Press, 1979), p. 7.

27. Ibid., p. 8.

28. G. B. White and E. J. Partenheimer, "Economic Impacts of Erosion and Sedimentation Control Plans: Case Studies of Pennsylvania Dairy Farms," *Journal of Soil and Water Conservation* 35, no. 2 (March-April 1980): 76.

29. James Lake, *Environmental Impact of Land Use on Water Quality: Final Report on the Black Creek Project* (Summary), (Fort Wayne, Ind.: Allen County Soil and Water Conservation Districts, 1977), p. 38.

30. Herbert Hoover and Marc Wiitala, *Operator and Landlord Participation in Soil Erosion Control in the Maple Creek Watershed in Northeast Nebraska,* ESCS Staff Report NRED 80-4. (Washington, D.C.: Economics, Statistics, and Cooperatives Service, USDA, 1980).

31. David Sheridan, *Desertification of the United States* (Washington, D.C.: Council on Environmental Equality, 1981), p. 82.

32. Ibid., p. 81.

Chapter 4

1. R. M. Davis, "Prime and Unique Farmlands," *Land Inventory and Monitoring Memorandum* #3 (Washington, D.C.: SCS, October 15, 1975).

2. Raymond I. Dideriksen, Allen R. Hidlebaugh, and Keith O. Schmude, "Trends in Agricultural Land Use," *Farmland, Food and the Future* (Ankeny, Iowa: Soil Conservation Society of America, 1979), p. 15.

3. Ibid., p. 21.

4. News release, U.S. National Alcohol Fuels Commission, Washington, D.C., September 5, 1980.

Chapter 5

1. Dr. Charles Benbrook, Council on Environmental Quality, personal correspondence. (Calculated from the NALS conversion data using 100 bushels per acre as the average corn yield.)

2. Ibid. (Calculated from conversion rates accepted in the NALS.)

3. John Cozart, "America's Vital Farmland Is Vanishing," *Tulsa Tribune,* July 23, 1980; George L. Baker, "Resource Crisis Perils State Farm Output," *Fresno Bee,* November 9, 1980; and M. Rupert Cutler, "Crops Die Where Cities Sprawl," *Chicago Tribune,* July 6, 1980.

4. Pierre Crosson, "The Long-Term Adequacy of Agricultural Land in the United States: An Overview" (Manuscript prepared for publication of an RFF book in 1981, Washington, D.C.: Resources for the Future, 1981).

5. Don Paarlberg, *Farm and Food Policy: Issues of the 1980's* (Lincoln: University of Nebraska Press, 1980), p. 151.

6. Charles E. Little, *Land & Food: The Preservation of U.S. Farmland* (Washington, D.C.: American Land Forum, 1979), p. 10.

7. Pierre Crosson, "Agricultural Land Use: A Technological and Energy Perspective," *Farmland, Food and the Future,* (Ankeny, Iowa: Soil Conservation Society of America, 1979), p. 106.

8. Robert E. Coughlin, "Agricultural Land Conversion in the Urban Fringe," *Farmland, Food and the Future,* p. 29.

9. Raymond I. Dideriksen, A. R. Hidlebaugh, and. Keith O. Schmude, "Trends in Agricultural Land Use," *Farmland, Food and the Future,* p. 21.

10. Frederick J. Napolitano, "Statement of the National Association of Home Builders" (Testimony before the Senate Agriculture Committee's Subcommittee on Environment, Soil Conservation and Forestry, September 16, 1980), p. 3.

11. Soil Conservation Service, *Basic Statistics: Status of Land Disturbed by Surface Mining in the United States, as of July 1, 1977* (Washington, D.C.: USDA, 1977).

12. *Farmland, Food and the Future*, p. 23.

13. J. Dixon Esseks, "Nonurban Competition for Farmland," *Farmland, Food and the Future*, p. 57.

14. Ibid.

15. Ibid., p. 58.

16. National Association of Counties, "Farmland and Coal: The Battle Over Stripmining," *Aglands Exchange* 2, no. 1 (September-October 1980): 7.

17. Ibid., p. 8.

18. *Farmland, Food and the Future*, p. 58.

19. U.S. Department of Energy, *Environmental Analysis of Synthetic Liquid Fuels.* (Washington, D.C.: DOE, 1979), Table 4, p. 16.

20. W. Wendell Fletcher, *Farmland and Energy: Conflicts in the Making* (Washington, D.C.: National Agricultural Lands Study, 1980), p. 22.

21. Office of Surface Mining Reclamation and Enforcement, *Final Environmental Impact Statement, Section 501(b) Regulations.* (Washington, D.C.: U.S. Department of the Interior, 1979), p. BIII-34. See also: Janet M. Smith, David Ostendorf, and Mike Schechtman, *Who's Mining the Farm?* (Herrin, Ill.: Illinois South Project, 1979).

22. U.S. Department of Energy, *Additions to Generating Capacity: 1978-1987* (Washington, D.C.: DOE, October, 1978), p. 1.

23. Sally Jacobsen, "Land Use Dispute in Illinois: Nuclear Power vs. Crops," *Bulletin of the Atomic Scientists* 29 (1973).

24. Jack Doyle, *Lines Across the Land.* (Washington, D.C.: Environmental Policy Institute, 1979), p. 163.

25. *Farmland, Food and the Future*, p. 60.

26. Keith O. Schmude, "A Perspective on Prime Farmland," *Journal of Soil and Water Conservation* 32, no. 5, (September-October 1977): 242.

27. *Farmland, Food and the Future,* p. 60.

28. Louise B. Young, *Power over People.* (New York: Oxford University Press, 1973).

29. Thomas W. Smith et al., *Transmission Lines: Environmental and Public Policy Considerations.* (Madison: University of Wisconsin Institute of Environmental Studies, 1977), p. 48.

30. *Lines Across the Land*, p. 168.

31. "Study Reveals Staggering Agricultural Land Losses," *Journal of Soil and Water Conservation* 35, no. 2 (March-April 1980): 101.

32. M. Rupert Cutler, "Crops Die Where Cities Sprawl," *Chicago Tribune*, July 6, 1980.

33. Kenneth A. Cook, "The National Agricultural Lands Study: In Which Reasonable Men May Differ," *Journal of Soil and Water Conservation* 35, no. 5 (September-October 1980): 249.

34. M. L. Cotner, M. D. Skold, and O. Krause, *Farmland: Will There Be Enough?* (Washington, D.C.: USDA, 1975), p. 3.

35. National Association of Conservation Districts, "Harris Poll Finds Widespread Support for Soil and Water Conservation," *Tuesday Letter*, January 22, 1980, p. 1.

36. Pierre Crosson, "The Long-Term Adequacy of Agricultural Land in the U.S.: An Overview" (See Note 4).

37. William M. Johnson, "What Has Been Happening in Land Use in America and What Are the Projections?" (Speech presented at the Combined Western Sectional Meetings of the American Society of Animal Science and Canadian Society of Animal Science, Pullman, Washington, July 20, 1976).

38. Robert E. Coughlin et al., *Saving the Garden: The Preservation of Farmland and Other Environmentally Valuable Land* (Washington, D.C.: National Science Foundation [RANN], 1977), p. 71.

39. Calvin L. Beale, "Internal Migration in the United States Since 1970," (Testimony before the House Select Committee on Population, February 8, 1978), p. 2.

40. W. Wendell Fletcher and Charles E. Little, "Land Impacts of Rural Population Growth" (Manuscript prepared for the Council on Environmental Quality, February 15, 1980), p. 4.

41. Herman Kahn and Anthony J. Werner, *The Year 2000: A Framework for Speculation on the Next Thirty-Three Years* (New York: Macmillan Co., 1967), p. 62.

42. U.S. Bureau of the Census, "Population Profile of the U.S.: 1979," *Current Population Reports* (Washington, D.C.: Bureau of the Census, 1980), p. 20.

43. "Land Impacts of Rural Population Growth," p. 6.

44. Ibid., p. 9.

45. Ibid., p. 17.

46. National Association of Conservation Districts, "Agricultural Land Survey Project." NACD polled local districts about the extent of land conversion over the past five years. The information was the estimate of local district officials and was not based on field measurements or surveys.

47. "Land Impacts of Rural Population Growth," p. 21.

48. Ibid., p. 49.

49. Ibid., p. 53.

50. Ibid., p. 54.

51. Steven E. Kraft, "Macro and Micro Approaches to the Study of Soil Loss," *Journal of Soil and Water Conservation* 33, no. 5 (1978): 238.

52. *Saving the Garden*, p. 71.

53. Edward Thompson, Jr., "'Right to Farm' Laws Examined," *Aglands Exchange* 2, no. 2 (November-December 1980): 1.

54. Aldo Leopold, *A Sand County Almanac* (New York: Ballantine Books, 1966), p. 263.

55. Charles E. Little, "Demise of the Agricultural Land Protection Act: Some Optimism in Defeat," *Journal of Soil and Water Conservation* 35, no. 2 (March-April 1980): 99.

56. Ibid.

57. James M. Jeffords, "Protecting Farmland: Minimizing the Federal Role," *Journal of Soil and Water Conservation* 34, no. 4 (July-August 1979): 158.

58. Robert Gray, "The National Agricultural Lands Study," *Journal of Soil and Water Conservation* 35, no. 4 (July-August 1980): 171.

59. National Agricultural Lands Study, *Final Report: 1981* (Washington, D.C.: NALS, 1981).

60. *Farmland and Energy* (See Note 20).

61. National Association of Conservation Districts, *Soil Degradation: Effects on Agricultural Productivity*, NALS Interim Report No. 4, (Washington, D.C.: NALS, 1980).

Chapter 6

1. Harry O. Buckman and Nyle C. Brady, *The Nature and Properties of Soils*, 7th ed. (London: Macmillan & Co., 1971), p. 41.

2. J. H. Stallings, *Soil Conservation* (Englewood Cliffs, N.J.: Prentice-Hall, 1959), pp. 197-99.

3. See several articles in *Soil Erosion: Prediction and Control* (Ankeny, Iowa: Soil Conservation Society of America, 1977) for examples of research relating soil characteristics to erodibility.

4. *Soil Conservation*, pp. 51-70.

5. *Climate and Food: Climatic Fluctuation and U.S. Agricultural Production* (Washington, D.C.: National Academy of Sciences, 1976), p. 50.

6. Ibid., p. 50.

7. *Soil Conservation*, p. 79.

8. Raymond I. Dideriksen, "Resource Inventory—Sheet, Rill and Wind Erosion" (Paper presented at Summer Meeting, American Society of Agricultural Engineers, San Antonio, Tex., June 18, 1980).

9. Prime farmland is the land best suited for producing food, feed,

forage, fiber, and oilseed crops. One of the factors considered in classifying land as prime farmland is that the K factor (erodibility) of the soil must be under established standards, so that prime farmlands are, in general, less erodible than non-prime lands. (Raymond I. Dideriksen and R. Neil Sampson, "Important Farmlands: A National View," *Journal of Soil and Water Conservation* 31 (1976): 195-97).

10. "Resource Inventory," p. 12.

11. Ibid., p. 11.

12. Charles Benbrook and Allen R. Hidlebaugh, *The Economic and Environmental Consequences of Agricultural Land Conversion*, NALS Technical Paper 14 (To be published in 1981 by the National Agricultural Lands Study), April 2, 1981 Draft, p. 82.

13. *Control of Water Pollution from Cropland*, Report #EPA-600/ 1-75-026a (Washington, D.C.: Environmental Protection Agency, 1975), p. 21.

14. David Pimentel et al., "Land Degradation: Effects on Food and Energy Resources," *Science*, October 1976, p. 150.

15. Council on Agricultural Science and Technology, *Land Resource Use and Protection* (Report to the Senate Committee for Agriculture and Forestry, Washington, D.C.: U.S. Senate, 1975).

16. A. F. Bartsch, "Biological Aspects of Stream Pollution," *Biology of Water Pollution* (Washington, D.C.: U.S. Department of the Interior, 1967), p. 13.

17. "Land Degradation," p. 150.

18. *The Nature and Properties of Soils*, pp. 297-99.

19. Walter H. Wischmeier, "Relation of Field Plot Runoff to Management and Physical Factors," *Soil Science Soc. of America Proc.* 30(2) (1966): 272-77.

20. Donald E. McCormack, Keith Young, and Leon Kimberlin, "Current Data for Determining Soil Loss Tolerance" (Staff paper, Soil Conservation Service, Washington, D.C., 1979).

21. W. D. Shrader and G. W. Langdale, "Effect of Soil Erosion on Soil Productivity" (In review for special ASA publication, "Determinants of Soil Loss Tolerance," Madison, Wis.: American Society of Agronomy, scheduled for publication in 1981).

22. G. W. Langdale et al., "Corn Yield Reduction of Eroded Southern Piedmont Soils" (1978).

23. G. J. Buntley and F. F. Bell, "Yield Estimates for the Major Crops Grown on Soils in West Tennessee," Ag. Experiment Station Bulletin No. 561 (University of Tennessee, 1976).

24. D. Harper Simms, *The Soil Conservation Service* (New York: Praeger, 1970), p. 7.

25. Engelstad et al., "The Effect of Soil Thickness on Corn Yield," *Soil Science Soc. of America Proc.* 25 (1961): 497-99.

26. *Soil Conservation*, p. 204.

27. Leon Kimberlin, Allen R. Hidlebaugh, and A. Grunewald, "The Potential Wind Erosion Problem in the US," *Transactions of the Amer. Soc. of Agricultural Engineers* 20, no. 5 (1977).

28. For an explanation of how the USDA estimates were developed, see National Association of Conservation Districts, *Soil Degradation: Effects on Agricultural Productivity* (Washington, D.C.: NALS, 1980), p. 26.

29. USDA, *RCA Appraisal 1980: Review Draft, Part II* (Washington, D.C.: USDA, 1980), p. 3-39.

30. R. Neil Sampson, "The Ethical Dimension of Farmland Protection," *Farmland, Food and the Future* (Ankeny, Iowa, Soil Conservation Society of America, 1979), p. 91.

Chapter 7

1. Harry O. Buckman and Nyle C. Brady, *The Nature and Properties of Soils*, 7th ed. (New York: Macmillan Co., 1969), p. 151.

2. E. J. Russell, *Soil Conditions and Plant Growth* (London: Longmans, Green and Co., 1956), p. 282.

3. *The Nature and Properties of Soils*, p. 11.

4. J. I. Rodale, *Pay Dirt: Farming and Gardening with Composts* (Emmaus, Pa.: Rodale Press, 1945), p. 3.

5. *Soil Conditions*, p. 283.

6. Samuel L. Tisdale and Werner L. Nelson, *Soil Fertility and Fertilizers* (New York: Macmillan Co., 1956), p. 250.

7. Gene Logsdon, *The Gardener's Guide to Better Soil* (Emmaus, Pa.: Rodale Press, 1975), p. 88.

8. National Association of Conservation Districts, *A Summary: Non-Federal Natural Resources of the United States* (Washington, D.C.: NACD, 1979), p. 3.

9. W. B. Voorhies, "Soil Tilth Determination Under Row Cropping in the Northern Corn Belt: Influence of Tillage and Wheel Traffic," *Journal of Soil and Water Conservation* 34, no. 4 (July-August 1979): 186.

10. W. B. Voorhees, "Soil Compaction," *Crops and Soils* 29, no. 6: 7.

11. N. J. Rosenberg, *Advances in Agronomy*, 16(1964): 181-96.

12. Paul D. Hebblethwaite and Michael McGowan, "The Effects of Soil Compaction on the Emergence, Growth and Yield of Sugar Beet and Peas, *Journal of the Science of Food and Agriculture* 31, no. 11 (November 1980): 1,131.

13. F. N. Swader, "Soil Productivity and the Future of American Agriculture," *The Future of American Agriculture as a Strategic Resource* (Washington, D.C.: The Conservation Foundation, 1980), p. 96.

14. E. P. Adams, G. R. Blake, W. P. Martin, and D. H. Boelter,

Transactions, 7th Congress of the Soil Science Institute, Vol. 1. (Madison, Wis.: Soil Science Institute, 1961), pp. 607-15.

15. "Soil Compaction: A Pressing Problem," *New Farm* 1, no. 5 (July-August 1979): 11.

16. *The Future of American Agriculture*, p. 96.

17. Ibid.

18. John A. Mabbutt, "The Impact of Desertification as Revealed by Mapping," *Environmental Conservation* 5, no. 1 (Spring 1978): 45.

19. David Sheridan, *Desertification of the United States* (Washington, D.C.: Piedmont Environmental Council, 1981) June 30, 1980, p. 4.

20. John Wesley Powell, *Selected Prose of John Wesley Powell* (Boston: David R. Godine, 1970), p. 49.

21. *Desertification of the United States,* p. 9.

22. Bureau of Land Management, *Final Environmental Impact Statement—Proposed Domestic Livestock Program for the Challis Planning Unit* (Boise, Idaho: BLM State Office, 1977), p. 1.

23. U. S. Department of Agriculture, *San Joaquin Valley Basin Study* (Davis, Calif.: Soil Conservation Service, 1977), p. 55.

24. Ibid., p. 56.

25. Ibid.

26. *Desertification of the United States,* p. 77.

27. Ibid., p. 86.

28. USDA, *RCA Appraisal 1980: Review Draft, Part II* (Washington, D.C.: USDA, 1980), p. 3-82.

29. George L. Baker, "Salt Buildup Ravages Vital Valley Farmland," *Fresno Bee*, November 11, 1980.

30. Mass and Hoffman, "Crop Salt Tolerance," *Journal of the Irrigation and Drainage Division, ASCE* 103, no. IR2 (June 1977).

31. Department of the Interior, *Westwide Study Report on Critical Water Problems Facing the Eleven Western States* (Washington, D.C.: USDI, 1975), p. 116.

32. *RCA Appraisal: Part II*, p. 3-83.

33. Ross Howard and Michael Perley, *Acid Rain: The North American Forecast* (Toronto: House of Anasi Press, 1980), p. 13.

34. Ibid., p. 64.

35. National Academy of Sciences, *Agricultural Production Efficiency* (Washington, D.C.: NAS, 1975), p. 101.

36. Ibid.

37. Gerald Darby, "Problem Analysis for the RCA" (Staff paper, Soil Conservation Service, Washington, D.C., March 1979).

38. Ellis Darley et al., "Effects of Chronic Low-Level Air Pollutants on Agriculture" (Statement prepared for the Subcommittee on the Environment and the Atmosphere of the House Committee on Science and Technology, November 12, 1975).

39. A. A. Millecon, *A Survey and Assessment of Air Pollution Damage to California Vegetation, 1970-1974* (Sacramento, Calif.: Department of Food and Agriculture, 1976).

40. *Scientific American*, June 1976, p. 122.

41. *Acid Rain*, p. 61.

42. *RCA Appraisal: Part II*, p. 3-219.

Chapter 8

1. Don Paarlberg, *Farm and Food Policy: Issues of the 1980's* (Lincoln,: University of Nebraska Press, 1980), p. 163.

2. John Timmons, "Agriculture's Natural Resource Base: Demand and Supply Interactions, Problems, and Remedies," *Soil Conservation Policies: An Assessment* (Ankeny, Iowa: Soil Conservation Society of America, 1979), p. 53.

3. Kenneth D. Frederick, "Irrigation and the Adequacy of Agricultural Land" (Paper presented at Resources for the Future Conference on the Adequacy of Agricultural Land, Washington, D.C., June 19-20, 1980).

4. There are considerable differences in the various estimates of irrigated cropland in America. In 1974, the Census of Agriculture estimated 41.2 million acres, while in 1975 the U. S. Geological Survey estimated that 54 million acres were irrigated. These differences have caused much consternation since several important resource studies have been based on each of them. In an exhaustive study of the various estimates, Kenneth D. Frederick of Resources for the Future states that "the 1977 NRI provides the best estimates." It leaves one dubious of our ability to plan carefully for water resource use when the estimates of the amount of irrigated land vary so widely.

5. Lewis D. Walker, "An Assessment of Water Resources in the United States, 1975-2000," *Resource-Constrained Economies: The North American Dilemma* (Ankeny, Iowa: Soil Conservation Society of America, 1980), p. 78.

6. The year "1975" is shown in quotes because the data used by WRC as a basis for the Second National Water Assessment was "normalized" for that year by the WRC staff in order to best approximate what occurred in that year. It is not based on 12-month calendar-year water records.

7. Soil Conservation Service, "Proposed Program for Resource Area 3" (Draft RCA staff paper, October 31, 1980).

8. USDA, *RCA Appraisal 1980: Review Draft, Part II* (Washington, D.C.: USDA, 1980), p. 3-82.

9. Environmental Protection Agency, *National Water Quality Inventory: 1977 Report to Congress* (Washington, D.C., EPA Office of Water Planning and Standards, 1978), p. 2.

10. *RCA Appraisal: Part II,* pp. 3-54 and 3-77.

11. Ibid., p. 3-76.

12. Ibid., p. 3-81.

13. Ibid.

14. Department of the Interior, *Westwide Study Report on Critical Water Problems Facing the Eleven Western States* (Washington, D.C.: USDI, 1975), p. 177.

15. James Risser, "Intensive Irrigation in U.S. Threatens Water Supplies," *Des Moines Register*, Sept. 14, 1978.

16. *Resource-Constrained Economies*, p. 79.

17. "Irrigation and Adequacy," p. 18.

18. *Farm and Food Policy*, p. 167.

19. Wes Jackson, *New Roots for Agriculture* (San Francisco: Friends of the Earth, 1980), p. 30.

20. *Farm and Food Policy*, p. 168.

21. Richard Barnet, *The Lean Years* (New York: Simon & Schuster, 1980), p. 201.

22. *Resource-Constrained Economies*, p. 90.

23. "Irrigation and Adequacy," p. 20.

24. U. S. Water Resources Council, *The Nation's Water Resources 1975-2000, Second National Water Assessment, Vol. 1* (Washington, D.C.: WRC, 1978), p. 18.

25. Although water conservation has recently received a great deal of attention as a solution to the water shortages of the nation, water conservation and reuse do not increase the natural water supply. In most river basins of the West, water is used, reused, then reused again downstream. The water that escapes from one person's irrigation system is not "lost," but simply continues downstream to the next user. Conservation can be more significant in extending the life and utility of groundwater supplies where recharge is so limited that the water is essentially being mined out. For a good discussion of the potential for conservation, see *Irrigation Water Use and Management*, an Interagency Task Force Report done by USDI, USDA, and EPA. (Washington, D.C.: USGPO, June 1979), p. 6.

26. "Proposed Program for Resource Area 3," p. 3.

27. "Irrigation and Adequacy," p. 44.

28. Ibid., p. 45.

29. Ibid., p. 36.

Chapter 9

1. Pierre Crosson, "The Long-Term Adequacy of Agricultural Land in the United States: An Overview" (Manuscript prepared for publication of an RFF book in 1981, Washington, D.C.: Resources for the Future, 1981), p. 18.

2. Charles C. Geisler et al., "Sustained Land Productivity: Equity Consequences of Technological Alternatives" (Report prepared for the Office of Technology Assessment, U.S. Congress, November 1980), p. 1.

3. Yao-chi Lu, Phillip Cline, and Leroy Quance, *Prospects for Productivity Growth in U.S. Agriculture*, Agriculture Economic Report No. 435 (Washington, D.C.: USDA, 1979), p. 4.

4. Sandra S. Batie and Robert G. Healy, "American Agriculture as a Strategic Resource: The Past and the Future," *The Future of American Agriculture as a Strategic Resource* (Preliminary papers prepared for a Conservation Foundation Conference, Washington, D.C.: The Conservation Foundation, 1980), p. 16.

5. Ibid., p. 24.

6. Ibid., p. 30.

7. Ibid., p. 31.

8. L. C. Gray, O. E. Baker, F. J. Marschner, B. D. Weitz, W. R. Chapline, Ward Shephard, and Raphael Zon, "The Utilization of Our Lands for Crops, Pasture and Forests," *1923 Yearbook of Agriculture* (Washington, D.C.: USDA, 1923), p. 495.

9. *Prospects for Productivity Growth*, p. 1.

10. A. Dexter Hinckley, *Impact of Climatic Fluctuation on Major North American Food Crops* (Washington, D.C.: The Institute of Ecology, 1976), p. 3.

11. *The Future of American Agriculture*, p. 18.

12. *Prospects for Productivity Growth*, p. 2.

13. Earl Heady, "Technical Change and the Demand for Land" (Paper prepared for Resources for the Future Conference on the Adequacy of Agricultural Land, Washington, D.C., June 19-20, 1980).

14. *Impact of Climatic Fluctuation*, p. 7.

15. Robert H. Shaw, "Climate Change and the Future of American Agriculture," *The Future of American Agriculture*, p. 263.

16. Kenneth D. Frederick, "Irrigation and the Future of American Agriculture," *The Future of American Agriculture*, p. 157.

17. USDA, *RCA Appraisal 1980: Review Draft, Part I* (Washington, D.C.: USDA, 1980), p. 2-42.

18. *RCA Appraisal: Part II*, p. 3-18.

19. Ibid., p. 3-47.

20. Ibid., p. 3-257.

21. George Reiger, " Wetlands Laws: A Breakdown in Commonsense," *Field & Stream* 84(12) (April 1980): 22.

22. Vernon W. Ruttan, "Agricultural Research and the Future of American Agriculture," *The Future of American Agriculture*, p. 125.

23. Ibid., p. 136.

24. Ibid.

25. *Prospects for Productivity Growth*, p. 40.

26. Dr. John Patrick Jordan, Director, Colorado State University Experiment Station, personal communication, April 6, 1981.

27. Ibid.

28. Ibid.

29. Wes Jackson, *New Roots for Agriculture* (San Francisco: Friends of the Earth, 1980), p. 114.

30. Burtt Trueblood, personal communication, December 16, 1980.

31. *RCA Appraisal 1980: Part II*, p. 2-9.

32. *Prospects for Productivity Growth*, p. i.

33. *The Future of American Agriculture*, p. 138.

34. "Sustained Land Productivity: Equity Consequences," p. 6.

35. George E. Brown, Jr., "Agricultural Science: The Challenge of Change" (Paper delivered at the Joint Meeting of the National Agricultural Research and Extension Users Advisory Board and the Joint Council on Food and Agricultural Sciences, July 15, 1980).

Chapter 10

1. Harvard Business School, *Energy Future* (New York: Davidson House, 1979), p. 197.

2. Cited in Barry Commoner, *The Politics of Energy* (New York: Alfred A. Knopf, 1979), p. 2.

3. The White House, "Fact Sheet: S. 932, The Energy Security Act," Public Law 96-294 (Washington, D.C.: Office of the White House Press Secretary, 1980).

4. W. E. Larson, R. F. Holt, and C. W. Carlson, "Residues for Soil Conservation," *Crop Residue Management Systems*, Spec. Pub. No. 31. (Madison, Wis.: American Society of Agronomy, 1978).

5. W. E. Larson and F. J. Pierce, "Crop Residues: Availability for Removal and Requirements for Erosion Control" (Paper presented at Bio-Mass World Congress and Exposition, Atlanta, Ga., 1980).

6. R. Neil Sampson, "Energy Analysis as a Tool in Conservation

Planning," *Critical Conservation Choices: A Bicentennial Look* (Proceedings of the 31st Annual Meeting, Ankeny, Iowa, Soil Conservation Society of America, 1976), p. 88.

7. Thomas Grubisich, "Not the Same Old Manure," *Washington Post*, August 31, 1980.

8. Lester R. Brown, *Food or Fuel: New Competition for the World's Cropland*, Worldwatch Paper 35, (Washington, D.C.: Worldwatch Institute, 1980), p. 19.

9. Paul Clancy, "Farmer Makes Fuel to Survive," *Washington Star*, August 20, 1980. See also Roger Blobaum, "Energy Alternatives for Agriculture," *Resource-Constrained Economies: The North American Dilemma* (Ankeny, Iowa: Soil Conservation Society of America, 1979), p. 140.

10. Wes Jackson, Mari Peterson, and Charles Washburn. "Impacts on the Land in the New Age of Limits," *Land Report No. 9* (Winter 1980): 20.

11. Harry J. Prebluda and Roger Williams, Jr., "Perspectives on the Economic Analysis of Ethanol Production from Biomass," *Preprints*, Division of Petroleum Chemistry, Inc., American Chemical Society 24, no. 2 (March 1979): 485.

12. Kathryn A. Zeimetz, *Growing Energy: Land for Biomass Farms*, Agricultural Economic Report No. 425 (Washington, D.C.: U.S. Department of Agriculture, 1979), p. 13.

13. U.S. Forest Service, *A Report on the Nation's Renewable Resources: Review Draft* (Washington, D.C.: U.S. Department of Agriculture, 1979), p. 40.

14. Pierre Crosson, "Future Economic and Environmental Costs of Agricultural Land" (Paper presented at a Resources for the Future Conference on the Adequacy of Agricultural Land, Washington, D.C., June 19-20, 1980; to be published by RFF).

15. Cited in *Growing Energy*, p. 6.

16. U.S. Department of Energy, *The Report of the Alcohol Fuels Policy Review*, DOE/PE-0012 (Washington, D.C: DOE, 1979), p. 15.

17. *Food or Fuel*, p. 19.

18. Bob Bergland, Secretary of Agriculture. Testimony before the House Committee on Science and Technology, Washington, D.C., May 4, 1979.

19. Office of Technology Assessment, *Gasohol: A Technical Memorandum* (Washington, D.C.: Congress of the United States, 1979), p. 31.

20. *Alcohol Fuels Policy Review*, p. 13. (The calculation that 4.7 billion gallons of ethanol per year would be less than 2 percent of our petroleum use was provided by Neil Hazlett who deals in such matters at the Naval Research Laboratory, in a personal communication.)

21. *Land Report No. 9*, p. 20.

22. Jim Williams, Deputy Secretary of Agriculture. Statement before the Committee on Science and Technology, House of Representatives, Washington, D.C., Feb. 22, 1980.

23. Bob Bergland, Secretary of Agriculture. Statement before the U.S. National Alcohol Fuels Commission, Washington, D.C., June 19, 1980.

24. Martin E. Abel, "Growth in Demand for U.S. Crop and Animal Production by 2005" (Paper presented at a Resources for the Future Conference on the Adequacy of Agricultural Land, Washington, D.C., June 19-20, 1980; to be published by RFF).

25. Bob Bergland. Testimony (See Note 18).

26. Howard Tankersley, USDA/SCS, personal correspondence, 1980.

27. Lemoyne Hogan, "Jojoba: A New Crop for Arid Regions," *New Agricultural Crops* (Boulder, Colo.: Westview Press, 1979), p. 182.

28. Gary Nabhan, "New Crops for Desert Farming," *New Farm* 1, no. 3: 52.

29. Noel D. Vietmeyer, "Guayule: Domestic Rubber Rediscovered," *New Agricultural Crops*, p. 169.

30. Omer J. Kelley, "The Agricultural Research System: A New Evaluation" (Paper presented at the 35th Annual Meeting of the Soil Conservation Society of America, Dearborn, Mich., August 6, 1980).

31. W. P. Bemis, James W. Berry, and Charles W. Weber, "The Buffalo Gourd: A Potential Arid Land Crop," *New Agricultural Crops*, p. 77.

32. "New Crops for Desert Farming," p. 59.

33. L. H. Princen, "New Crop Developments for Industrial Oils," *Journal of the American Oil Chemists' Society* 56, no. 9: 845.

34. L. H. Princen, "Potential Wealth in New Crops: Research and Development," *Crop Resources* (New York: Academic Press, 1977), p. 9.

35. Ibid., p. 10.

36. L. H. Princen, *Alternate Industrial Feedstocks from Agriculture* (Peoria, Ill.: SEA-AR, USDA, 1977), p. 7. Also "Need for Renewable Coatings Raw Materials and What Could be Available Today," *Journal of Coatings Technology* 49, no. 635: 88-93.

Chapter 11

1. See Chapter 2 of the *RCA Appraisal 1980: Review Draft, Part II* for a complete explanation of the assumptions utilized by USDA in the RCA study.

2. J. S. McCorkle and Arnold Heerwagen, "Effects of Range Condition on Livestock Production," *Journal of Farm Managers* 4, no. 4: 242-48.

3. Irirangi C. Bloomfield, "Emptying the Horn of Plenty," *Journal of Soil and Water Conservation* 35, no. 6: 283.

4. Kenneth A. Cook, "The National Agricultural Lands Study: In Which Reasonable Men May Differ," *Journal of Soil and Water Conservation* 35, no. 5: 247.

5. National Agricultural Lands Study, "NALS Technical Paper VI:

Competing Demands for Agricultural Land in the Year 2000" (Draft paper dated July 24, 1980, p. vi-97. In publication).

6. Soil Conservation Service, "Response to Concern over Agricultural Land Supply Information in 1980 Draft," (Draft RCA Staff Paper, February 13, 1980).

7. Louis Harris and Associates, Inc., "Outline for Press Briefing on a Survey of the Public's Attitudes toward Soil, Water, and Renewable Resources Conservation Policy" (Washington, D.C., January 17, 1980), p. 3.

Chapter 12

1. Aldo Leopold, *A Sand County Almanac* (New York: Ballantine Books, 1966), p. 243.

2. J. H. Stallings, *Soil Conservation* (Englewood Cliffs, N.J.: Prentice-Hall, 1959), p. 160.

3. See Harold A. Hughes, *Conservation Farming* (Moline, Ill.: The John Deere Company, 1980) for a well-illustrated broad overview of conservation issues and practices.

4. Ibid., p. 6.

5. John N. Cole, "Learning a Hard Lesson in Aroostook County," *Country Journal*, November 1980, p. 44.

6. William Lockeretz, Georgia Shearer, Robert Klepper, and Susan Sweeney, "Field Crop Production on Organic Farms in the Midwest," *Journal of Soil and Water Conservation* 33, no. 3 (May-June 1978): 130.

7. USDA, *Report and Recommendations on Organic Farming* (Washington, D.C.: USDA, 1980), p. xii.

8. *Journal of Soil and Water Conservation* 33 no. 3 (May-June 1978): 130.

9. The study was initiated, according to Secretary of Agriculture Bob Bergland's foreword, because of the growing number of requests for alternative farming systems, from both large-scale and small-scale producers.

10. Garth Youngberg, "Organic Farming: A Look at Opportunities and Obstacles," *Journal of Soil and Water Conservation* 35, no. 6 (November-December 1980): 261.

11. Ibid., p. 262.

12. Council on Agricultural Science and Technology, "Comparison of Conventional and Organic Farming Published," *Journal of Soil and Water Conservation* 35, no. 6 (November-December 1980): 263.

13. John James Ingalls, "In Praise of Blue Grass," *Grass: The Yearbook of Agriculture, 1948* (Washington, D.C.: Government Printing Office, 1948), p. 7.

14. P. V. Cardon, "A Permanent Agriculture," *Grass: The Yearbook of Agriculture 1948* (Washington, D.C.: Government Printing Office, 1948), p. 1.

15. Louis Bromfield was a Pulitzer-winning American author who

settled on a farm near Lucas, Ohio, in 1933 and wrote about his new understanding of the land and the farm as a living, dynamic, productive system. His two books *Pleasant Valley* (1945) and *Malabar Farm* (1948) are probably the most well known of several books that dealt, in part, with his feeling for man's relationship to the land.

16. Channing Cope, *Front Porch Farmer* (Atlanta, Ga.: Turner E. Smith & Co., 1949), p. xxvii.

17. Ibid., p. xxvi.

18. Ibid., p. xx.

19. Ibid., p. xi.

Chapter 13

1. David G. Unger, "Evolution of Institutional Arrangements: A Federal View," *Soil Conservation Policies: An Assessment* (Ankeny, Iowa: Soil Conservation Society of America, 1979), p. 26.

2. South Carolina Association of Soil Conservation District Supervisors, *Keepers of the Land: A History of Soil and Water Conservation Districts in South Carolina* (Columbia, S.C.: SCASCD, 1972), p. 6.

3. Ibid., p. 6.

4. H. H. Bennett. Statement before the Subcommittee of House Committee on Public Lands, Washington, D.C., March 20, 1935.

5. Rita S. Dallavalle and Leo V. Mayer, *Soil Conservation in the United States: The Federal Role* (Washington, D.C.: Congressional Research Service, 1980), p. 8.

6. The number of conservation districts has varied around 3,000 for over a decade. After the period of district formation, there have been both additions and subtractions as multi-county districts have split up into smaller units and, conversely, as small districts merged. A major change was in 1974, when Nebraska combined 83 single-county districts into 24 Natural Resource Districts.

7. George R. Bagley, "Evolution of Institutional Arrangements: A Nongovernmental View," *Soil Conservation Policies*, p. 33.

8. *Keepers of the Land*, p. 40.

9. *Soil Conservation in the United States*, p. 10.

10. Ibid., p. 12.

11. R. Burnell Held and Marion Clawson, *Soil Conservation in Perspective* (Baltimore: Johns Hopkins Press, 1965), p. 51.

12. Ibid., p. 52.

13. *Soil Conservation in the United States*, p. 13.

14. General Accounting Office, *To Protect Tomorrow's Food Supply, Soil Conservation Needs Priority Attention* (Washington, D.C.: General Accounting Office, 1977), p. 43.

15. Ibid., p. 29.

16. D. Harper Simms, *The Soil Conservation Service* (New York: Praeger, 1970), p. 138.

17. *Soil Conservation Policies*, p. 12.

18. Ibid., p. 13.

19. *Soil Conservation in Perspective*, p. 80.

20. *Soil Conservation Policies*, p. 13.

21. *To Protect Tomorrow's Food Supply*, p. 53.

22. *Soil Conservation Policies*, p. 37.

23. National Association of Conservation Districts, *The Future of Districts: Strengthening Local Self-Government in Conservation and Natural Resource Development* (Washington, D.C.: NACD, n.d.), p. 18.

24. *Soil Conservation Policies*, p. 38.

25. Ibid., p. 39.

26. Norman A. Berg, "The National Outlook," *Proceedings: Soil Conservation Service Conservationist's Conference, Huron, S. D., March 21-23, 1972* (Washington, D.C.: SCS, 1972), p. 3.

27. National Association of Conservation Districts, *RCA Notes*, No. 7, July 25, 1980.

28. *Des Moines Sunday Register*, November 16, 1980, p. A1.

29. Lyle Bauer, "1980: Year of Decision," *The 1980 Proceedings and Thirty-Fourth Annual Convention, Houston, Texas* (League City, Tex.: NACD, 1980), p. 46.

30. *To Protect Tomorrow's Food Supply*, p. 8.

31. Philip M. Glick, "Soil Conservation: Highlights of Political and Legal Arrangements," *Soil Conservation Policies*, p. 23.

32. Ibid., pp. 23-24.

33. Aldo Leopold, *A Sand County Almanac* (New York: Ballantine Books, 1966), p. 244.

34. Charles McLaughlin, "Statement before the Subcommittee on Environment, Soil Conservation and Forestry, Ames, Iowa, August 14, 1980," *Soil Conservation* (Washington, D.C.: Government Printing Office, 1980), p. 111.

35. *Soil Conservation in Perspective*, p. 74.

Chapter 14

1. Shirley Foster Fields, *Where Have the Farmlands Gone?* (Washington, D.C.: National Agricultural Lands Study, 1979), p. 8.

2. For a running account of the emergence of the farmland issue, it is instructive to read the sections on "natural resources" in the Annual Reports of the President's Council on Environmental Quality, beginning in 1975

and proceeding to 1980. The reports are for sale by the Superintendent of Documents, Government Printing Office, Washington, D.C. 20402.

3. Wendell Berry, *The Unsettling of America: Culture and Agriculture* (New York: Avon Books, 1978), p. 22.

4. Ibid., p. 13.

5. Charles E. Little, *Land & Food: The Preservation of U.S. Farmland* (Washington, D.C.: American Land Forum, 1979), p. 47.

6. Christopher Derrick, *The Delicate Creation: Towards a Theology of the Environment* (Old Greenwich, Conn.: The Devon-Adair Co., 1972), p. 72.

7. Lance Gay, "Block Urges Preservation of Farm Land," *Washington Star*, February 10, 1981.

8. Don Reeves, "A Quaker Farmer Comments on Pope's Land Statement," *Catholic Rural Life* 29, no. 3 (March 1980): 21.

9. Aldo Leopold, *A Sand County Almanac* (New York, N.Y.: Ballantine Books, 1966), p. 238.

10. Piedmont Environmental Council, *Toward a New Land Use Ethic* (Warrenton, Va.: Piedmont Environmental Council, 1981).

11. R. Neil Sampson, "Protecting Farmland: The Ethical Dimension," *Farmland, Food and the Future* (Ankeny, Iowa: Soil Conservation Society of America, 1979), p. 94.

12. E. F. Schumacher, *Small is Beautiful: Economics as Though People Mattered* (New York, N.Y.: Harper & Row, 1973), p. 45.

13. Wayne H. Davis, "The Land Must Live," *The New Food Chain* (Emmaus, Pa.: Rodale Press, 1973), p. 47.

14. *Land & Food*, p. 10.

15. Calculated at 100 bushels of corn per acre, 60 pounds per bushel, 2,000 kilocalories per pound. 12 million kilocalories per acre $\div$ 730,000 kilocalories per person per year (2,000 kilocalories/day x 365) = 16.4

16. Robert Cahn, *Footprints on the Planet* (New York, N.Y.: Universe Books, 1978), p. 9.

17. Richard C. Collins, "Developing the Needed New Land Use Ethic," *Land Issues and Problems, No. 18* (Blacksburg, Va.: Virginia Polytechnic Institute, 1976), p. 3.

18. Ibid.

19. See Chaplin B. Barnes, "A New Land Use Ethic," *Journal of Soil and Water Conservation* 35, no. 2, (March-April, 1980): 61-62, for a short explanation of the land use ethic developed by the Piedmont Environmental Council.

20. National Association of Conservation Districts, *Position Paper on the Resources Conservation Act (RCA)* (Washington, D.C.: NACD, 1980).

21. James M. Jeffords, "Protecting Farmland: Minimizing the Federal Role," *Journal of Soil and Water Conservation* 34, no. 4 (July-August 1979): 158-59.

Chapter 15

1. Agricultural Stabilization and Conservation Service, *National Summary Evaluation of the Agricultural Conservation Program: Phase 1* (Washington, D.C.: ASCS, 1980), p. viii.

2. Charles Benbrook. Testimony on Soil Conservation Policy Needs before the Conservation, Credit and Rural Development Subcommittee of the House Agriculture Committee, Washington, D.C., March 12, 1981.

3. Economic Research Service, *The Land Utilization Program: 1934 to 1964,* Agricultural Economic Report No. 85 (Washington, D.C.: Government Printing Office, 1965).

4. Lindsay Grant, "Speculators in the Cornfield," *Journal of Soil and Water Conservation* 34, no. 2 (March-April 1979): 51.

5. Tom Barlow, "How Federal Policies are Encouraging Soil Erosion," *Catholic Rural Life* 29, no. 3 (March 1980): 7.

6. Tom Barlow, "Cross Compliance Urged to Protect Conserving Farmers" (Testimony before the House Agriculture Committee, Washington, D.C., March 12, 1981).

7. Harold D. Guither, "How Farmers View Agricultural and Food Policy Issues" (Testimony before the House Agriculture Committee, Washington, D.C., March 10, 1981).

8. Charles E. Little, "Middleground Approaches to the Preservation of Farmland" (Discussion paper prepared for the Council on Environmental Quality, 1980), p. 5. For a published version that did not focus on the costs of development rights, but rather on the advantages of farmland conservancies, see Charles E. Little, "Farmland Conservancies: A Middleground Approach to Agricultural Land Preservation," *Journal of Soil and Water Conservation* 35, no. 5 (September-October 1980): 204-11.

Appendix D

1. Walter H. Wischmeier and Dwight D. Smith, *Predicting Rainfall Erosion Losses — A Guide to Conservation Planning*, Agriculture Handbook 537 (Washington, D.C.: USDA, 1978) This handbook gives full details on how to use the Universal Soil Loss Equation. In addition, several state handbooks have been prepared that give farmers the actual coefficients to use for their particular soils. One example is: Robert D. Walker and Robert A. Pope, *Estimating Your Soil Erosion Losses with the Universal Soil Loss Equation (USLE)* (Urbana: Cooperative Extension Service, College of Agriculture, University of Illinois, revised May, 1980).

2. E. L. Skidmore and R. Woodruff, *Wind Erosion Forces in the U.S. and Their Use in Predicting Soil Loss*, Agriculture Handbook 346 (Washington, D.C.: USDA, 1968).

3. *Predicting Rainfall Erosion Losses*, p. 7.

4. National Association of Conservation Districts, *A Summary: Non-Federal Natural Resources of the United States* (Washington, D.C.: NADC, 1979), p.6.

Index